D0781779

AutoCAD Workbook for Architects and Engineers

Shannon R Kyles

Professor of Architecture
Mohawk College, Ontario, Canada

Blackwell Publishing

Blackwell Publishing was acquired by John Wiley & Sons in February 2007. Blackwell's publishing programme has been merged with Wiley's global Scientific, Technical, and Medical business to form Wiley-Blackwell.

Registered office
John Wiley & Sons Ltd, The Atrium, Southern Gate, Chichester, West Sussex, PO19 8SQ, United Kingdom

Editorial office
9600 Garsington Road, Oxford, OX4 2DQ, United Kingdom

For details of our global editorial offices, for customer services and for information about how to apply for permission to reuse the copyright material in this book please see our website at www.wiley.com/wiley-blackwell.

Library of Congress Cataloging-in-Publication Data

Kyles, Shannon.
AutoCAD workbook for architects / S.R. Kyles.
   p. cm.
   Includes index.
   ISBN-13: 978-1-4051-8096-2 (pbk. : alk. paper)
   ISBN-10: 1-4051-8096-X (pbk. : alk. paper)   1. Architectural drawing–Computer-aided design–Handbooks, manuals, etc.   2. AutoCAD–Handbooks, manuals, etc.   I. Title.
NA2728.K95 2008
720.28′40285536–dc22

                                        2007033120

A catalogue record for this book is available from the British Library.

Set in Times New Roman 10/13pt by S.R. Kyles, Canada
Printed in Singapore by C.O.S. Printers Pte Ltd
1 2008

Contents

Acknowledgments

The exercises in this book went through many years of student testing both at Mohawk College and at McMaster University in Hamilton, Ontario, Canada. Since the student edition appeared in January 1993, there have been many positive suggestions and much constructive criticism. I would like to thank all my students over the past 27 years of teaching CAD for working with me on the development of new projects, and for proofing tutorials and exercises.

I would like to thank Brian McKibbin, Diego Gomez, and Trevor Garwood Jones for their generosity in letting me use their architectural designs to provide students with up-to-date architectural work. I would also like to thank Robert Deeks and Henry Brink for their help in updating the mechanical engineering examples, and Ken Mercer for checking the accuracy of the wood-related drawings.

Finally I would like to thank Catriona Dixon and Madeleine Metcalfe from Blackwell Publishing for their efforts in polishing this text.

Shanon R. Kyles
Canada

December 2007

Introduction

AutoCAD is a very popular, flexible software system that allows the user to create both 2-dimensional and 3-dimensional models and drawings. This book offers a series of exercises to help you learn the 2D drawing techniques of AutoCAD. Most of the 2D commands in Releases 2006, 2007 and 2008 have not changed since Release 2000i. These exercises can be used on all releases from 2000i through 2008.

For those who are familiar with computers, learning AutoCAD will be easy, simply because you are aware of the typical response structure and the format of your system.

Using this Book

System Prompts and User Responses

All commands listed within the text of this book are in the command font. CIRCLE, LINE, ZOOM, MIRROR, etc. When shown as in this font are commands.

In this book, the system command information will be shown in this style:

```
Command:
Specify first point:
Specify second point or [Undo]:
```

The user responses (what you should type in) will be shown in bold:

```
Command:LINE
Specify first point:0,0
Specify second point or [Undo]:5,3
```

The Enter or Return Key

At the end of each command or entry on the command line, use the Enter key (symbolized by ↵) to signal the end of:

- a command entry:

```
Command:LINE↵
```

- a coordinate entry:

```
Specify first point:2,4↵
```

- a value:

```
New fillet radius.0000:3↵
```

- text:

```
Text:All Holes 2.00R Unless Noted↵
```

Please note that the ↵ will not be shown at the end of every entry after Chapter 1; it will only be used when the user should press ↵ rather than entering any other response.

Disks and File Storage

Most of the drawings in this book are small, so a floppy disk may be used for file storage if your computer still has a floppy drive. Zip drives, Jazz drives, mass storage units and CD writers are better suited to storing larger files. Always make three copies of your files. Save your file every half hour on the permanent computer drive, either C: or D:. If you have been working for more than three hours, also save onto an external file system like a flash drive. Before you sign off from the system, e-mail a copy of the file to yourself as well.

Starting AutoCAD

Your Windows environment should have been set up so that a double-click on the AutoCAD application icon will bring up AutoCAD. You will automatically be placed in the Drawing Editor.

When you first start up, there may be palettes overlaying your screen. Click on the "X" to remove them. On releases 2007 and 2008 a dashboard may come up instead of toolbars. Pick AutoCAD Classic from the Workspaces toolbar.

Unfortunately, there is no way of guaranteeing how the last user has left the screen, so you may not have the necessary toolbars showing. If your screen is not the same as shown below, you can either open toolbars in the View pull-down menu (very cumbersome and difficult) or right-click any toolbar on screen and pick from the list a toolbar that you want.

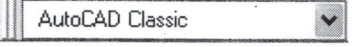

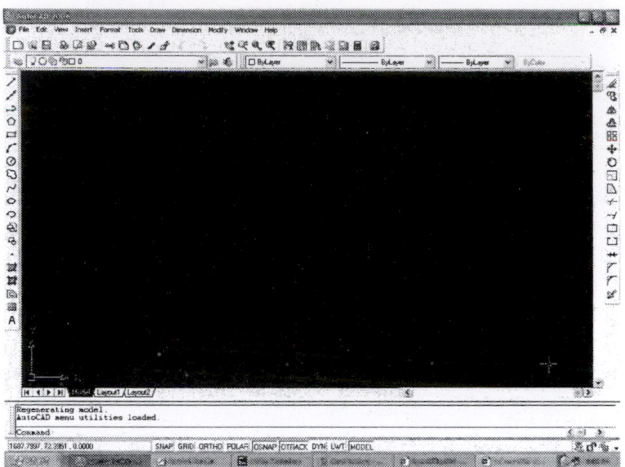

The Windows Drawing Editor

The initial Windows screen contains the menu bar, the status bar, the drawing window or graphics area, and several toolbars. Toolbars contain icons that represent commands.

The menu bar (along the top) contains the pull-down menus. The status bar (along the bottom) displays the cursor coordinates and the status modes such as GRID and SNAP. Mode names are always visible in the status bar as selectable buttons. Click the buttons to toggle the modes. The command line in Windows is "floating," that is, it may be dragged to any location on the screen. The command line is where your commands will be written out. Keep reading this to see where you are.

Keyboard and Mouse Functions

There are many different kinds of pointing devices or mice on the market. Some have two or three buttons, others have as many as 20. Two buttons are adequate for most operations. A central roller on the top of the mouse will help with display commands.

In releases after 2004, the roller ball on the mouse will both ZOOM and PAN your file.

The Pick Button

On all mice there is a point or command indicator or *pick button*; on a two-button mouse, it is usually on the left side of the device.

The pick button is used to indicate the command you want to access either from the on-screen menu or from the digitizer tablet. It is also used to indicate point positions.

The Enter Button

The button on the right of the mouse will often have the function of the ↵ key (Enter or Return) on the keyboard. This signals the end of a command. There is also a right-click facility that accesses the functions associated with each command. This can be turned off under the Tools pull-down menu by selecting Options and then User Preferences.

Function Buttons

Many people who have used AutoCAD for a few years still make use of the function keys on the top of the keyboard (F6, F7, F8, etc). Move the mouse around the screen noting the movement of the crosshairs.

F6	toggles the Co-ordinate readout from absolute, to incremental to off and back
F7	toggles GRID on and off
F8	toggles ORTHO on and off
F9	toggles the SNAP on and off

Entering Commands and Coordinates

You can enter information either through the keyboard or through your mouse or pointing device. There are also toolbars and icons that help to access the information. You can enter a command by typing it in at the command prompt or you can use the pointing device to pick up commands from:

- the pull-down menus in the menu bar
- the icons on the toolbars.

Windows Toolbars

Toolbars are groups of icons or tools compiled according to application. Toolbars can be on-screen or not, and can be on the top or side of your screen or floating.

Accessing Toolbars

In Releases 2000 to 2005, Toolbars can be accessed through the View pull-down menu. Pick View, then Toolbars, then the toolbar that you need. In Releases 2006, 2007 and 2008, simply right-click the two parallel lines on the end of the toolbar, and a list of possible toolbars will appear. Pick the one you want.

To remove a toolbar from your screen, click on the X icon on the top right of the toolbar.

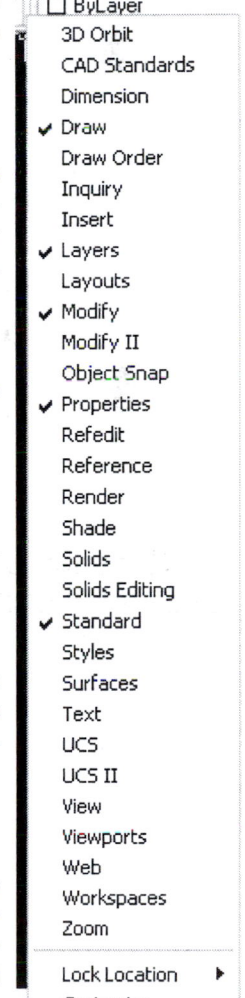

Using Windows Toolbars

Toolbars contain tools that represent commands. When you move the pointing device over a tool, Tooltips display the name of the tool below the cursor. Pick that tool to invoke the command.

Placing Toolbars

The Standard toolbar is visible by default. It carries frequently used tools such as Zoom, Redraw, and Undo. A ***docked toolbar*** attaches to any edge of the graphics window. A ***floating toolbar*** can lie anywhere on the application screen, and it can be resized and does not overlap with the drawing window.

If none of your toolbars appear on screen, exit AutoCAD and open the software again. Your start-up file may not have been properly loaded.

To Dock a Toolbar

1. Position the cursor on the toolbar, and press the pick button on the pointing device.

2. Drag the toolbar to a dock location at the top, bottom, or either side of the drawing window.

3. When the outline of the toolbar appears in the docking area, release the pick button.

To place the toolbar in a docking region without docking it, hold down the Ctrl key as you drag.

Placing theFirst Toolbar

If your screen comes up with no toolbars, type in the word 'toolbar' preceeded by a dash as shown below. Then type in 'Standard' and your Standard toolbar will appear. Dock it, as explained above, then right-click the two parallel lines on the end and place the other toolbars as required. If this doesn't work, exit AutoCAD and reload it.

```
Command:-toolbar
Enter toolbar name or [All]:Standard
```

The Windows Command Window

Like the toolbars, the Windows command line or response area can be moved and docked. By default the command window is docked at the bottom of your screen.

You can resize the command window vertically and horizontally, both with the pointing device and with the splitter bar located at the top edge of the window when docked on the bottom and on the bottom edge when docked at the top. Resizing and docking the command window can help you to create more space for your drawings on-screen. It can also help you see your commands to find out where you may have gone wrong.

Scroll Bars

In most Windows applications there are scroll bars that advance the file you are viewing. Each scroll bar has arrows that indicate a move up or down. To access an area not displayed, click on the up or down arrow until the information is displayed or pick the box within the scroll bar and move it quickly up and down the screen.

Scroll bars can be either vertical or horizontal. In Windows, the scroll bars on the top and bottom move the file across the screen in the same way that PAN does.

Opening or Accessing Drawings

Once you have accessed the Drawing Editor, you can start drawing and later save your work under a specified name in a specified directory. If you have a drawing started in AutoCAD Release 2000i, 2004 or some earlier version, you can use OPEN to find it and then work on it.

Opening Existing Drawings

The command line equivalent is OPEN.

In Windows, under File Name double-click the file name in the list of files. Use the scroll bars to access other files. To access other directories, pick the down arrow beside the words ' Look in:'. You can also type in the drawing name by picking the long white box beside File Name:, then typing in the name of the file. If you prefer to type in both the directory and the name, type that into the File Name box.

Once your file is open, any changes can be saved to the same directory with the SAVE command.

Starting a New File

If you would like to start a new file, access the same File menu and choose New.

Before the new file is created, you can choose a default drawing file environment and/or enter the name of the file that you wish to create.

```
Command: _new
Enter template file name or [. (For none)] <acad.dwt>:
Enter template file name or [. (For none)] <acadiso.dwt>:
```

The .dwt extension stands for drawing template. acad.dwt is imperial, acadiso.dwt is metric.

The default file environment can be either the acad.dwt standard file or a prototype file that contains all the settings for a specific application. Once you are familiar with AutoCAD, you can save drawing templates that contain plotter information, layer information, groups, blocks, linetypes, and other standard information so that you don't need to set up your file from scratch each time.

Recovering Files

If you have a problem with retrieving a file using OPEN, you may need to RECOVER the file. Usually these problems are caused by either bad diskettes or removing the floppy disk from the drive before AutoCAD has completely exited from the file. If you need to restore a file, simply type in RECOVER at the command prompt. Theoretically, the OPEN command should automatically repair any damaged files, but if this doesn't work, try RECOVER.

Saving Files

Computers have a tendency to lose information at the worst possible times. It is suggested that when you are using AutoCAD you save your files at least every hour.

The first time you save a drawing, you will be prompted for the name of the file before it is saved. If you have already entered the name of the current file under the New option under File, then AutoCAD simply saves the file under the given name and directory and you will not be prompted for a name.

To save a named file, use SAVE. Use the icon or

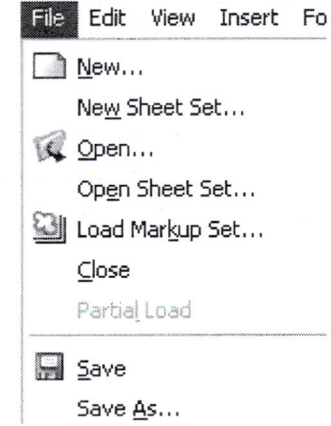

1. Type in the word SAVE at the command prompt

2. From the File menu, choose SAVE. In the Save Drawing As dialog box, enter the new drawing name. Then choose OK.

Choose SAVE every subsequent time you would like to save the drawing, and the drawing will automatically be saved under this specified file name.

If you specified a directory and file when you signed on, use SAVE to save the file under this name.

To save the file under a new name or on a different directory, choose Save As from the File pull-down menu.

Notes

To save a file to be read on an earlier release of AutoCAD, choose Save As, then under the Files of Type box, pick Release 2000.

To change the directory, double-click on the directory listing that you want. The line reading 'Look In:' must reflect the directory chosen.

You can save a file as a different release of AutoCAD by specifying the file type. Specify the release you need under the Files of Type box in the Save As dialog box.

Changing the Drawing Name or Directory

If you want to change the drawing name or directory, use Save As. If you have been addressing C: while creating your drawing, you can save the file onto a disk before exiting the file by using Save As, then pick A: or B: for the directory or drive.

Exiting AutoCAD

Once you have saved the file, you can exit AutoCAD either by clicking on the X at the top right or by picking Exit from the File menu.

The command line equivalent is QUIT.

The large red X will exit you from AutoCAD. The smaller black x will exit you from the current drawing.

Do not remove your floppy disk from the drive before you have completely exited from AutoCAD.

Options Dialog Box

In previous releases and in many other Windows programs the Options dialog box is called Preferences. The Options set up your screen display, the drawing environment, and the system. If you find the color of the screen difficult to work with, change it under Tools, Options, Display, Color. You can also set right-click preferences here.

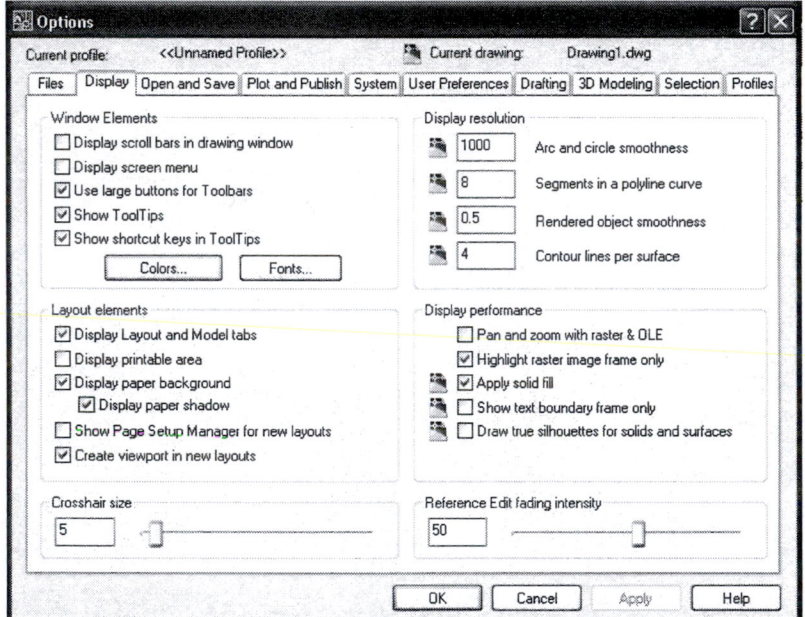

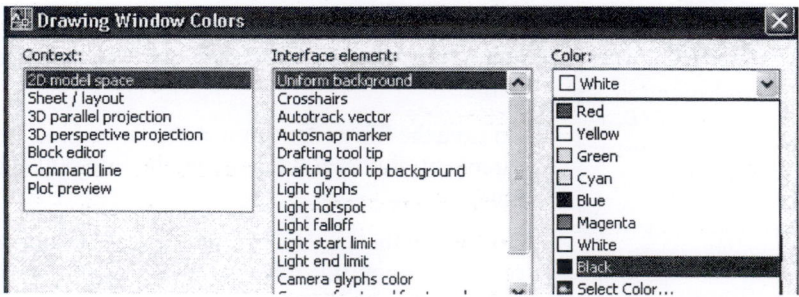

1

Introductory Geometry and Setting Up

On completion of this chapter you should be able to:
1. Change the screen LIMITS, SNAP, and GRID
2. Use coordinate entry methods and on-screen picking
3. Create simple geometry using LINE, CIRCLE, ARC, and FILLET
4. Use the Display commands ZOOM and PAN
5. Set up the UNITS.

This book is about how to use AutoCAD to make drawings. The information is relevant to all AutoCAD releases from 2000 to 2008. Commands not available before a particular release are noted.

Starting a Drawing in Metric or Imperial

Once you have entered the Drawing Editor, AutoCAD establishes a default working environment. There may also be some 'floating palettes' on the screen (Sheet Sets, Tool Palettes).

Click on the X at the top right corner of each palette to clear these off your screen so it looks like Figure 1.1. Make sure your workspace is AutoCAD Classic (Releases 2007 and 2008).

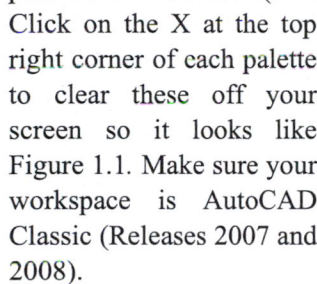

Move your cursor across the screen to the right of the drawing area. If the num-bers in the coordinate readout are under 100, you have opened an imperial file. If they are over 500, you have opened in metric.

Figure 1.1

Changing Imperial and Metric

It is best to start off immediately with the units that you want to use. If you have started in the wrong units, open a new file with acad.dwt (imperial) or acadiso.dwt (metric).

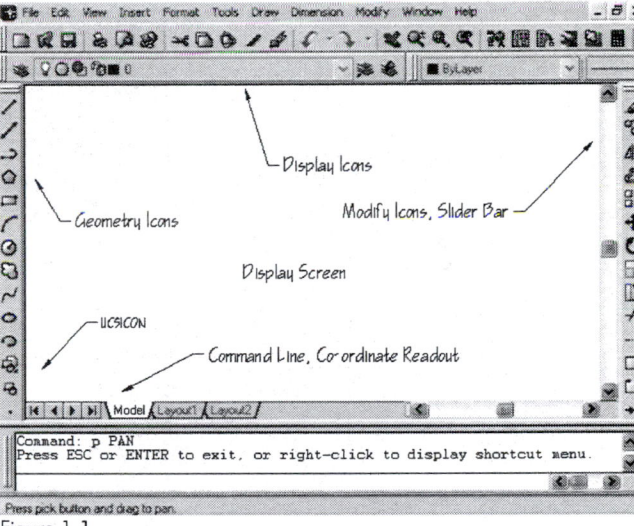

The command STARTUP can also be used. This will prompt for either imperial or metric without the other options.

```
Command:STARTUP
   Enter new value for startup <0>:1
Command:NEW
```

Pick either imperial or metric from the dialog box. If you don't start in the right units, your dimensioning, area, and volume calculations will be difficult.

In AutoCAD it is suggested that you draw everything at full scale or 1:1 scale, and plot the drawing at the required scale factor later.

The UNITS Command

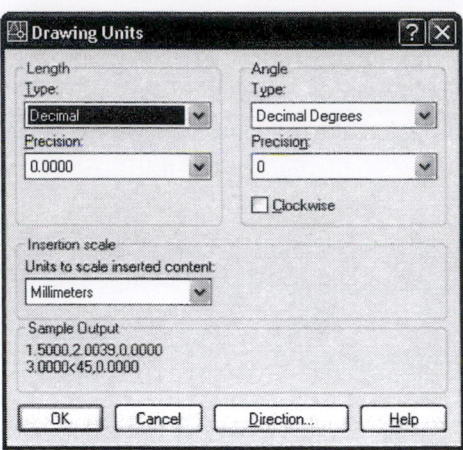

Figure 1.2

Once you have chosen your deired units from the startup menu, you then use the UNITS command to set your readout only.The type of units chosen determines how AutoCAD interprets coordinate and angle command entries. The 'Insertion scale' area indicates again your base units.

AutoCAD offers various types of units of measure for use on your drawings. Before setting up the parameters of the drawing, first set up the units so that the readout displays the required units. Decimal mode may be used for metric units as well as for imperial units. Be sure you have set up your file correctly for the units that you require.

The decimal unit type will display one millimeter for one unit. Specify the number of decimal places for your readout using precision, as in Figure 1.2.

The engineering and architectural modes assume that one drawing unit equals one inch. Again set your precision, as shown in Figure 1.3. Fractional and scientific settings will give a readout in those specific units. Again, the UNITS command only sets the readout. If you are setting your UNITS in inches, but your 'Insertion scale' is millimeters, then you will have problems later.

The UNITS command can be accessed either through the command line or through the UNITS dialog box from the Format pull-down menu at the top of your screen.

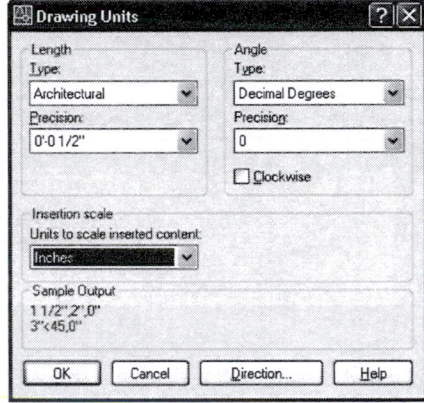

Figure 1.3

Starting to Draw

AutoCAD uses Cartesian coordinates for point entry. The points are set around a determined origin at $X0$, $Y0$, $Z0$. In this case X is 21'10". Y is 5'0", and Z is 0.0000. All points to the right of 0,0 have a positive X value; all points to the left have a negative X. All points above 0,0 have a positive Y value; all points below have negative Y.

Moving the cursor around the screen you will notice that the 0,0 position defaults to the bottom left corner of your screen. See Figures 1.4a and Figure 1.4b.

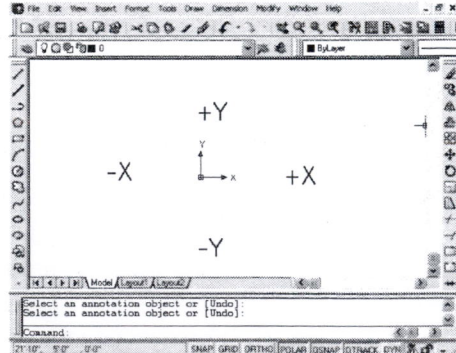

Figure 1.4a

218.2813, -148.6999, 0.0000

Figure 1.4b

Choosing the Origin

The origin or 0,0 should be the most easily accessible point on the design. If a large percentage of the dimensions on a drawing stem from one point, it should be made the origin. The coordinate readout on the bottom of the screen is there to help you find your position. The placement of the origin is important to establish a base for your readouts. It will be more important later when merging files.

To move 0,0 from the bottom of the screen use the PAN command, as in Figure 1.5.

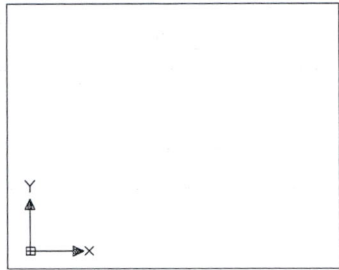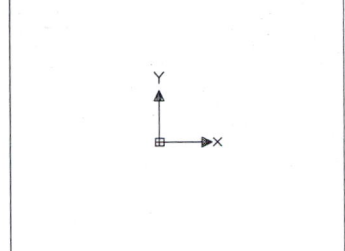

Figure 1.5

Often you can press down on the roller ball of your mouse to get PAN. The command line equivalent is PAN or just P.

`Command: PAN`
`   (drag the icon across the screen to where you want it)`

In architectural drawings the origin is often at the bottom left corner, as in Figure 1.6.

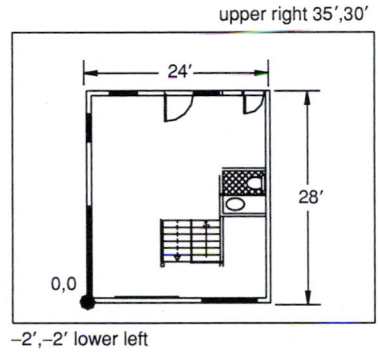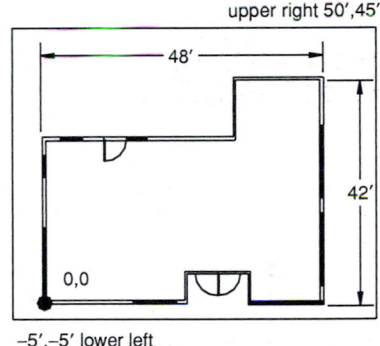

Figure 1.6

In mechanical applications it is often in the center, as in Figure 1.7.

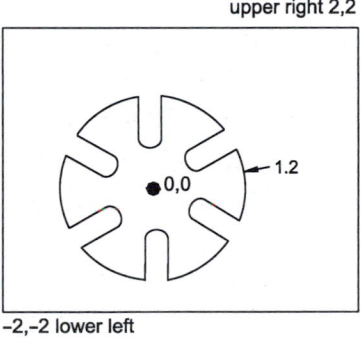

Figure 1.7

Using PAN to get Started

The easiest way to start a file is by using 0,0 as the starting point. Use PAN to move the origin or 0,0 to the center of the screen. Then draw your first object using 0,0 as the first point.

The PAN command is as follows:

Toolbar Within the Standard toolbar choose

Pull-down menu From the View menu, choose Pan....... or hold down the central roller on the mouse and move it.

The command line equivalent is PAN or P.

Once centered, draw a circle, then use ZOOM All to fit it to your screen, as in Figure 1.8. The same can be done using LINE. The CIRCLE command is explained further on page 11, but the commands below will show how it works.

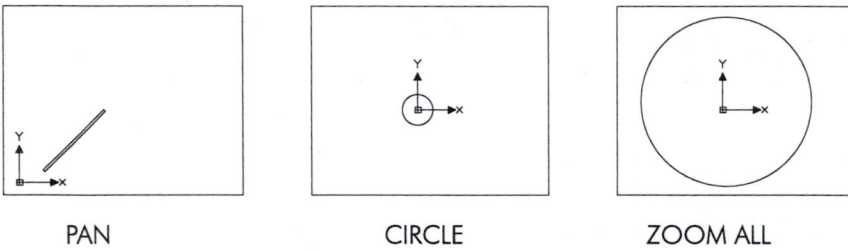

| PAN | CIRCLE | ZOOM ALL |

Figure 1.8

```
Command:PAN (move your 0,0 to the center of the screen as in
   Figure 1.5)
Command:CIRCLE
Specify center point for circle or [3P/2P/Ttr (tan tan
   radius)]:0,0
Specify radius of circle or [Diameter]<1'-0">:5
Command:ZOOM
Specify corner of window, enter a scale factor (nX or nXP),
   or [All/Center/Dynamic/Extents/Previous/Scale
   /Window/Object]<real time>:ALL
```

The LIMITS Command

LIMITS sets a flexible general size for your drawing. LIMITS sets the size of your screen and the area covered by the screen grid. Unlike drawing on paper, you can change the LIMITS size at any time. It simply gives you a place to start and helps provide a visual size that you can identify with.

Toolbar The LIMITS command is not found on the icon menus.

Pull-down menu From the Format menu, choose Drawing Limits.

Drawing Limits

The command line equivalent is LIMITS.

Setting LIMITS does not limit your model; it merely lets you determine how big the finished product might be. You can reset the LIMITS at any time simply by picking new points on the screen. ZOOM All allows you to view the size you have chosen.

A Sample Set Up

A house that is 40' x 36'.

The following commands will center the first line on your screen without LIMITS.

```
Command:LINE
LINE Specify first point:0,0
Specify next point or [Undo]:40',0
Specify next point or [Undo]:↵
Command:ZOOM
Specify corner of window, enter a scale factor (nX or nXP),
   or [All/Center/Dynamic/Extents/Previous/Scale
   /Window/Object]<real time>:ALL
```

You can also draw this using LIMITS.

```
Command:LIMITS
Reset Model space limits
Specify lower left corner or [ON / OFF]
   <0'-0",0'-0">:-5',-5'
Specify upper right corner <12.0000,9.0000>:45',40'
Command:ZOOM
Specify corner of window, enter ..../Object]<real time>:ALL
Command:LINE
LINE Specify first point:0,0
Specify next point or [Undo]:40',0
Specify next point or [Undo]:↵
```

Setting LIMITS, SNAP and GRID

LIMITS sets an overall size for your design. SNAP sets an increment that the cursor will move by. GRID sets a visual aid to help you place objects, and is often set to twice the SNAP value. The grid will extend over the area given by the LIMITS command.

To find GRID and SNAP:

> **Toolbar** Click SNAP and GRID on the status bar at the bottom of the screen to turn them off or on. **Right-click** either GRID or SNAP to change the spacing or angle.

The command line equivalent is SNAP or GRID.

```
Command:LIMITS
Reset Model space limits
Specify lower left corner or [ON/OFF]<0.0000,0.0000>:-5,-40
Specify upper right corner<12.0000,9.0000>:240,180
Command:ZOOM
Specify corner.....Extents/Left/Previous/Scale/Window]<real
   time>:ALL
Command:SNAP
Specify snap spacing (X) or
   [ON/OFF/Aspect/Rotate/Style/Type]<1.0000>:5
Command:GRID
Specify grid spacing or [ON/OFF/Snap/Aspect]<0>:10
```

Entry of Points

All parts of geometry are entered by means of points. Lines have two points each. circles have a center point and a point determining the radius. Arcs have a center point, a radius point, a start point, and an end point.

There are three ways of entering points:

- · by coordinates: absolute values, relative values, or polar values
- · picking them on the screen, with or without SNAP or DYNamic
- · relative to existing geometry

In this chapter we will look only at the first two methods of point entry. The LINE command will be used to illustrate coordinate entries.

The *LINE* Command

Find LINE as follows:

Toolbar From the Draw toolbar, choose the Line button.

Pull-down menu From the Draw menu choose <u>L</u>ine.

The command line equivalent is LINE or the command alias L.

```
Command:LINE or L
```

To create a LINE, you will need to know where it starts and where it ends. Pick two or more points on the screen or enter the coordinates. Terminate the command by pressing the Enter key (↵) .

Coordinate Entry using Absolute, Relative, and Polar Values

The coordinates of an item, the X and Y values, can be entered either relative to the origin (the absolute value of the line) or relative to the last point entered (the incremental value).

Absolute Value Entries

In this method, the origin of the model or drawing does not change: the objects are placed relative to the origin. To enter the absolute value of an item, type in the X value, then the Y value, separated by a comma. You will need to enter two sets of values to draw a line. Press the enter key ↵ to signal the end of the coordinate entry and you should get the line shown in Figure 1.9.

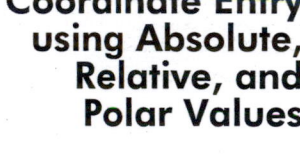

The DYNamic function is very useful but confusing at first. Turn it off for coordinate entry by clicking the icon.

```
Command:LINE
Specify first point:0,0
Specify next point or [Undo]:4,0
```

0,0 ——————— 4,0

DYNamic off

DYNamic on

Figure 1.9

This will draw a line from the absolute position of 0,0 to the absolute position of 4,0.

Relative Value Entries

To enter an incremental or relative value, type the @ symbol (Shift-2) before the number. @ means 'from the last point.'

```
Command:LINE
Specify first point:2,3
Specify next point or [Undo]:@4,0
```

This will draw a line from the absolute position of 2,3 to a position 4 units in positive X from this point.

Try these two examples:

Absolute
```
Command:LINE
Specify first point:0,0
Specify next point or[Undo]:4,0
Specify next point or[Undo]:4,4
Specify next point or [Undo]:0,4
Specify next point or [Undo]:0,0
Specify next point or [Undo]:↵
```

Relative
```
Command:LINE
Specify first point:5,5
Specify next[Undo]:@4,0
Specify next[Undo]:@0,4
Specify next[Undo]:@-4,0
Specify next[Undo]:@0,-4
Specify next [Undo]  :↵
```

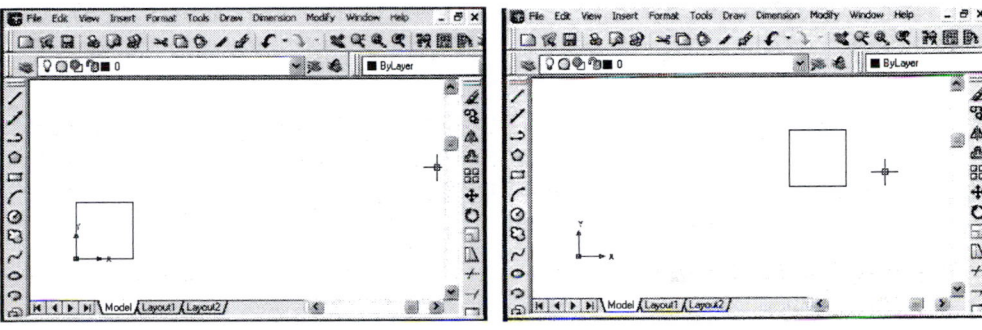

Figure 1.10

The example on the left in Figure 1.10 is a four-unit square starting at 0,0. The example on the right is a four-unit square starting at 5,5. Both squares are created relative to the origin, 0,0.

To draw a line from point 5,6 to point 8.3,6 use either of the following:

Absolute
```
Command:LINE
Specify first point:5,6
Specify next point or[Undo]:8.3,6
```

Relative
```
Command:LINE
Specify first point:5,6
Specify next[Undo]:@3.3,0
```

In choosing between the absolute and the incremental method, the deciding factor is what you know. If you know that the final point is going to be 8.3,6, use the absolute value. If you know that the line is going to be 3.3 units in positive X from the last point, then enter the incremental coordinates. AutoCAD will do the calculations.

Polar Value Entries

Polar coordinates allow you to enter an item, relative to the last item, at a specified length and angle. Angles are normally calculated counterclockwise from the positive *X* direction, as shown in Figure 1.11.

```
Command:LINE
Specify first point:3,4
Specify next point or [Undo] :@4<45
```

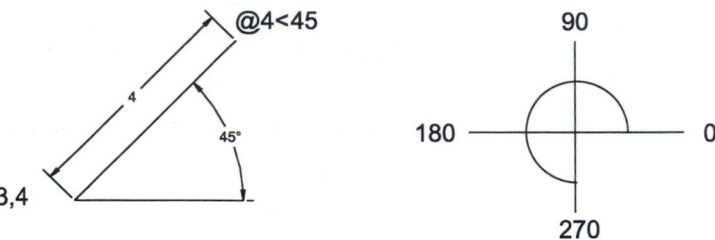

Figure 1.11

Where: @ = relative to the last point
        4 = the length of the line
        < = angle
      45 = the angle that the line will be drawn at; all angles are calculated
              counterclockwise

Try this example:

```
Command:LINE
Specify first point:6,0
Next point:@2<0
Next point:@3<90
Next point:@2<0
Next point:@1<270
Next point:@2<0
Next point:@2<90
Next point:@6<150
Next point:@1<210
Next point:C (for close)
```

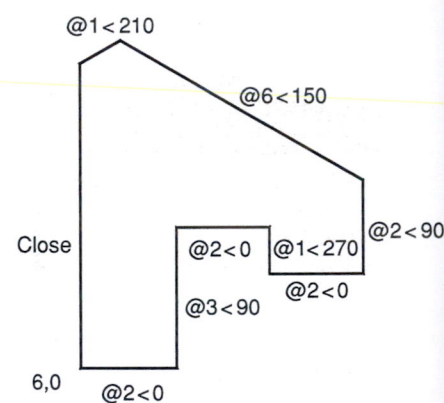

Figure 1.12

As noted above, angles are calculated counter-clockwise from the furthest point in positive *X*.

Coordinate Entry using SNAP, ORTHO, POLAR, and DYNAMIC

Functions that can help you enter your drawing information are found at the bottom of your screen. The lefthand button on the mouse will enter a point every time you press it while in a draw command. You can make your digitizing or picking of points much easier and more accurate by using functions such as SNAP, POLAR and ORTHO. The function bar is shown in Figure 1.13.

| SNAP | GRID | ORTHO | POLAR | OSNAP | OTRACK | DUCS | DYN | LWT | MODEL |

Figure 1.13

SNAP

With SNAP you can draw lines, arcs or circles at preset integers. SNAP can also be set at an angle or on an isometric. GRID follows the SNAP settings. More advanced SNAP functions are found in Chapter 2. If you set the SNAP to .25, all entries will be rounded to the nearest .25 interval as shown in Figure 1.14.

SNAP on

SNAP off

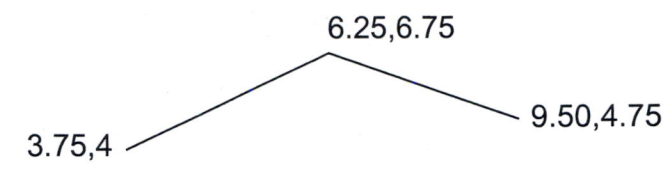

6.25,6.75

3.75,4

9.50,4.75

Figure 1.14

The toggle turns SNAP off and on, you need the command to change the size.

```
Command:SNAP
Specify snap spacing (X) or
   [ON/OFF/Aspect/Rotate/Style/Type]<1.0000>:.25
```

Try repeating the examples on pages 7 and 8 using the mouse and setting the SNAP value to 1. If the coordinate readout does not move, press F6. Remember that it is a three-way toggle; off, absolute, and incremental. If you set SNAP to 1, all the points you digitize or pick from the screen will be accurate to one-unit integers. You cannot be accurate without using SNAP and/or POLAR.

ORTHO

With the ORTHO option (F8 or the ORTHO button), lines can only be drawn vertically or horizontally. Draw a LINE across your screen. Keep adding segments to the line and turn ORTHO on. You will notice that the cursor only goes vertically and horizontally. By turning ORTHO off, you will be able to draw diagonal lines again.

The GRID (F7 or the GRID button) gives you a visual display of distance.

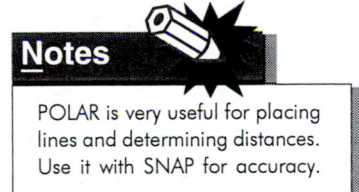

Notes

POLAR is very useful for placing lines and determining distances. Use it with SNAP for accuracy.

POLAR

Now turn ORTHO off and POLAR on. As shown in Figure 1.15, POLAR allows lines to be drawn vertically or horizontally at a given distance and it gives the incremental distance from the last entered point. The button is on when it looks pushed in.

Polar: 13.0309 < 0°   Polar: 6.1133 < 90°

POLAR

Figure 1.15

```
Command:LINE
Specify first point:(pick a point) (with POLAR on move your
   cursor to the right)
Specify next point or [Undo]:60
Specify next point or [Undo]:50(move your cursor up)
```

The line will be drawn 60 units in positive *X* and 50 units in positive *Y*. POLAR will allow length entry for lines going 0, 90, 180 or 270 degrees. Leave these lines on screen.

DYNamic

The DYNamic function displays the command line beside the cursor. DYNamic also gives you a dynamic angle readout. You can draw lines at anglesin a similar way to drawing using the POLAR option. Beware, however, because these angle readouts are not given in decimal places; lines drawn in using DYNamic are not always accurate. Change your angle precision in UNITS for more accuracy. Place one line beside the otheras i n Figure 1.16. Zoom in to see how accurate they are.

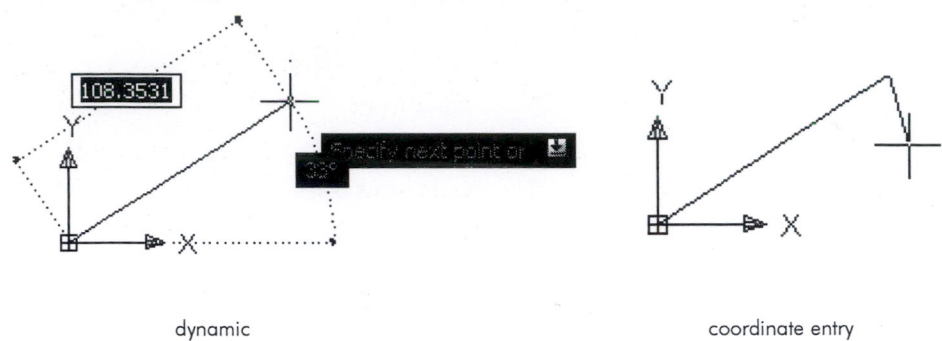

dynamic   coordinate entry

Figure 1.16

```
Command:LINE                          Command:LINE
Specify first point:0,0               Specify first point:0,0
Specify next point:100                Specify next point:@100<33
```

The second line in Figure 1.16 is accurate, the first is not. The line put in using the DYNamic angle is approximately 33 degrees, but is actually placed whereever your cursor is between 33 and 34 degrees.

The DYNamic readout is useful for those who like to read their commands on the display screen rather than on the command line. In Figure 1.17, the ERASE command is being used and the user is selecting the objects to be erased.

```
Command:ERASE
Select objects:(pick 1)
Select objects: ↵
```

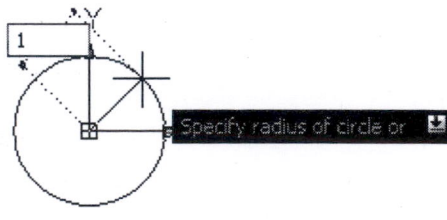

Figure 1.17

With geometry commands, you are prompted for each point entry or value.

The value box will turn from blue to white as you enter the value, as shown in the radius option of Figure 1.18.

```
Command:Circle
Specify center point
   for circle or
   [3P/2P/Ttr (tan
   tan radius)]:0,0
```

Figure 1.18

Geometry Commands

All geometry commands are similar in the way they are entered. The user picks the command LINE, CIRCLE, ARC, etc., and then AutoCAD will prompt for the points or distances needed to create that geometry.

The LINE Command

The LINE command is as simple as the above examples indicate. With either a pick on the screen or a coordinate position you can tell AutoCAD where each point should be. Any combination of points is accepted. Use the enter key ↵ to exit the command.

> **Toolbar** From the Draw toolbar, choose
>
> **Pull-down menu** From the Draw menu, choose Line.

The command line equivalent is LINE or L.

```
Command:LINE
Specify first point:(pick a point)
Specify next point or (Undo):@3<250
Specify next point or (Undo):(pick another point)
Specify next point or (Undo):↵ (Enter)
```

When drawing lines, you are creating objects that are described by two points: a beginning and an end. Any number of points can be entered in the LINE command with each point joined to the last by a separate line. If you have entered five or six points in a single command, any of the lines can be erased.

LINE Options

C will close the string of lines with a line from the last point to the first point.

U can be entered within the command line to undo the last entered point.

The CIRCLE Command

When drawing a CIRCLE you are also describing an object that has two points; a center and a radius. An ARC has four points: a center, a radius, a start, and an end.

The CIRCLE command will prompt you for the information needed to complete the circle.

> **Toolbar** From the Draw toolbar, choose .
>
> **Pull-down menu** From the Draw menu, choose Circle – Center, Radius.

The command line equivalent is CIRCLE or C.

Figure 1.19 shows a circle with a radius of 4.

```
Command:CIRCLE
Specify center point for circle or
   [3P/2P/Ttr (tan tan radius)]:(pick a
   point or type 0,0)
Specify radius of circle or
   [Diameter]<1'-0">:(pick another point
   or type in a radius value, for
   example 4)
```

0,0   4,0

Figure 1.19

Where: 3P = a circle fitted through three points
2P = a circle fitted through two points
Ttr = a circle that is tangent on its diameter to two selected objects
indicated with a specified radius

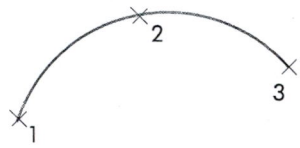

Notes

Turn OSNAP off if your cursor keeps going to another object on ..the screen.

Options appear when you type in C or CIRCLE. The default is to have a circle defined by the radius. Type in D then space if you prefer to enter a diameter. When picking CIRCLE from either the screen menu or the pull-down menu, you will be prompted for one of the options listed above.

The ARC Command

Arcs are also created by using options to control how the ARC is entered. The default is to define the first, then the second, then the third or final point of an arc.

Toolbar From the Draw toolbar choose

Pull-down menu From the Draw menu choose Arc – 3 point

The command line equivalent is ARC or A.

The default arc is created through three contiguous points. Pick three points on the screen after initiating the ARC command as shown in Figure 1.20. The points can be entered with SNAP on, by coordinate entry, or by using existing objects with OSNAP, as shown in Chapter 2.

Figure 1.20

```
Command:ARC
ARC Specify start point of arc or [Center]:(pick point - 1)
Specify second point of arc or [Center/End]:(pick point - 2)
Specify end point of arc:(pick a third point - 3)
```

There are many variations on the ARC command.

You can access all ARC options through the Draw pull-down menu under Arc. Should you want to enter the options at the command line, simply type in the option that you want. Only the first letter of the option is required: e for end, r for radius, etc. Put in the arc using the Start End Radius formula, as shown in Figure 1.21.

```
Command:ARC
ARC Specify start point of arc or
   [Center]:(pick 1)
Specify second point of arc
or [Center/End]:E
Specify end point of arc:(pick 2)
Specify center point of arc or
   [Angle/Direction/Radius]:R
Specify radius of arc <0.00>:(pick
   3)
```

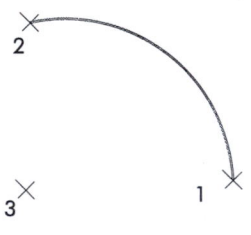

Figure 1.21

ARCs will be calculated counterclockwise.

The FILLET Command

The FILLET command provides an easy way to place an arc between two existing objects, usually lines. FILLET can also be used with radius 0 to clean up corners and connect lines to an apex.

Toolbar From the Modify toolbar choose

Pull-down menu From the Modify menu choose Fillet.

The command line equivalent is FILLET or F.

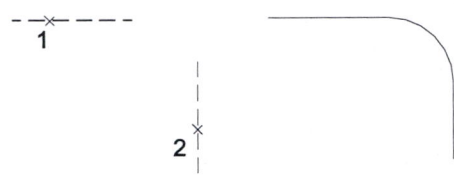

Figure 1.22

```
Command:FILLET
Current settings: Mode = Trim,Radius = <5.0000>
Select first object or
   [Undo/Polyline/Radius/Trim/Multiple]:(pick 1)
Select second object:(pick 2)
```

In Figure 1.22 the default radius of 5 was used. To change the radius choose the option R. Always press the enter key ↵ after letter options. In the example below the option R to change the radius, has been used to create a FILLET with the new radius value of 12. Note that R must be followed by a space. R indicates that you would like to enter a new radius value, the actual number is entered on the next line.

```
Command:FILLET
Current settings: Mode = Trim,Radius = <5.0000>
Select first object or
   [Undo/Polyline/Radius/Trim/Multiple]:R↵
Specify fillet radius:12
Select first object or
   [Undo/Polyline/Radius/Trim/Multiple]:(pick 1)
Select second object:(pick 2)
```

View Commands

The View menu and toolbar offer commands which will change the display of the model or drawing relative to the screen. Commands from the View menu will not change the coordinates or position of the model or the database. They only change the way you look at it. The following commands appear in the View menu:

ZOOM	=	magnifies a section of the screen
PAN	=	moves the model across the screen without changing the magnification factor (zoom)
REDRAW	=	updates the view and erase marks
REGEN	=	recomputes the file

If your scroll bars are not displayed, go to the Tools pull-down menu and choose Options, and then the Display tab.

The ZOOM Command

ZOOM is accessed by typing it on the command line, using the slider bars on the top and bottom of the screen, or using the pull-down menus or the standard toolbar. For the ZOOM options on the toolbar, hold the Zoom Out button down and the options will be shown.

ZOOM is a transparent command which means it can be used with the icons below within a command string.

Toolbar From the standard toolbar choose ZOOM All

ZOOM Window

ZOOM Previous

The command line equivalent is ZOOM or just Z.

```
Command:ZOOM (or Z)
Specify corner of window, enter a scale factor (nX or nXP), or
[All/Center/Dynamic/Extents/Previous/Scale/Window/Object]<re
    al time>:
```

The options for the Zoom command are:

All (A)	=	expands or shrinks the model or drawing to fit onto the screen relative to your limits
Center (C)	=	centers the model on the screen; you mustenter a magnification factor or ''height''
Dynamic (D)	=	creates a dynamic display of the item for zooming
Extents (E)	=	expands or shrinks the model or drawing to fit all of the objects on screen
Previous (P)	=	returns you to the Previous zoom factor
Realtime	=	zooms interactively to a logical extent. Activate with either the button or a right-click
Window (W)	=	describes by two diagonal points a rectangle around the area you want to view
Scale nX	=	specifies a percentage of the existing size
Scale nXP	=	specifies a size relative to paper space

Many mice have a roller ball that acts as a ZOOM function. Roll the roller ball to change the Zoom factor, or press down on it and move it across the screen to PAN. In Release 2008 the ZOOM function on the roller ball sometimes malfunctions. Use ZOOM All or ZOOM Extents to center your drawing on the screen.

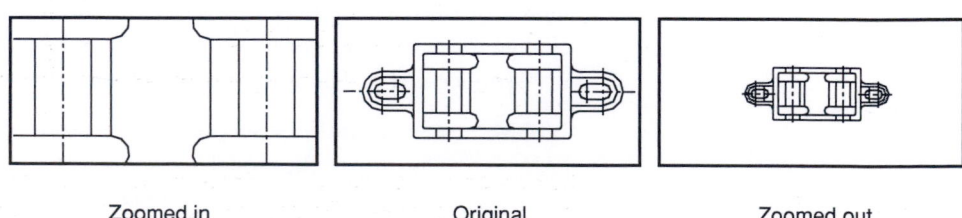

| Zoomed in | Original | Zoomed out |

Figure 1.23

In Figure 1.23 the object is centered on the screen using ZOOM ALL. If your object gets lost use ZOOM ALL to get it back on screen.

ZOOM Window

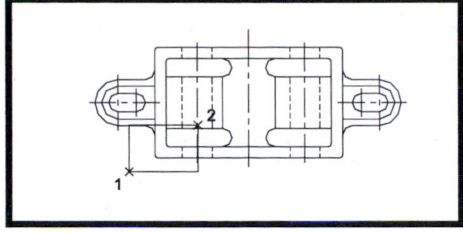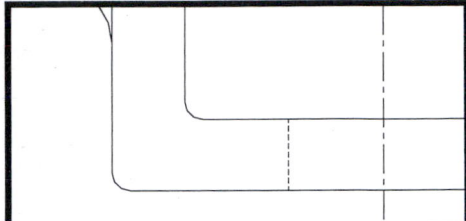

Figure 1.24

```
Command:ZOOM (or Z)
Specify corner of window, enter a scale factor (nX or nXP),
    or [All/Center/Dynamic/Extents/Previous/Scale
    /Window]<real time>:W
Specify first corner:(pick 1)
Specify opposite corner:(pick 2)
```

ZOOM Extents

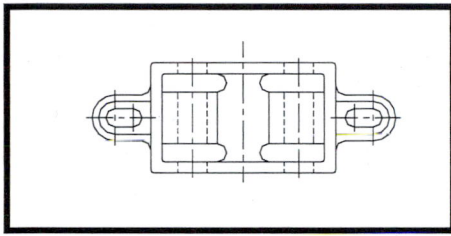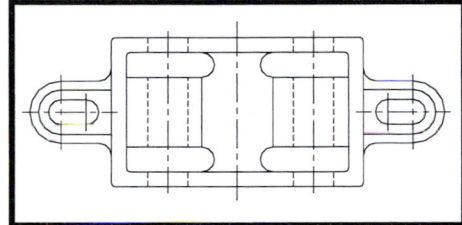

Figure 1.25

```
Command:Z
Specify corner of window, enter a scale factor (nX or nXP),
    or [All/Center/Dynamic/Extents/Previous/Scale
    /Window]<real time>:E
```

ZOOM with scale

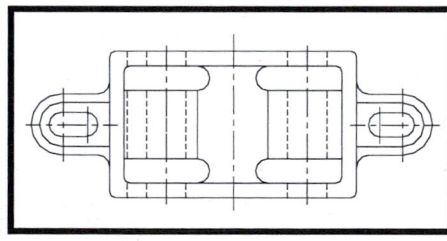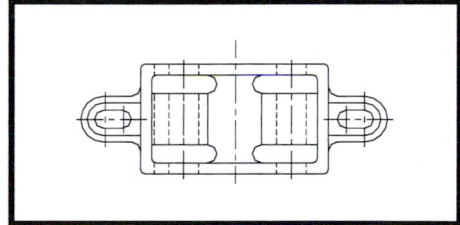

Figure 1.26

Scale works like this: .8x will display an image at 80% of its current size; .5x will display an image at half the current size; and 2x will display an image twice the size of the current size.

```
Command:Z
Specify corner of window, enter a scale factor (nX or nXP),
    or[All/Center/Dynamic/Extents/Previous/Scale
    /Window]<real time>:.5x
```

Zooming In and Out

Zooming in doubles the size of the image, zooming out reduces the image by half.

Zoom Limits shows the screen limits. If you have a center rolling wheel on your mouse you can zoom in and out quickly, but ZOOM Window is usually more efficient.

ZOOM All

To display the entire drawing, use ZOOM All or ZOOM Extents.

```
Command:ZOOM
Specify corner of window, enter a scale factor (nX or nXP),
    or[All/Center/Dynamic/Extents/Previous/Scale/Window]<real
    time>:A
```

The PAN Command

To move the view across the screen without changing the display size, use PAN as shown in Figure 1.27.

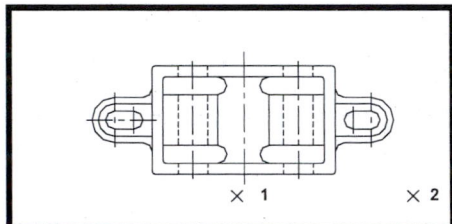 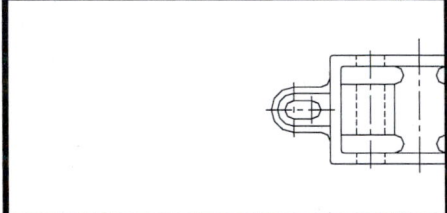

Figure 1.27

```
Command:PAN (or P)
Press ESC or ENTER to exit, if you have a roller on your
    mouse, just press down and move it to PAN.
```

PAN and ZOOM

The PAN command moves the model across the screen, while ZOOM magnifies the model within the screen. The database, i.e.the 0,0,0 point and associated coordinate points, remain the same.

The roller ball can be very useful in placing the image on the screen, but be sure to experiment with ZOOM All and ZOOM with a window (pick two points around what you want to look at) and you may find these useful as well.

Windows Scroll Bars

Windows scrollbars can be used instead of the PAN command. To move the drawing up, pick the down arrow on the vertical scroll bar. To move the drawing to the left, pick the right arrow on the scroll bar. If you have a smaller screen, you may want to have the scroll bars not displayed. To remove them from the screen, choose the Tools menu, then Options, then Display. Remove the check mark from the box beside Display Scroll Bars.

REGEN

While your data are always available, to save memory, they are not always completely generated. The REGEN command (RE) is used to update arc and circle displays to make the objects look more rounded. If your arcs are choppy or squared, use REGEN to update the screen to the current magnification factor and display a superior image.

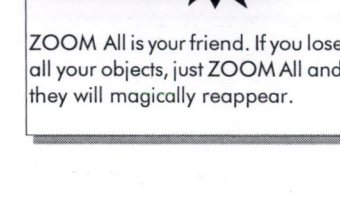

Notes

ZOOM All is your friend. If you lose all your objects, just ZOOM All and they will magically reappear.

Alternate Units

It is important to have the correct nomenclature when entering geometry.

Architectural

```
Command:LINE
Specify first point:2,3 (X 2 inches, Y 3 inches)
Specify next point or (Undo :@4',0 (4 feet in X, none in Y)
Specify next point or (Undo):@0,2'3 (none in X, 2 feet 3
    inches in Y)
Specify next point or (Undo):@3'4-1/2",0 (3 feet 4 1/2
    inches in X, none in Y)
```

Fractional

With fractional units the information is entered as a fraction with a slash (/). For mixed numbers a hyphen (-) must be added with the slash (/).

```
Command:LINE
Specify first point:1-1/2,2-3/4
Specify next point or (Undo):@3/4,0
```

Surveying

The surveyor's compass rose is much the same as a ship's compass. It is divided into four parts with the top being north, the left being west, etc. Angles are expressed in 90 degree quadrants.

The quadrant between north and east, for example, starts at 0 degrees due east and progresses 90 degrees to due north. To express 25 decimal degrees using AutoCAD's default origin for angles, enter N25d0'0"E. You may omit null minutes and/or seconds and enter N25dE.

When entering this measurement, do not use spaces.

```
Command:L
Specify first point:0,0
Specify next point or (Undo):@38<S44d14'9"W
```

Measuring Angles

AutoCAD's default setting for angles is zero degrees at due east. You may change this zero-degree reference point to due north, due west, or due south. These are the only four positions offered by the UNITS command. To orient the zero reference at an angle other than those specified, you can change the user coordinate system (UCS) as explained later.

Untranslated Angles: If the UNITS command is set to a nondecimal angular mode (e.g. radians), an angle can be preceded by a 'less than' symbol < to enter a measurement counterclockwise from 3 o'clock.

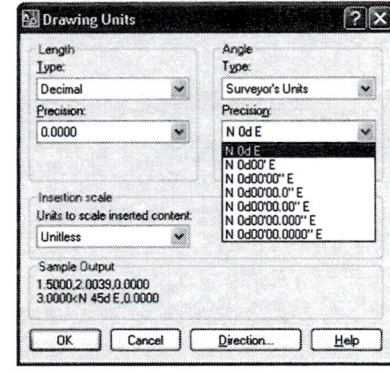

Figure 1.28

If an angle measurement direction or origin has been changed, enter < before an angle measurement to have the angle measured counterclockwise from 3 o'clock.

Change the precision of your angle readouts to show up on the POLAR and DYNamic readouts, as shown in Figure 1.28.

Open AutoCAD. If AutoCAD is already running, pick File from the pull-down menu (top left), then New to start a new file in metric by using **acadiso.dwt**.

Step 1

Use the Tools pull-down menu and Drafting Settings or the following typed commands to set your LIMITS and SNAP. Then pick ZOOM All from the View pull-down menu, use the icon shown, or type it in.

The bold type in the following commands is the user response or what *you* type in. Use the Enter Key ↵ to enter each command or value.

```
Command:LIMITS↵
Reset model space limits
Specify lower left corner or [ON/OFF]<0.0000,0.0000>:-5,-5↵
Specify upper right corner<12.0000,9.0000>:100,60↵
Command:(from the View pull-down menu, pick ZOOM All)
Command:SNAP
Specify snap spacing (X) or
   [ON/OFF/Aspect/Rotate/Style/Type]<1.0000>:5
```

Step 2

Start by drawing a series of lines using absolute coordinates. Turn DYNamic off.

```
Command:L↵
Specify start point:0,0↵
Specify next point or [undo]:75,0↵
Specify next point or [undo]:75,50↵
Specify next point or [undo]:0,50↵
Specify next point or [undo]:c↵
Command:ZOOM
Specify corner of ......Scale/Window]<real time>:Extents
Command:ZOOM
Specify corner of ......Scale/Window]<real time>:.8x
```

Step 3

Make sure that ORTHO, and OTRACK are off, but that POLAR and SNAP are on.

| SNAP | GRID | ORTHO | POLAR | OSNAP | OTRACK | DUCS | DYN | LWT | MODEL |

Notes

When entering coordinates, make sure your DYN button is turned off! Its on the bottom line of your screen

Then draw in the next lines using POLAR to complete an inside rectangle. Move your cursor directly up, down, right or left from the last entered point.

```
Command:L↵
Specify start point:5,5↵
Specify next point or
   [undo]:(move your cursor to the right)65↵
Specify next point or [undo]:(move your cursor up)40↵
Specify next point or [undo]:(move your cursor left)65↵
Specify next point or [undo]:c↵
```

Step 4

Enter a line at an angle from 10,10.

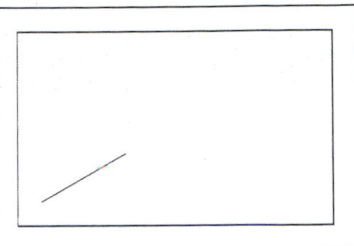

```
Command:L
Specify start point:10,10
Specify next point or [undo]:@20<30
Specify next point or [undo]:↵
```

Then another from the opposite direction.

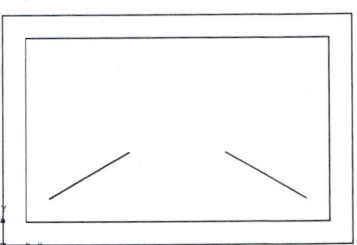

```
Command:L
Specify start point:50,10
Specify next point or [undo]:@20<150
Specify next point or [undo]:↵
```

Step 5

Notice that the ends of the diagonal lines are not near the snap points. ZOOM into the end of the line on the left to start adding circles. Use either the icon or type the letter Z.

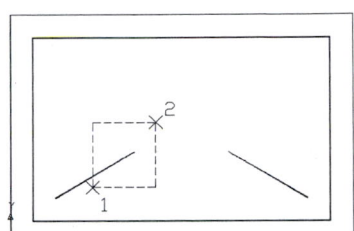

```
Command:Z
Specify corner of window, enter a
  scale factor (nX or nXP), or
[All/Center/Dynamic/Extents/Previo
  us/Scale/Window]<real time>:(pick 1, pick 2)
```

Turn DYNamic mode on.

```
OTRACK DUCS DYN LWT MODEL
```

Make sure SNAP is on too. Use the CIRCLE command with DYNamic.

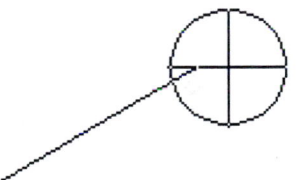

```
Command:CIRCLE
Specify center point for circle
  or [3P/2P/Ttr (tan tan
  radius)]:(pick the snap point
  to the right of the end of the
  diagonal line)
Specify radius of circle or
  [Diameter]<1.0000>:2.7
```

The radius of the circle will snap to the closest snap point. Your radius entry at the command line will override this. When asked for the radius, use ↵ to accept the default.

```
Command:CIRCLE
Specify center point for circle or [3P/2P/Ttr (tan tan
  radius)]:(pick the snap point shown)
Specify radius of circle or [Diameter]<2.7000>:↵
```

Step 6   Now use ZOOM All to place your total image within your screen again.

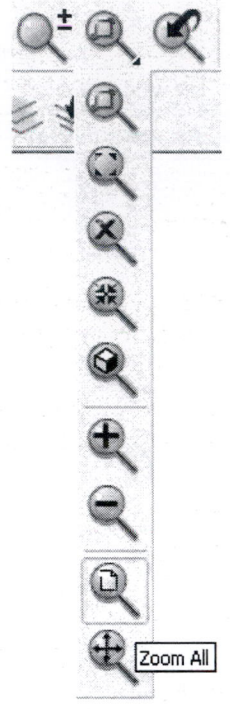

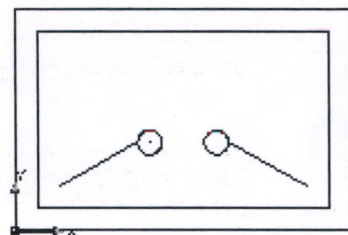

Then use CIRCLE with the Ttr option to add a circle in the center.

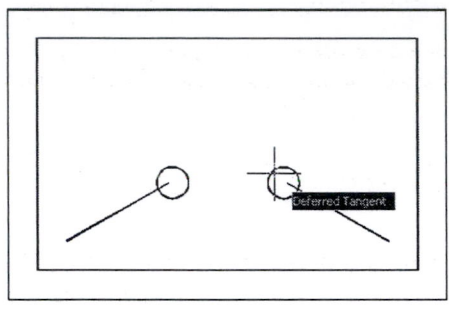
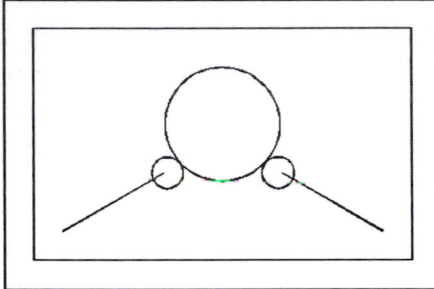

```
Command:CIRCLE
Specify center point for circle or [3P/2P/Ttr (tan tan
    radius)]:Ttr
Specify point on object for first tangent of circle:(pick
    the first circle)
Specify point on object for first tangent of circle:(pick
    the other circle)
Specify radius of circle <2.7000>:10
```

Step 7   Finally add some FILLETs to the inside rectangle.

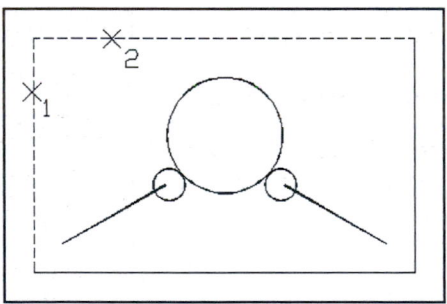
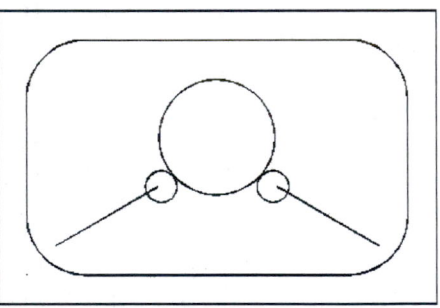

```
Command:FILLET
Current settings: Mode = Trim,Radius = <5.0000>
Select first object or
    [Undo/Polyline/Radius/Trim/Multiple]:r
Specify fillet radius:10
Select first object or [Undo/...Multiple]:(pick 1)
Select first object or [Undo/...Multiple]:(pick 2)
```

Continue to fillet the edge until your drawing looks like the one on the right above.

In this tutorial we will create a drawing in imperial measurements without using LIMITS or SNAP. Make sure your status bar looks like this: only MODEL and POLAR on:

| SNAP | GRID | ORTHO | POLAR | OSNAP | OTRACK | DUCS | DYN | LWT | MODEL |

Step 1

LIMITS, GRID and SNAP are very useful for placing items that are rectilinear with regular integers. For many applications, they are simply not needed. Here is how to get started without them.

First open a new file and make sure that the units are imperial. Pick either File, then New, or pick the icon from the Standard toolbar which should be on the top left of your screen.

```
Command:NEW
```

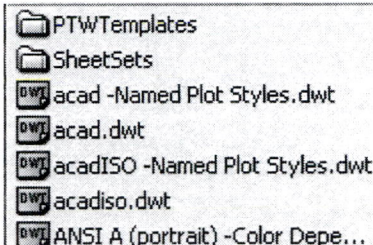

Then choose **acad.dwt** from the list of template options.

Another way to get started is to use STARTUP.

```
Command:STARTUP
Enter new value for
   startup <0>:1
Command:NEW
(Choose imperial)
```

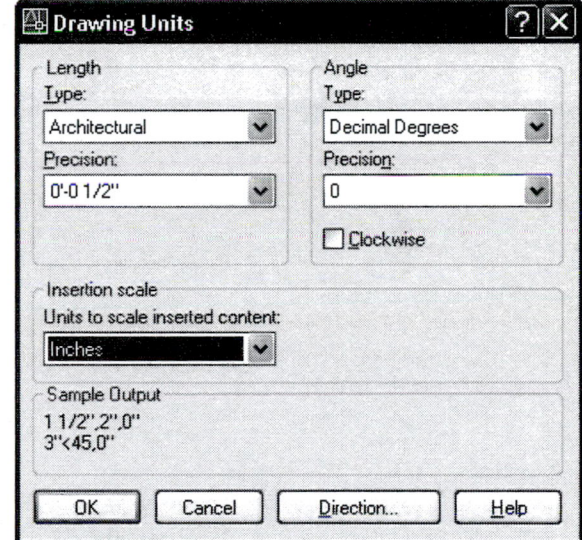

Step 2

Now set your UNITS to architectural.

```
Command:UNITS
```

Set your units to architectural.

Choose the ½″ under Precision. There is no need to change the Angle readout.

Choose OK.

Step 3

Now draw a CIRCLE, then use ZOOM to place it on the page.

```
Command:CIRCLE
Specify center point for circle or [3P/2P/Ttr (tan tan
   radius)]:0,0
Specify Radius of circle or [Diameter]:4'
```

Don't forget the foot symbol. AutoCAD defaults to inches, so if you only enter 4 as your value it will be accepted as four inches not four feet.

Step 4 You will not be able to see the object on screen, so use ZOOM to place it.

Use the roller on your mouse, the icons, or simply Z from your keyboard to start the ZOOM command.

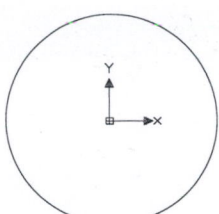

```
Command:Z
Specify corner of window, enter a
  scale factor (nX or nXP),or
  [All/Center/Dynamic/Extents/Previous
  /Scale/Window]<realtime>:A
Command:Z
Specify corner  of  window,  enter  a
  scale factor (nX or nXP), or
[All.....ow]<real time>:.5X
```

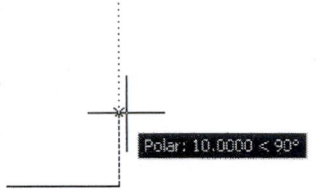

This centers your circle on the screen. Let's assume this circle is a patio table, now let's make chairs.

Step 5 Use the LINE command to create a small rectangular chair. Use POLAR to make the chair 24" × 18". Just move the cursor directly right, then up, and type in the value.

```
Command:LINE
Specify first point:(pick a point to start)
Specify next point or [Undo]:24 (move the cursor right)
Specify next point or [Undo]:18 (move the cursor up)
Specify next point or [Undo]:24 (move the cursor left)
Specify next point or (Undo):c
```

Step 6 Now use CIRCLE and ARC to create another type of chair.

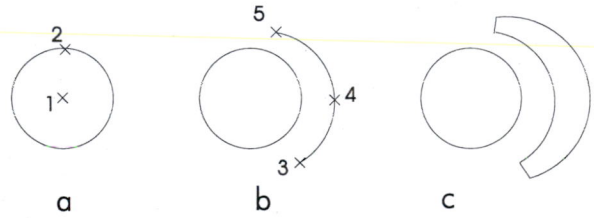

a   b   c

```
Command:CIRCLE
Specify center point for circle or [3P/2P/Ttr (tan tan
  radius)]:(pick 1)
Specify radius of circle or [Diameter]<1'-0">:(pick 2)
```

```
Command:ARC
ARC Specify start point of arc or
  [Center]:(pick 3)
Specify second point of arc or
  [Center/End]:(pick 4)
Specify end point of arc:(pick 5)
Command: (from the View pull-down
  menu, pick ZOOM All)
```

Step 7 The only way to make progress is to practice. Use ZOOM to zoom in and out, creating more chairs.

Exercise 1 Units Practice 1

Open a NEW file and set the UNITS accordingly.
PAN the origin (0,0) onto your screen and draw in the first line starting at 0,0.
Use ZOOM All or your roller ball to adjust the size of your image.

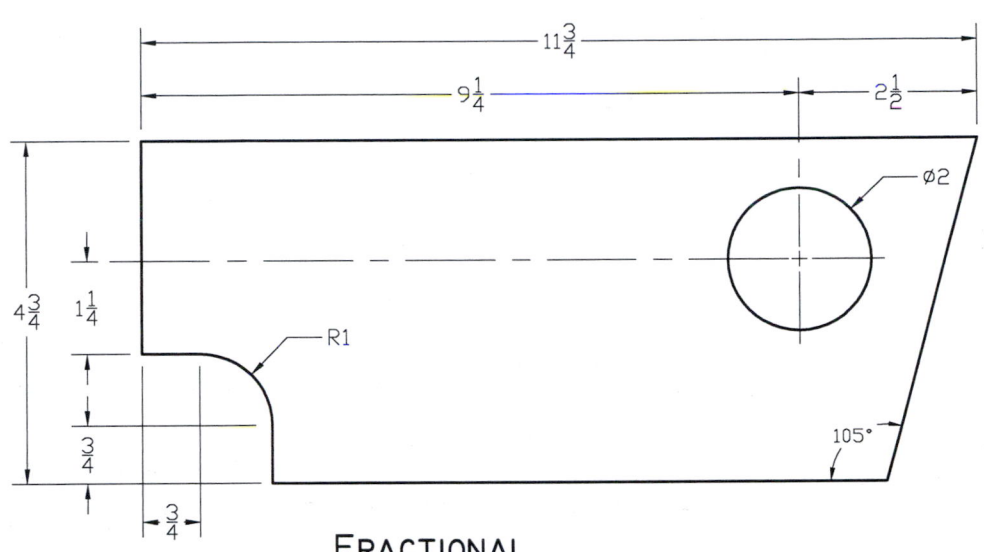

FRACTIONAL

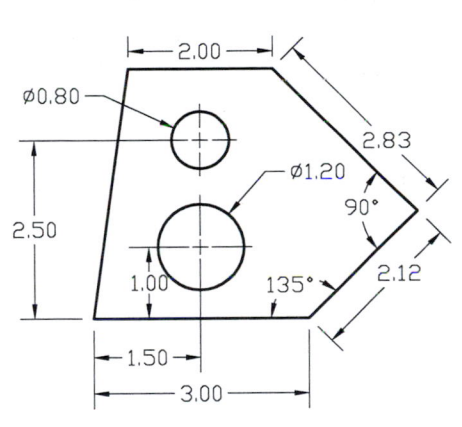

DECIMAL

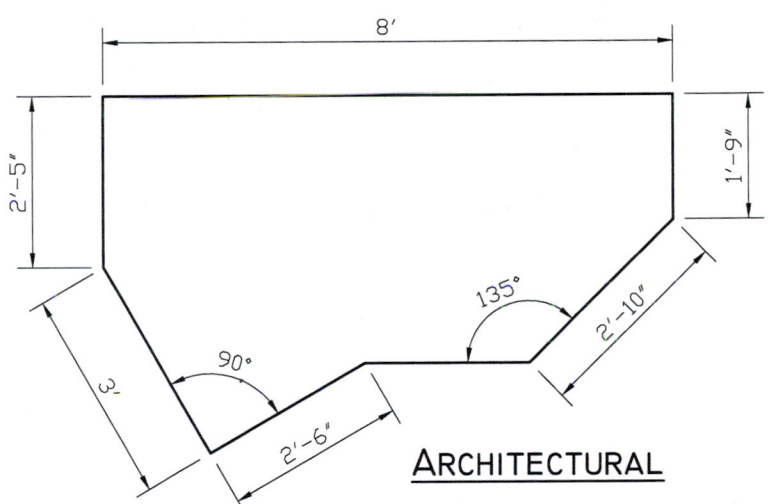

ARCHITECTURAL

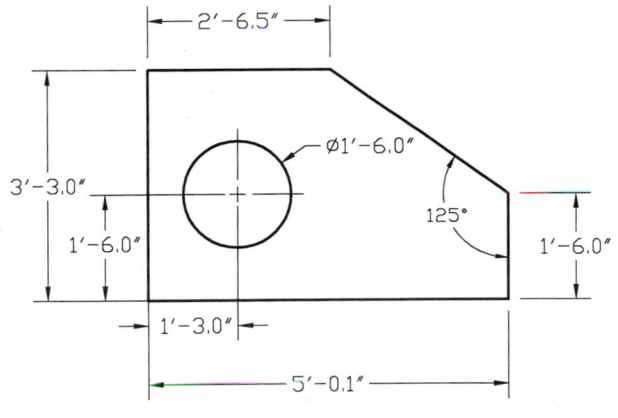

ENGINEERING

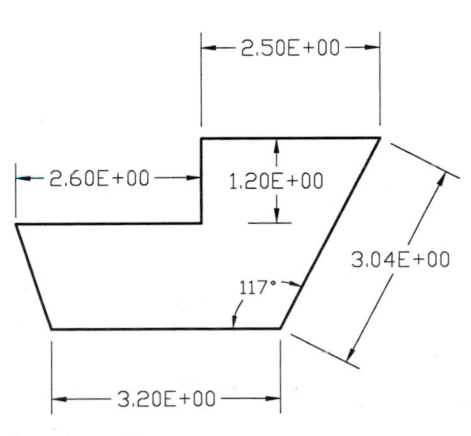

SCIENTIFIC

Exercise 1 Units Practice 2

Open a file in imperial units - acad.dwt - and set your UNITS.
Use PAN to move the object onto your screen and start drawing from 0,0.

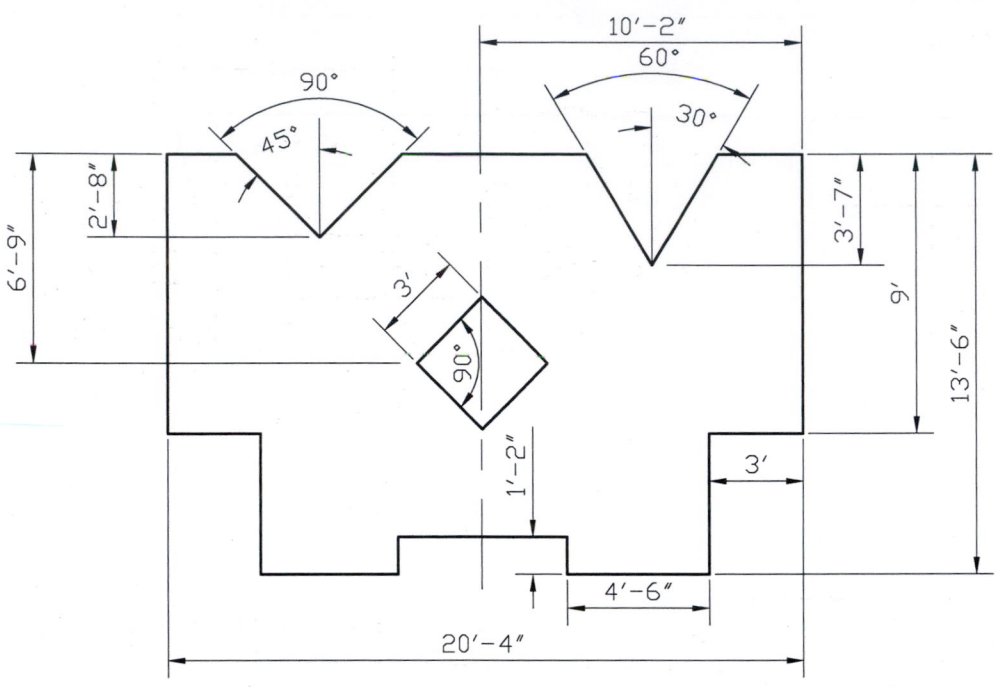

ARCHITECTURAL

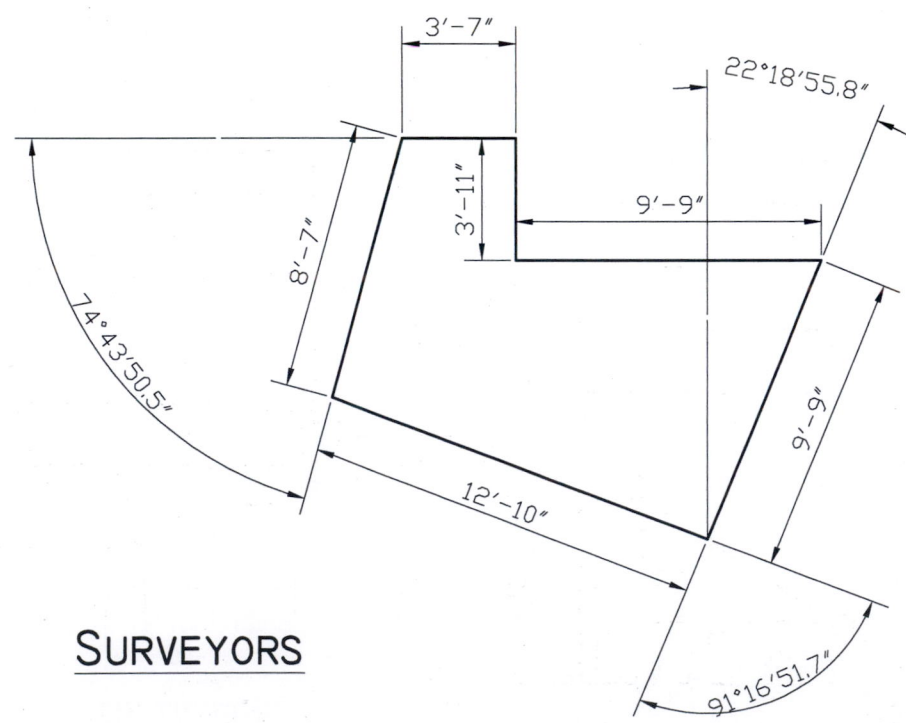

SURVEYORS

Exercise 1 Practice

LIMITS may be used to get started, or put in the first line and ZOOM ALL.
GRID and SNAP make these much easier. Don't forget POLAR.
These objects are drawn individually as practice. They are not scaled to the page.

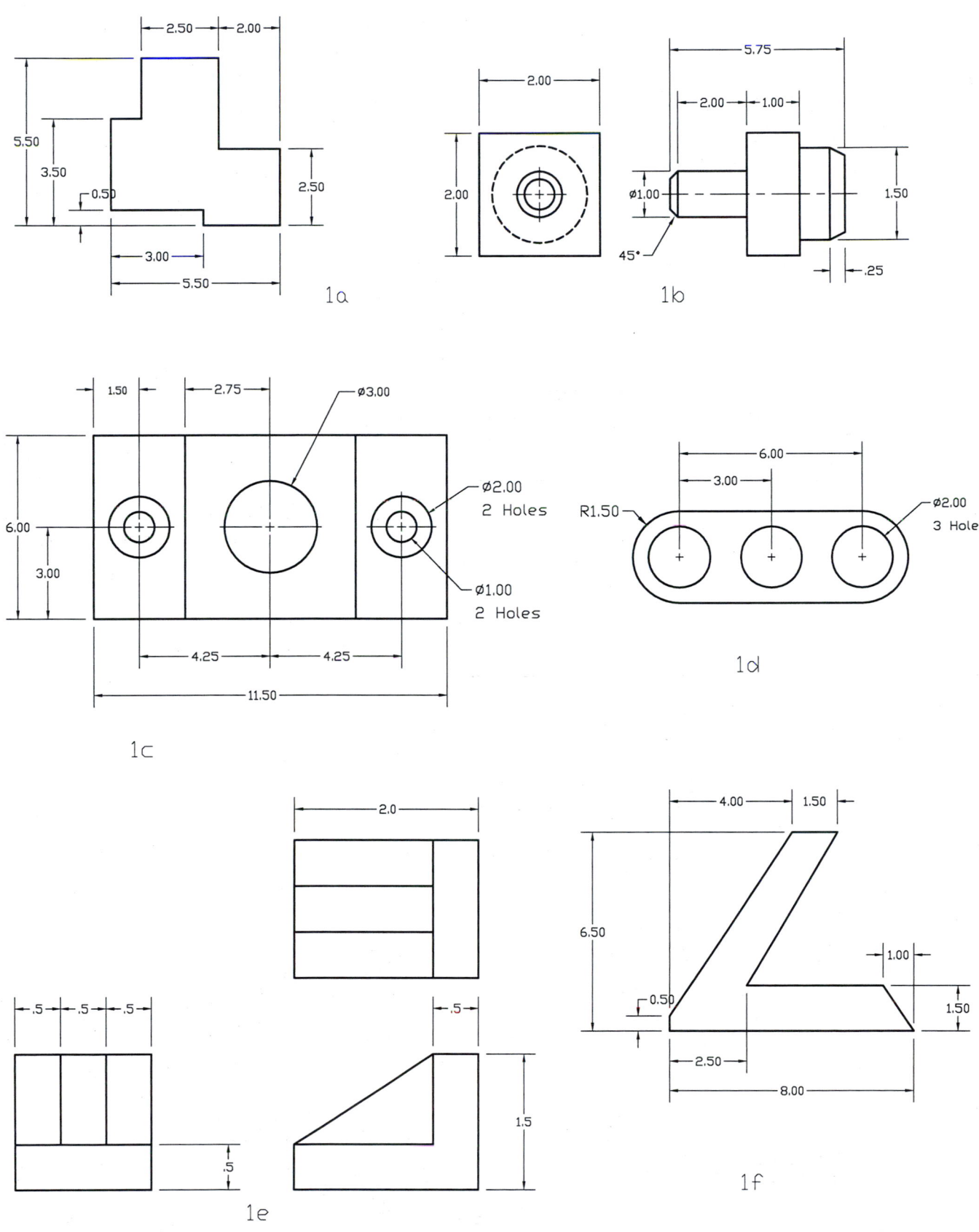

Exercise 1 Architectural

Open an imperial file - acad.dwt - and draw the objects as shown.
Do not attempt to dimension the drawing.

```
| Command:UNITS, Architectural, Precision 0'-0"
  Command:LIMITS
  Reset model space limits
  Specify lower left corner or [ON/OFF]<0,0>: -2',-2'
  Specify upper right corner<12,9>: 25',20'
  Command:SNAP
  Specify snap spacing (X) or [ON/OFF/Aspect/Rotate/Style/Type]<1>: 3
  Command:GRID
  Specify grid spacing or [ON/OFF/Snap/Aspect]<1>: 6
```

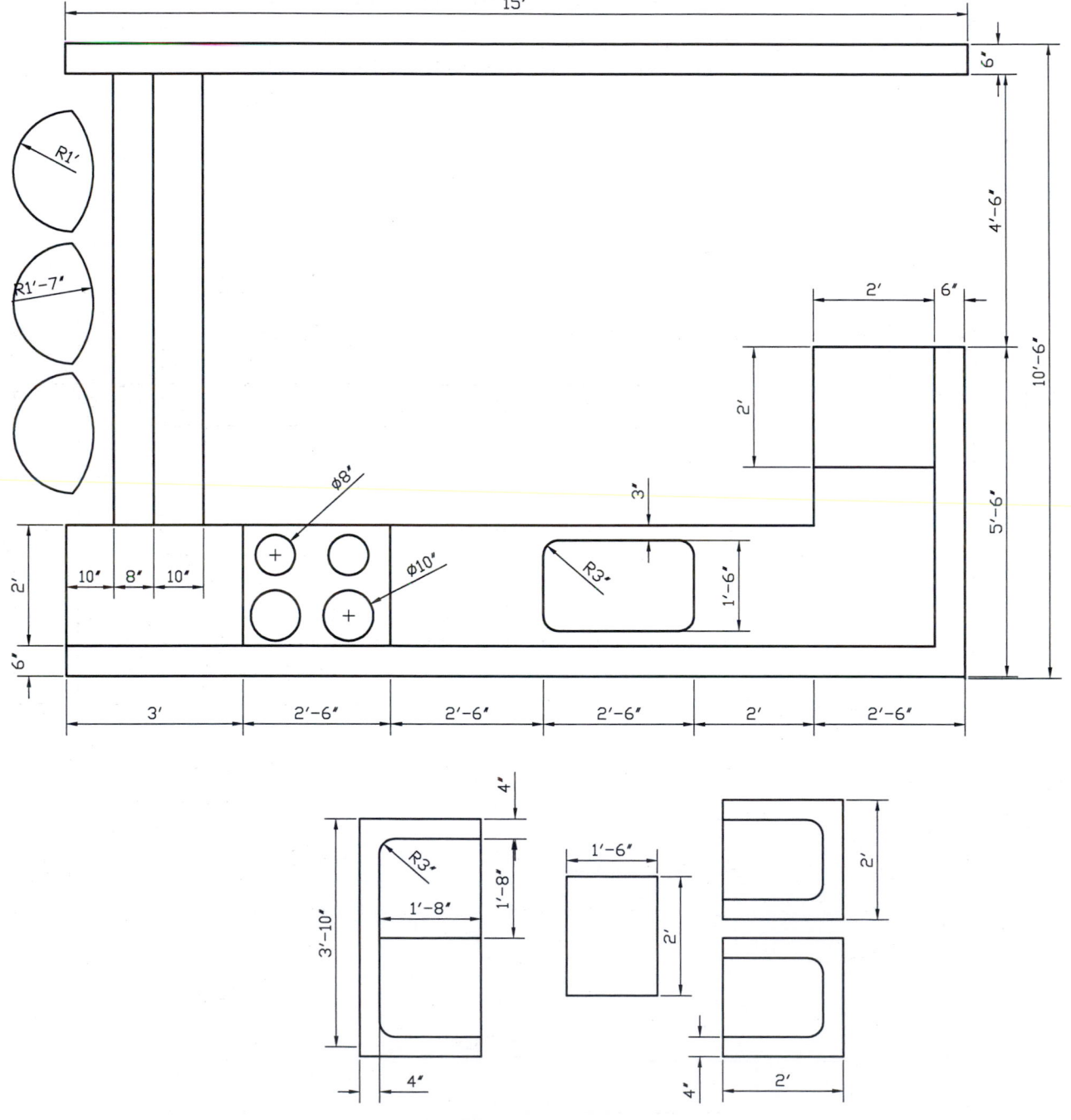

Kitchen Layout

Exercise 1 Mechanical

Open a NEW file in metric, acadiso.dwt, and draw the object below.
The GRID and SNAP will help you get started.
Some people find LIMITS handy.
Draw the LINEs first, then add the CIRCLEs, ARCs, and FILLETs.

```
Command:UNITS, Decimal, Precision 0
Command:LIMITS
Reset model space limits
Specify lower left corner or [ON/OFF]<0,0>:-10,-40
Specify upper right corner<12,9>:220,110
Command:SNAP
Specify snap spacing (X) or [ON/OFF/Aspect/Rotate/Style/Type]<10>:5
Command:GRID
Specify grid spacing or [ON/OFF/Snap/Aspect]<10>:10
Command:  (from the View Pull Down menu, pick ZOOM All)
```

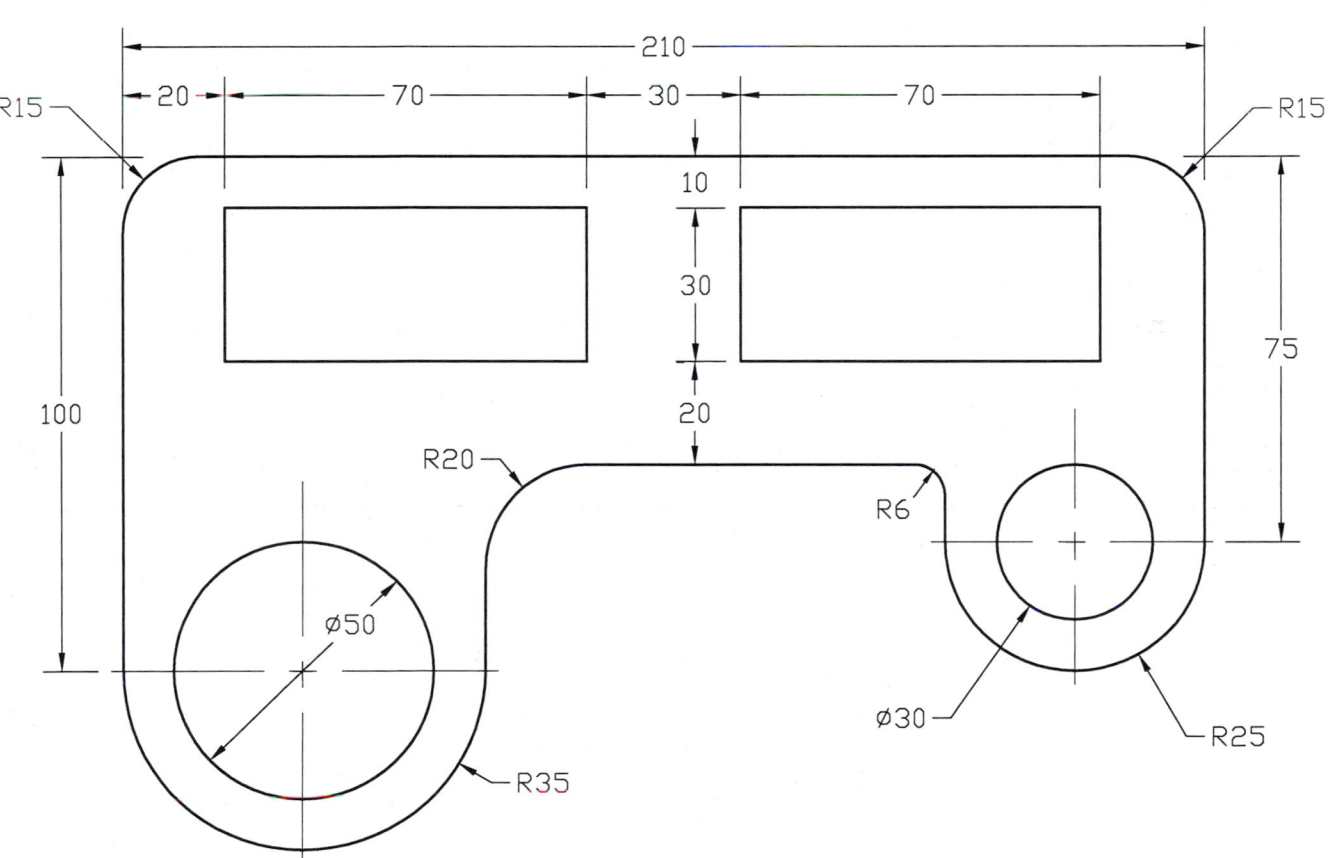

Template

Exercise 1 Woodwork

Start a NEW file in metric - acadiso.dwt - and set it up as shown.
Either use LIMITS as shown below, or put in a line from 0,0 to 580,0 and ZOOM ALL.

```
Command:UNITS, Decimal, Precision 0
Command:LIMITS
Reset model space limits
Specify lower left corner or [ON/OFF]<0,0>:-10,-10
Specify upper right corner<550,340>:1000,800
Command:SNAP
Specify snap spacing (X) or [ON/OFF/Aspect/Rotate/Style/Type]<10>: 5
Command:GRID
Specify grid spacing or [ON/OFF/Snap/Aspect]<10>: 10
Command:  (from the View Pull Down menu, pick ZOOM All)
```

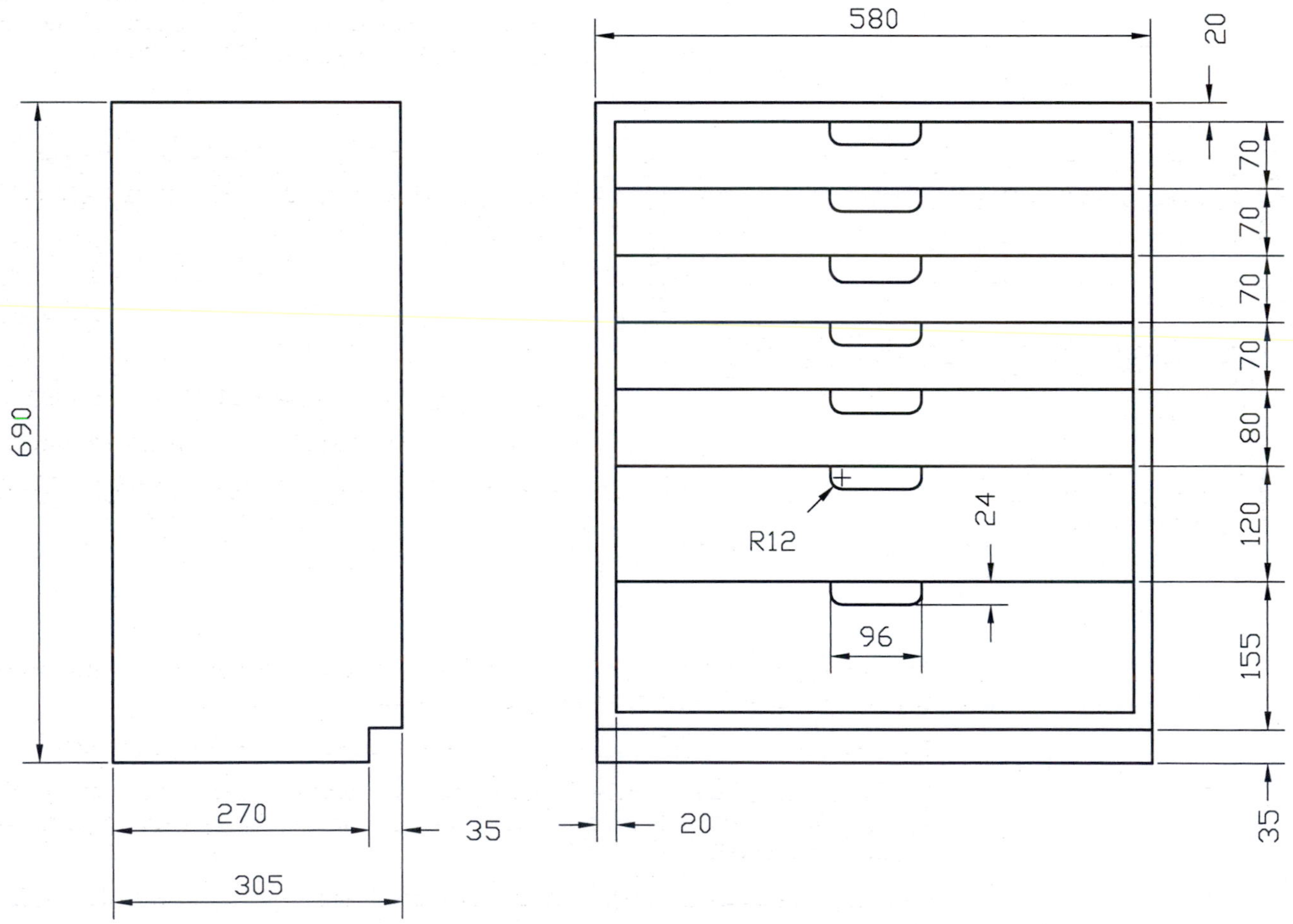

Paper Cabinet

Help Files, OSNAP, OTRACK, BREAK, TRIM, and ERASE

On completion of this chapter you should be able to:

1. Retrieve on-line documentation or Help files
2. Use the commands GRID and SNAP effectively
3. Use OSNAP and OTRACK modes
4. Use TRIM and BREAK to erase portions of objects
5. Use ERASE to remove whole objects.

Once you have learned how to sign on to the system and have located all the menus, you can learn the system from on-screen documentation. The Help files have two main functions.

The first function serves as an index of commands. When looking for a command that will change the magnification of the data on the screen, you may be able to spot it by using the following (Note: There is no 'Ask Me' in Release 2007 or 2008).

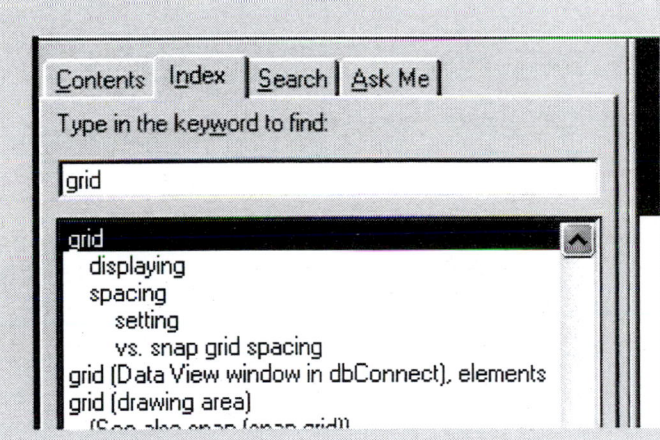

Pull-down menu Choose Help from the pull-down menu, then choose Help or AutoCAD Help, or use ? as follows:

    Command:?

Choose Index from the tabs listed along the top of the folders, then type in the word that you would like help with, or scroll through the list.

If you have just opened a new file in AutoCAD, you can get to the Help files simply by pressing the space bar before entering a command.

Use the X at the top right of the dialog box to remove the box from your screen.

The Help index gives you a listing of the various commands which are available on the system. Once you have found the correct command, you can retrieve information about how it works.

AutoCAD responds to any command with a series of prompts. The Help files explain those prompts.

On any version of AutoCAD, the Help files explain the syntax or command line sequence, and the prompts and options of each command.

Understanding Command Strings

Command strings are generally very similar in construction. Once you understand how they work, your AutoCAD skills should improve very quickly.

Option and Default Brackets

The AutoCAD commands are set up so that you can understand what the options are and accept the defaults within the commands. Once you understand how the commands are set up, it should be easier to figure out how to use them. Remember *you* are in control.

Defaults

The information offered in angle brackets, < >, is the default. This is what the command will do if you do not specify something else.

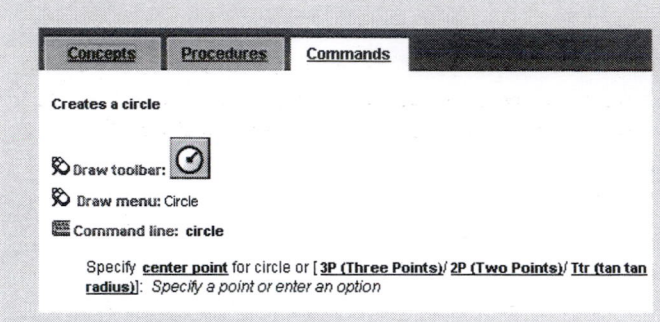

Once you have found the command in the Help files, use Display to open the page on it.

The Concepts and Proceedures will show how the command is executed. The Command tag shows you the command syntax.

Click on the X in the top right corner to exit.

In the case of the CIRCLE command, the previous radius is used by default if no other is specified. The square brackets offer you construction and diameter options.

If you have asked AutoCAD to construct a circle for you, you will then be prompted for the information needed to construct that circle.

```
Command:CIRCLE
Specify center point for circle or [3P/2P/Ttr (tan tan
    radius)]: (pick a point)
Specify radius of circle or [Diameter] <5.00>:
```

The first line says that by default you start the circle by indicating the center point.

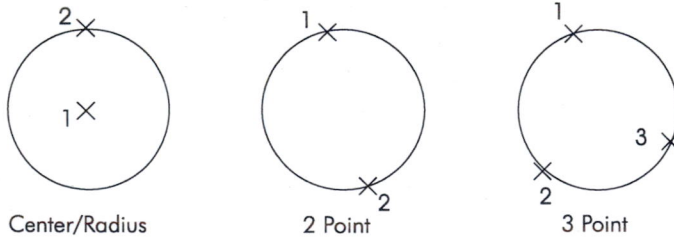

Center/Radius   2 Point   3 Point

Figure 2.1

Then you can specify the radius or, with the use of 'D' to indicate diameter, you can indicate a diameter.

If you would like to specify another means by which you would like the circle constructed, choose one of the three options within the square brackets.

Figure 2.1 illustrates three different ways that AutoCAD can construct a circle without reference to any other objects. The first circle on the left is the default. 2 Point and 3 Point are the other two circle types.

The AutoCAD prompts change according to the options you choose. Most options require a numeric value. For example, to change the diameter, type in d for diameter, then type in the diameter that you need, 1.5, 2.75, etc. You can also choose the value on screen by picking a point.

Option Brackets

The square brackets offer the various options available with each command. In the CIRCLE command, many of the options are shown in the first line; the Diameter option is shown in the second line.

```
Command:CIRCLE
Specify center point for circle or [3P/2P/Ttr (tan tan
   radius)]:(pick a point)
Specify radius of circle or [Diameter] <5.00>:
```

With the ARC command, some options are shown in the first prompt, but others show up with each option chosen. There are many ways of calculating arcs, and these are found in the Draw pull-down menu and illustrated in the Help files, but you generally only use one or two.

In Figure 2.2, three different arcs are generated. The first is calculated through three points. The second is calculated through the start, end and radius. To enter this, type in C to place the center of the arc or E to place the end instead of the default.

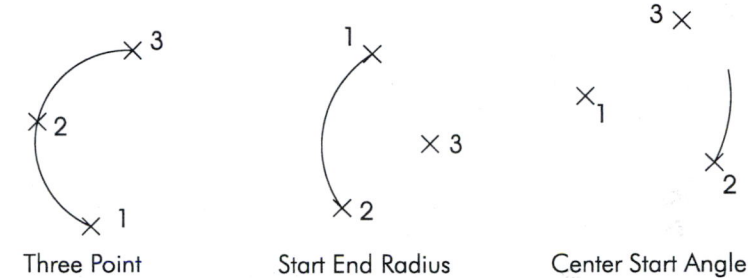

Three Point   Start End Radius   Center Start Angle

Figure 2.2

<table>
<tr><td>3 Points</td></tr>
<tr><td>Start, Center, End</td></tr>
<tr><td>Start, Center, Angle</td></tr>
<tr><td>Start, Center, Length</td></tr>
<tr><td>Start, End, Angle</td></tr>
<tr><td>Start, End, Direction</td></tr>
<tr><td>Start, End, Radius</td></tr>
<tr><td>Center, Start, End</td></tr>
<tr><td>Center, Start, Angle</td></tr>
<tr><td>Center, Start, Length</td></tr>
<tr><td>Continue</td></tr>
</table>

```
Command:ARC
ARC Specify start point of arc or [Center]:(pick 1)
Specify second point of arc or [Center/End]:(pick 2)
Specify end point of arc:(pick 3)
```

The third is calculated through the center point, then the start point, which will give the radius, and then the angle through which it is drawn, in this case 135 degrees.

```
Command:ARC
_arc Specify start point of arc or [Center]:_c
Specify center point of arc:(pick 1)
Specify start point of arc:(pick 2)
Specify end point of arc or [Angle/chord Length]:_a
Specify included angle:135
```

Picking the arc from the Draw pull-down menu will bring up the list of arcs shown on the left. If you pick an arc from this menu, you will be prompted for each component of the arc in turn.

In some cases you are prompted to pick a point through which the arc can be generated, such as the start, end or center of the arc.

In other cases you are prompted for a value. In the example above, 135, the angle of the arc, is a value. When specifying an option such as a diameter or angle, you first choose the option and use ↵ to have AutoCAD accept this option. You will then be prompted for the required value. Values can be entered on screen with a 'pick' as well as by entering the numeric value at the command line.

Object SNAPs

There are three ways to enter points:

PICKING Pick on screen, using SNAP to be accurate. (Chapter 1)

COORDINATE ENTRY You can enter absolute, relative, or polar coordinates in any order.

ENTITY SELECTION, OBJECT SNAP, OR OSNAP This allows you to use existing objects to create your file. Accessing points on existing objects is called using OSNAPs.

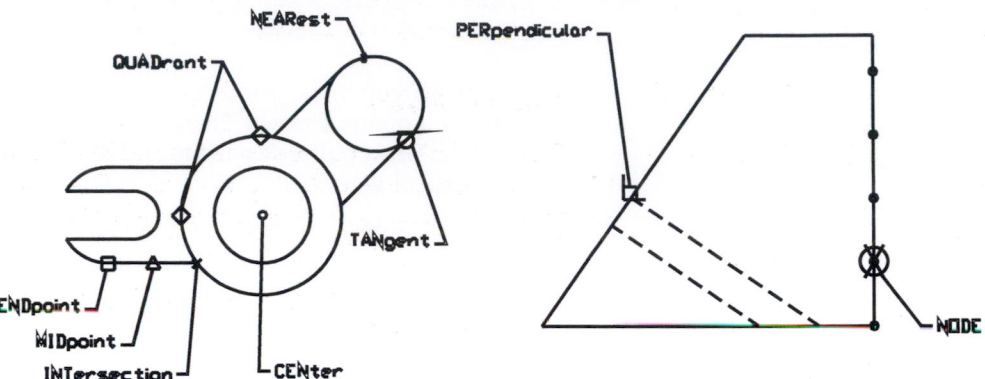

Figure 2.3

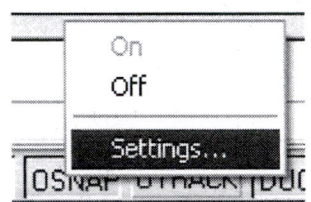

OSNAP: *Object SNAP*

Object Snaps allow you to specify precise points on objects in order to create or edit objects. AutoCAD has 16 Object Snap modes that allow you to specify precise points on objects. The capitalized letters are those needed when typing in the option.

APPint	snaps to a real or imaginary intersection of two objects (apparent ntersection)
CENter	snaps to the center of an arc or circle
ENDpoint	snaps to the closest end of any object
EXTention	snaps to a point along the extension of a line or arc
FROM	establishes a temporary reference point from the parameters of an existing objects
INSertion	snaps to the insertion point of a block
INTersection	snaps to the intersection of two items
MIDpoint	snaps to the midpoint of a selected item
NEARest	snaps to a point on an object nearest to the digitized point
NODE	snaps to a point created by POINT, DIVIDE, or MEASURE
NONE	turns the Object Snap mode off
PARallel	draws a line parallel to an existing line
PERpendicular	snaps to a 90 degree angle to an existing line
QUADrant	snaps to the 0, 90, 180, or 270 degree point on an arc, circle, or ellipse
Quick	snaps to the first snap point found
TANgent	snaps to the tangent of an arc or circle

Accessing OSNAPs

OSNAPs can be accessed in one of four ways:

1. There may be a designated button on your mouse. Often if you hold the Shift button on your keyboard down and use the right-click button on your mouse, an OSNAP menu will appear where your cursor is.
2. Type in the first few letters of the OSNAP mode.
3. Choose Object Snap from the Drafting Settings under the Tools menu.
4. Use the OSNAP buttons.

Using OSNAPs

Figure 2.4 illustrates the use of some OSNAPs within the command line. When entering the OSNAP, type in the first three or four letters, or Shift right-click and access the OSNAP list.

First, draw three circles as shown in Figure 2.5. The size doesn't matter.

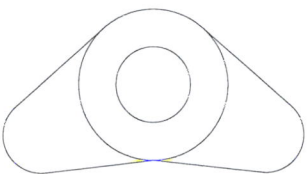

Figure 2.4

```
Command:CIRCLE
Specify center point for circle or
  [3P/2P/Ttr (tan tan radius)]:0,0
  Specify radius of circle or
  [Diameter]:3
Command:C
Specify center point for circle or
  [3P/2P/Ttr (tan tan radius)]:9,4
Specify radius of circle or
  [Diameter]:6
Command:C
Specify center point for circle or
  [3P/2P/Ttr (tan tan radius)]:18,0
  Specify Radius of Circle or
  [Diameter]:3
```

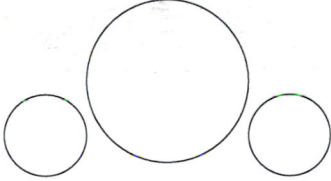

Figure 2.5

Now create lines tangent to the arcs. The TANgent object snap creates a tangent to the identified object from the last object to a circle or arc. For TANgent type in TAN.

```
Command:LINE
Specify first point:TANgent to (pick 1)
Specify next point or [Undo]:
TANgent to (pick 2)
Specify next point or [Undo]:↵
Command:LINE
Specify first point:TANgent to (pick 3)
Specify next point or [Undo]:
TANgent to (pick 4)
Specify next point or [Undo]:↵
```

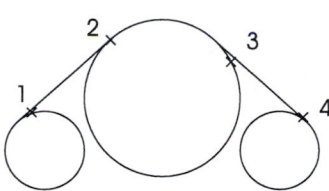

Figure 2.6

AutoCAD will calculate the tangent for you; just pick which side of the object the tangent should be on, as shown in Figure 2.6. You need to secure a tangent connection for both sides of the line. To set the OSNAP to TANgent, right-click the OSNAP button, set tangent, and make sure the OSNAP button is on.

Now use the QUADrant option to attach the other two lines as shown in Figure 2.7. QUADrant takes the top, the bottom, or the far right or left of a specified circle or arc. Type in QUAD for QUADrant.

```
Command:LINE
Specify first point:TANgent to (pick 1)
Specify next point or [Undo]:
QUADrant of (pick 2)
Specify next point or [Undo]:↵
Command:LINE
Specify first point:
QUADrant of  (pick 3)
Specify next point or [Undo]:
TANgent to (pick 4)
Specify next point or [Undo]:↵
```

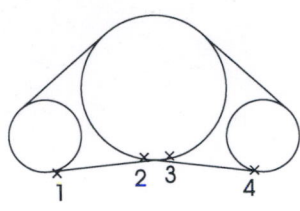

Figure 2.7

Now use CENter to place a circle concentric with the first as shown in Figure 2.8, CENter takes the center of an arc or circle. Only items with a defined radius can have a center point.

```
Command:CIRCLE
Specify center point for circle or
   [3P/2P/Ttr (tan tan radius)]:CENter
   of (pick 5)
Specify Radius of Circle or
   [Diameter]:3
```

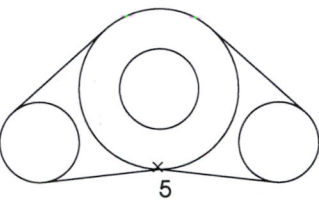

Figure 2.8

Now use the TRIM command (see page 72) to trim off the extra portions of the circle, as shown in Figure 2.9. The TRIM command will erase a portion of an object between two edges identified by intersections with other objects.

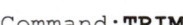

```
Command:TRIM
Current settings: Projection=UCS
   Edge=None
Select cutting edges...
Select objects:(pick 1)
Select objects:(pick 2)
Select objects:(pick 3)
Select objects:(pick 4)
Select objects:↵ (enter - no more cutting edges are needed)
<Select object to trim> or
   shift-select to extend [Fence/
   Project/Edge/eRase/Undo]:(pick 5,6)
```

Figure 2.9

Your final image should look like Figure 2.10.

The TRIM command will not work unless the TANgent option has been properly used. If the lines don't intersect the circles, they will not be trimmed.

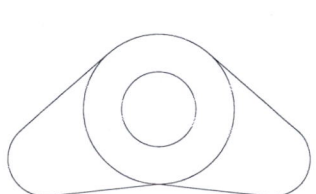

Figure 2.10

This example has shown how to use OSNAPs within a command string. You can type in the first few letters of the OSNAP option, use Shift right-click to get a list on screen or get the OSNAP options from a pull-down menu or floating toolbar. You can also set up your OSNAP so that the options automatically appear.

Running OSNAPs

You can set up OSNAP so that the options are automatically turned on.

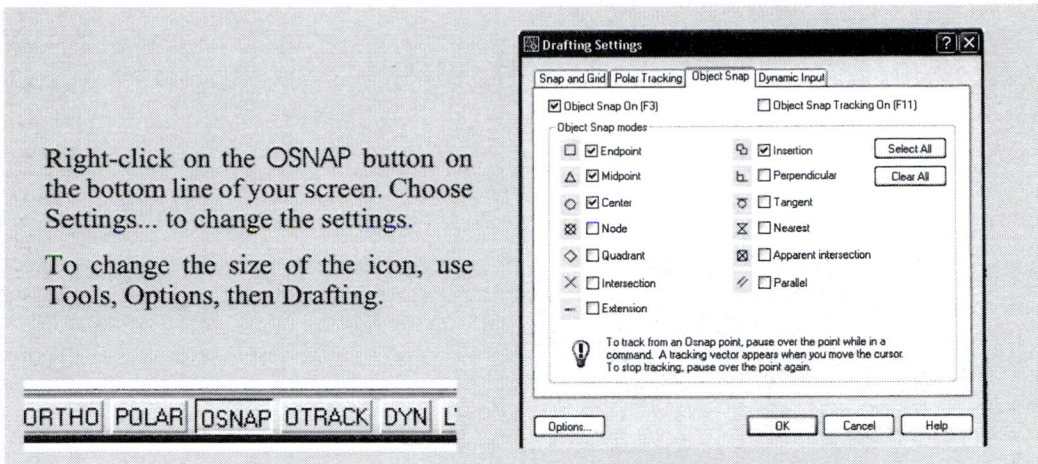

Right-click on the OSNAP button on the bottom line of your screen. Choose Settings... to change the settings.

To change the size of the icon, use Tools, Options, then Drafting.

In this example, ENDpoint, MIDpoint and CENter are turned on. These are the usual choices. We have seen how CENter works. ENDpoint takes the closest end point. All lines are made up of two end points. MIDpoint takes the middle point of an object. These options are illustrated in the Tutorials at the end of this chapter.

OTRACK

The OTRACK command is useful for lining objects up relative to other objects.

Figure 2.11

OTRACK is a toggle switch found on the function bar at the bottom of your screen, as shown in Figure 2.11.

Using **OTRACK**

With OSNAP on, OTRACK will track the ENDpoint, MIDpoint or CENter of an object and line up a new point exactly across from or beneath it. OTRACK works vertically or horizontally. First draw in the lines as shown in Figure 2.12 (A).

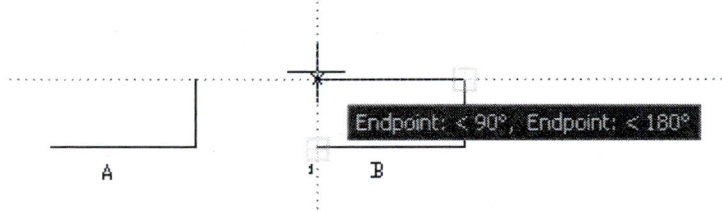

Figure 2.12

```
Command:LINE
Specify first point:(pick 1)
Specify next point or [Undo]:(pick 2)
Specify next point or [Undo]:↵
```

Make a line from the top right corner, lining it up with point 1 as shown on Figure 2.12 (A). Then draw your cursor across the endpoint of point 1 without picking it. Move it up level to your beginning point. When you have found the position that lines up, pick that point. AutoCAD will create a line using the *X* value of point 1 and the *Y* value of your start point in Figure 2.12 B. It's a lot easier to do than to explain.

BREAK, TRIM, and ERASE

The BREAK command removes a portion of a single object between two specified points. The TRIM command trims a series of objects according to a cutting line or cutting lines. The ERASE command removes an entire object from the file.

The *BREAK* Command

BREAK is found under the Modify menu in both the Toolbar and the pull-down menu. The object will be broken from the point chosen on the object or by another point specified after entering F to indicate a different first break point. In Figure 2.13, the line indicated in the first pick is cut from the center of the circle to the intersection of the line and the circle.

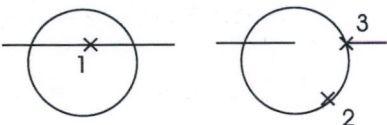

Figure 2.13

```
Command:BREAK
Select object:(pick 1)
Enter second point (or F for first point):F
Enter first point:CENter of (pick 2)
Enter second point:INTersection of (pick 3)
```

The *TRIM* Command

TRIM is found under the Modify menu in both the Toolbar and the pull-down menu. Pick either one or a series of cutting lines, and then choose the objects to trim.

In Figure 2.14, the lines are all trimmed to the circle; the circle was indicated first.

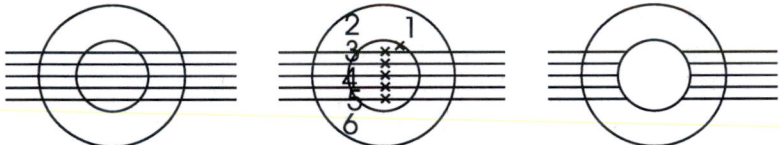

Figure 2.14

```
Command:TRIM (or TR)
Current settings: Projection=UCS Edge=None
Select cutting edges...
Select objects:(pick 1)
Select objects:↵ (no more cutting edges are needed)
<Select object to trim> or shift-select to extend [Fence/
   Project/Edge/eRase/Undo]:(pick 2,3,4,5,6)
```

The *ERASE* Command

ERASE is found under the Modify menu in both the Toolbar and the pull-down menu. You can also simply type in E. In Figure 2.15 the non-concentric circle is erased.

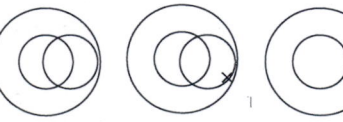

```
Command:ERASE (or E)
Select objects:(pick 1)
Select objects: ↵
```

Figure 2.15

GRIPS

GRIPS are used to identify a selection set or group of objects prior to a Modify command such as ERASE or MOVE (see Chapter 3). If you select a group of objects, blue grips will appear. This means you have selected the objects for editing. Now you can use ERASE to erase them.

```
Command:(pick 1) other corner
(pick 2)
Command:ERASE Erase 5 found
```

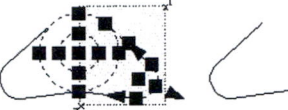

In Figure 2.16, five objects were selected by a crossing window, going from right to left. The objects were then erased. To undo an erase, use Undo or U.

Figure 2.16

```
Command:U Undo Erase
```

Use the Escape key on the top left of your keyboard to clear the grips if you have chosen them incorrectly.

ERASE with Window and Crossing

ERASE can also be used with the *Crossing* and the *Window* options to erase a group of objects as seen in Figure 2.17.

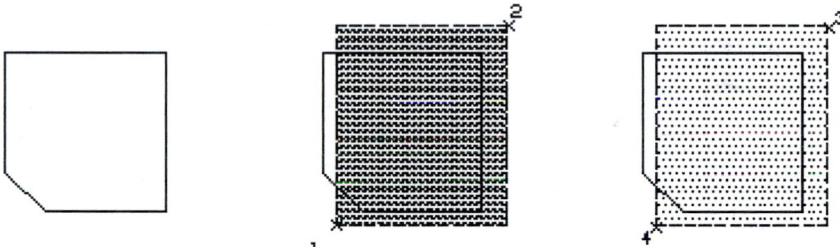

Figure 2.17

Window
```
Command:ERASE
Select objects:(pick 1, 2)
Select objects: ↵
```

Crossing
```
Command:ERASE
Select objects:(pick 3, 4)
Select objects: ↵
```

The *Window* selection set, as shown in picks 1 and 2, provides a blue-colored window. Any objects contained in this window will be erased. Since the diagonal line and the top line do not lie completely within the window, they will not be erased.

The *Crossing* selection set, as shown in picks 3 and 4, provides a green-coloured rectangle. Any object that touches or is within this rectangle will be affected by the command. In this example, every object except the vertical line on the left will be erased.

UNDO and REDO

The UNDO command will undo the previous command, regardless of what it is. If you erase something by mistake, trim something by mistake or enter some geometry that is not correct, use UNDO to return to correct it. You can UNDO as many commands as you have entered since the file was last opened.

```
Command:UNDO (or U)
Command:REDO
```

REDO will bring back a command that you have taken out with UNDO.

Step 1

Start a new file in imperial units. You are making a tricycle. No limits are needed, but you need to make sure you are in imperial. Choose acad.dwt. (You make the entries shown in **BOLD**).

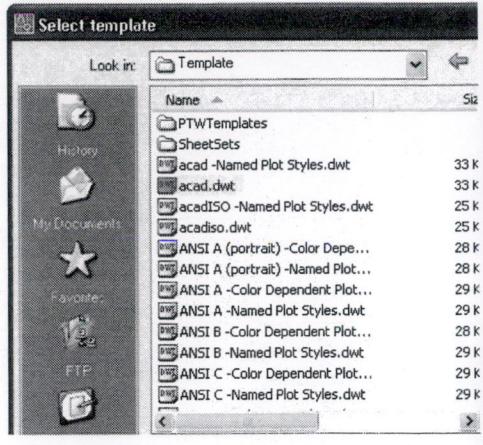

```
Command: NEW
(Choose acad.dwt)

Or

Command: STARTUP
Enter new value for startup
   <0>:1
Command: NEW (Choose Imperial)

Command: UNITS
Set units to architectural.
```

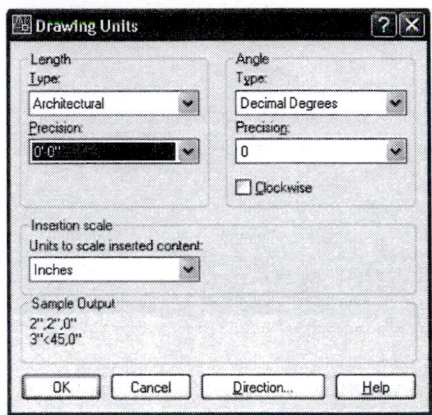

Step 2

Often with architectural designs the layout is set at regular intervals so LIMITS, GRID and SNAP are an advantage. With this tricycle elevation these are not needed. Simply start with a couple of circles, then ZOOM Extents.

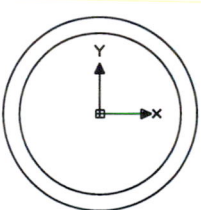

```
Command:CIRCLE
Specify center point for circle or [3P/2P/Ttr
   (tan tan radius)]:0,0
Specify radius of circle or [Diameter]:10
Command:CIRCLE
Specify center point for circle or [3P/2P/Ttr
   (tan tan radius)]:0,0
Specify radius of circle or [Diameter]:12

Command: (from the View pull-down menu, pick ZOOM Extents)
Command:Z
Specify corner of window, enter a scale factor (nX or
   nXP),or[All/Center/Dynamic/
   Extents/Previous/Scale/Window]<realtime>:E
Command:Z
Specify corner of window, .....Scale/Window]<real time>:.7X
```

You can also use your roller ball to ZOOM the objects into the middle of your screen.

Step 3

Make sure OSNAP is on, and draw an arc from the CENter of these two circles just up and to the right. When your cursor approaches the circle, a smaller circle should appear at the center indicating an OSNAP. If this doesn't show up, use Shift right-click for the menu, and pick Center as shown on the left, or simply type in CEN.

```
Command:ARC
ARC Specify start point of arc or [Center]:CEN of (pick 1)
Specify second point of arc or [Center/End]:(pick 2)
Specify end point of arc:(pick 3)
```

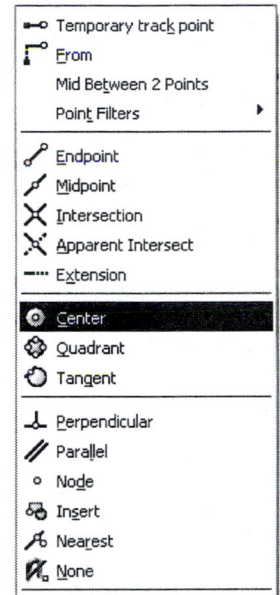

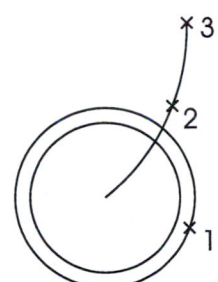

Next you want to place an arc starting from the same center point, just a little to the right of the first arc. **OSNAP**

After you have picked the first point, turn OSNAP off or you will have difficulty placing the next points. If the cursor 'picks up' another line, an OSNAP is assumed, and the second arc will be generated along the endpoints of the existing arc.

```
Command:ARC
ARC Specify start point of arc or
   [Center]:CEN of (pick 1)
Specify second point of arc or
   [Center/End]:(pick 2)
Specify end point of arc:(pick 3)
```

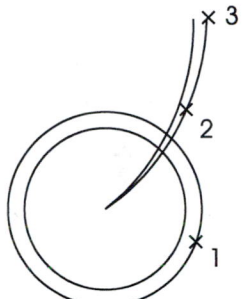

You should now have two arcs and two circles.

Step 4

Place a circle with a radius of 1′ at the same center point. Use Zoom Window to zoom in and see it. You can either use your roller ball on your mouse, the icon as shown, or type in Z then pick two points.

```
Command:Z
Specify corner of window, enter a
   scale ...ow]<realtime>:(pick 1)

corner:(pick 2)
```

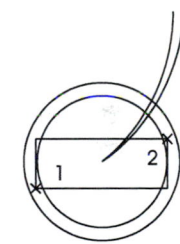

Specify opposite

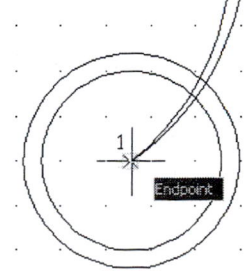

Now put the two circles in using the END of the two arcs. Turn your OSNAP on.

```
Command:CIRCLE
Specify center point for circle or
   [3P/2P/Ttr (tan tan radius)]:END of
   (pick 1)
Specify Radius of Circle or
   [Diameter]:3
```

```
Command:CIRCLE
Specify center point for circle or [3P/2P/Ttr (tan tan
   radius)]:END of (pick 1)
Specify Radius of Circle or [Diameter]:2
```

Step 5

Use TRIM to remove the ends, of the arcs as shown. The circle is your cutting edge.

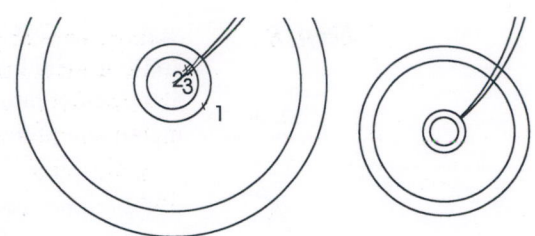

```
Command:TRIM
Current settings:
Select cutting edges...
Select objects:(pick 1)
Select objects:↵
<Select object to trim> or shift-select to extend [Fence/
   Project/Edge/eRase/Undo]:(pick 2,3)
<Select object to trim> or shift-select to extend [Fence/
   Project/Edge/eRase/Undo]:↵
```

Since you have now been using the file for at least a few minutes, you should save at this point.

Step 6

Use the line command with OSNAP, OTRACK and PO-LAR to place the rectangle at the top of the two arcs.

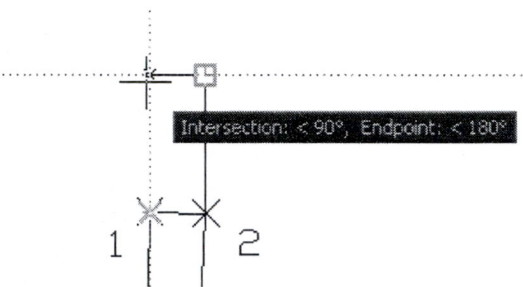

```
Command:LINE
Specify first point:END of (pick 1)
Specify next point or [Undo]:END of (pick 2)
Specify next point or [Undo]:4(move your cursor up)
Specify next point or [Undo]:(move your cursor over pick 1,
   then back up to a horizontal, then pick the point shown)
Specify next point or [Undo]:c
```

Step 7

Now we need to draw the back wheels. Draw in a line from the bottom QUADrant of the first wheel, 36 inches to the back. Make sure POLAR is on..

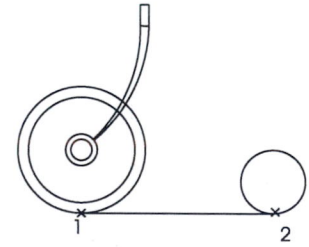

```
Command:LINE
Specify first point:QUAD of (pick
   1)
Specify next point or [Undo]:36
   (move your cursor to the right)
Specify next point or [Undo]:↵
```

```
Command:CIRCLE
 Specify center point for circle or [3P/2P/Ttr (tan tan
   radius)]:2P
Specify first end point on circle's diameter:END of (pick 1)
Enter second end point of circle's diameter:12(move your
   cursor straight up)
```

Once the CIRCLE has been drawn in, ERASE the line.

Step 8

Using a similar proceedure as for the front wheel, put in three more circles and two arcs. For the arcs, make the first from the center of the small circle to the MID-point of the vertical line, and the second to a point slightly higher using NEARest. TRIM all unwanted lines.

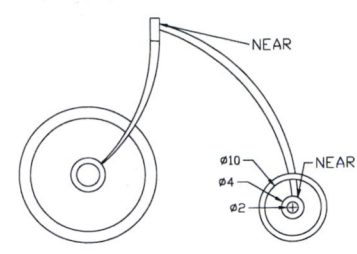

```
Command:CIRCLE
Specify ...radius)]:CEN of (the
  right circle)
Specify radius of circle or [Diameter]:D
Specify diameter:10
Command:ARC
ARC Specify start point of arc or [Center]:NEAR the line
Specify second point of arc or [Center/End]:(pick 2)
Specify end point of arc:NEAR the circle
```

Step 9

For the handle, use ARC and LINE. The ARC command will continue from the last entered point if you use ↲ when prompted for the first point.

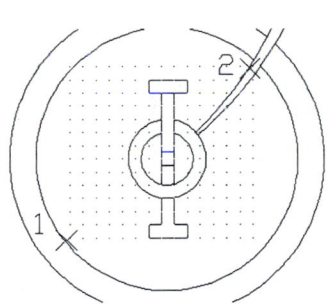

```
Command:ARC
ARC Specify .... [Center]:(pick 1)
Specify second ......End]:(pick 2)
Specify end point of arc:(pick 3)
Command:↲ (brings back ARC)
ARC Specify . of arc or [Center]:↲
  (continues from last point)
Specify second point of arc or [Center/End]:(pick 2)
Specify end point of arc:(pick 3)
```

Step 10

For the pedals, set up a GRID and SNAP, and draw them in without OSNAP. Use two points to pick the size of your limits.

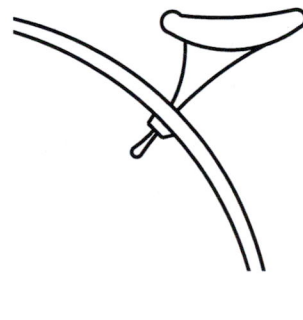

```
Command:LIMITS
Specify bottom left:(pick 1)
Specify upper right:(pick 2)
Command:SNAP
Specify snap spacing<>:.5
Command:GRID
Specify grid spacing<>:1
```

Finally use LINE and ARC to draw in a seat.

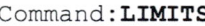

In this Tutorial we will practice using OSNAP and OTRACK to draw the mechanical piece shown below. Make sure ORTHO, OTRACK and POLAR are all on.

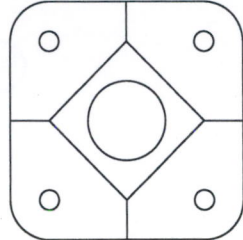

Step 1

For many applications, particularly mechanical, LIMITS, GRID and SNAP are not needed. Another way to set up your screen is simply to choose your units, metric or imperial, then draw in the first object. Use ZOOM to place it on the page.

```
Command:CIRCLE
Specify center point for circle or [3P/2P/Ttr (tan tan
    radius)]:0,0
Specify Radius of circle
    or [Diameter]:100
```

Now ZOOM the object on screen.

```
Command:Z
Specify corner of window, enter a scale factor (nX ornXP),or
    [All/Center/Dynamic/Extents/Previous/Scale/Window]
    <realtime>: A
Command:Z
Specify corner of window, enter a scale factor (nX or nXP), or
    [All/Center/Dynamic/Extents/Previous/Scale/Window]
    <real time>:.3X
```

You can also use your mouse roller ball to center the circle on the screen.

Step 2

Use the POLAR option to place the lines around the outside of the circle. Move your cursor along the direction indicated by the arrows, then type in the value.

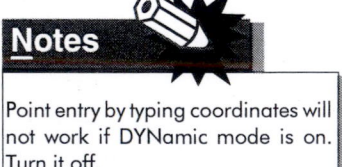

Notes

Point entry by typing coordinates will not work if DYNamic mode is on. Turn it off.

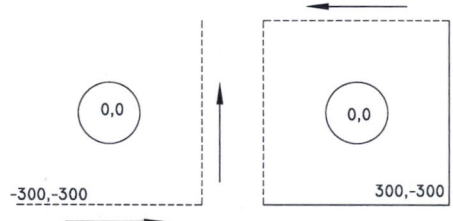

```
Command:LINE
Specify first point:-300,-300
Specify next point or [Undo]:600 right
Specify next point or [Undo]:600 up
Specify next point or [Undo]:600 left
Specify next point or (Undo):c
```

Step 3 Fillet the corners with a radius of 100.

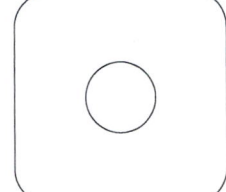

```
Command:FILLET(or F)
Current settings: Mode = Trim,
Radius = <5.0000>
Select first object or[Undo/Polyline/
   Radius/Trim/Multiple]:R
Specify fillet radius:100
Select first object or
   [Undo/Polyline/Radius/Trim/Multiple]:
   (pick 1)
Select second object:(pick 2)

Command:FILLET
Current settings: Mode = Trim,
Radius = <5.0000>
Select first object or[Undo/Polyline/
   Radius/Trim/Multiple]:(pick 3)
Select second object:(pick 4)
```

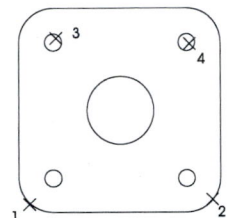

Continue until all four corners are done. By pressing the spacebar at the Command: prompt, the last entered command will be retreived. This could make your data entry a lot quicker.

Step 4

Place CIRCLEs in the four corners. As shown in the illustration, once the circle is identified, you can pick either the perimeter or the center of the circle to place the smaller ones.

For pick 1 place the cursor over the circle, then move the cursor to the center of the circle. The object snap picked up will be the CENter because the center mark is highlighted. The OSNAP button must be on.

```
Command:CIRCLE
Specify center point for circle or
   [3P/2P/Ttr (tan tan radius)]:CENter of (pick 1)
Specify Radius of Circle or [Diameter]:25
Command:CIRCLE
Specify center point for circle or
   [3P/2P/Ttr (tan tan radius)]:CENter of (pick 2)
Specify Radius of Circle or [Diameter]:25
```

Notes

If the CENter OSNAP does not come on, right-click the OSNAP button, choose settings.

If you type in CEN this will override the ENDpoint OSNAP, which is the system default.

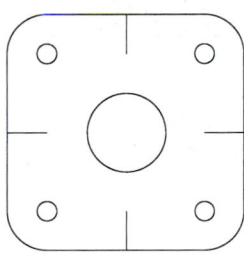

Step 5
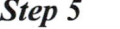

Now draw in the lines as shown in the illustration. To place these, use both OSNAP MIDpoint and the OTRACK of the center of the circles for placement as described below.

Make sure that CENter, MIDpoint and ENDpoint OSNAP is set and that the OSNAP button is turned on.

For the first point, pick the MID of the left vertical line.

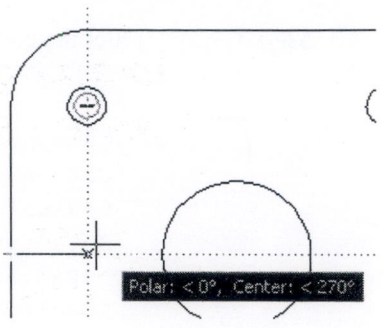

```
Command:LINE
Specify first point:MID of (pick the vertical line)
```

For the next point, drag your cursor over the circle as shown. Do not pick the circle, simply drag the cursor across it. Then drag it down to line up with the horizontal. This will give you a line with the *X* value of the center of the circle and the *Y* value of the MIDpoint of the vertical line, producing a straight line.

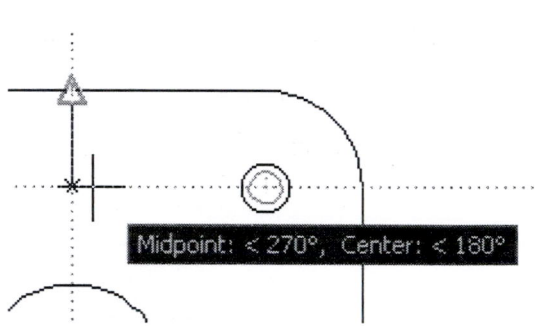

Draw in all four lines.In this example, the MIDpoint is taken from the top line and the end of the line is lined up with the center of the circle on the right.

Once you have done this for all four lines, you should be familiar with the way it works.

Step 6

Finally, access the ENDpoints of the small lines to finish the drawing.

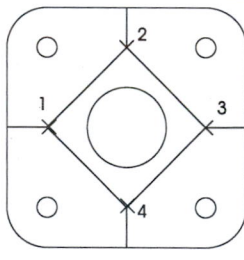

```
Command:LINE
Specify first point:END of (pick 1)
                              Specify next point or
  [Undo]:END of (pick 2)
Specify next point or [Undo]:END of (pick 3)
Specify next point or [Undo]:END of (pick 4)
Specify next point or [Undo]:c
```

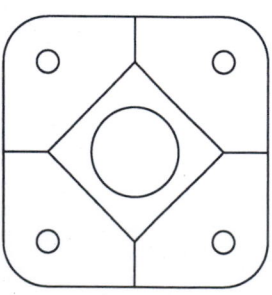

When completed, your object should look like this.

Exercises 2 Practice

DRAW all of these, just for practice.

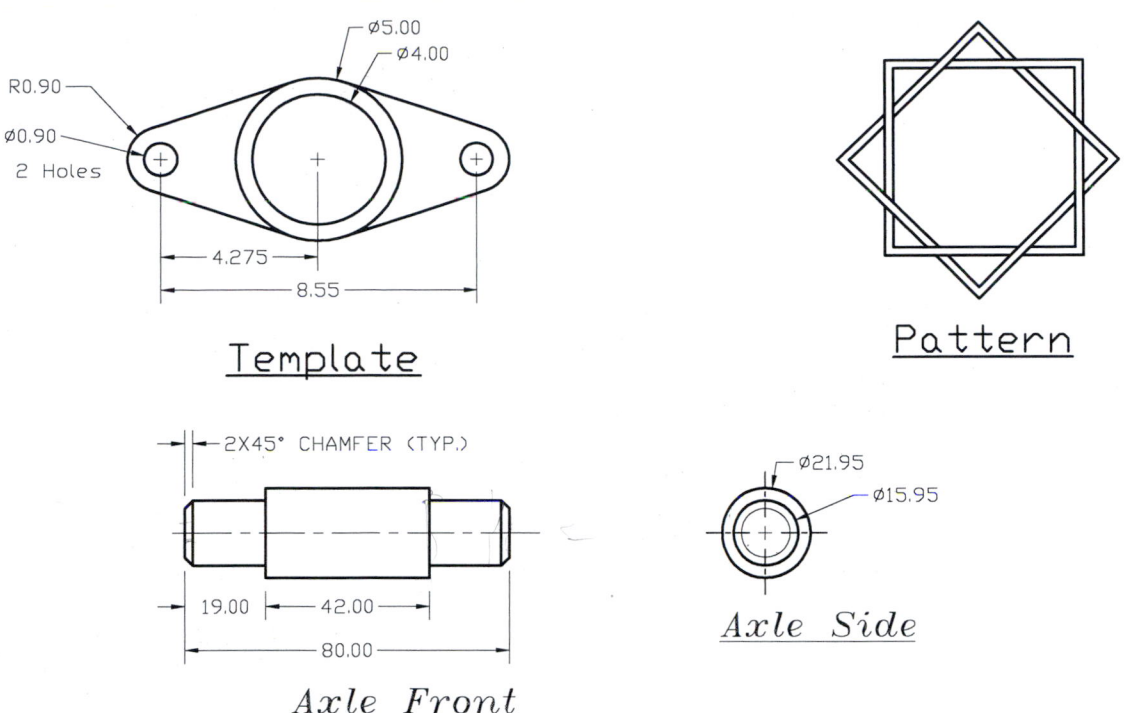

Template

Pattern

Axle Front

Axle Side

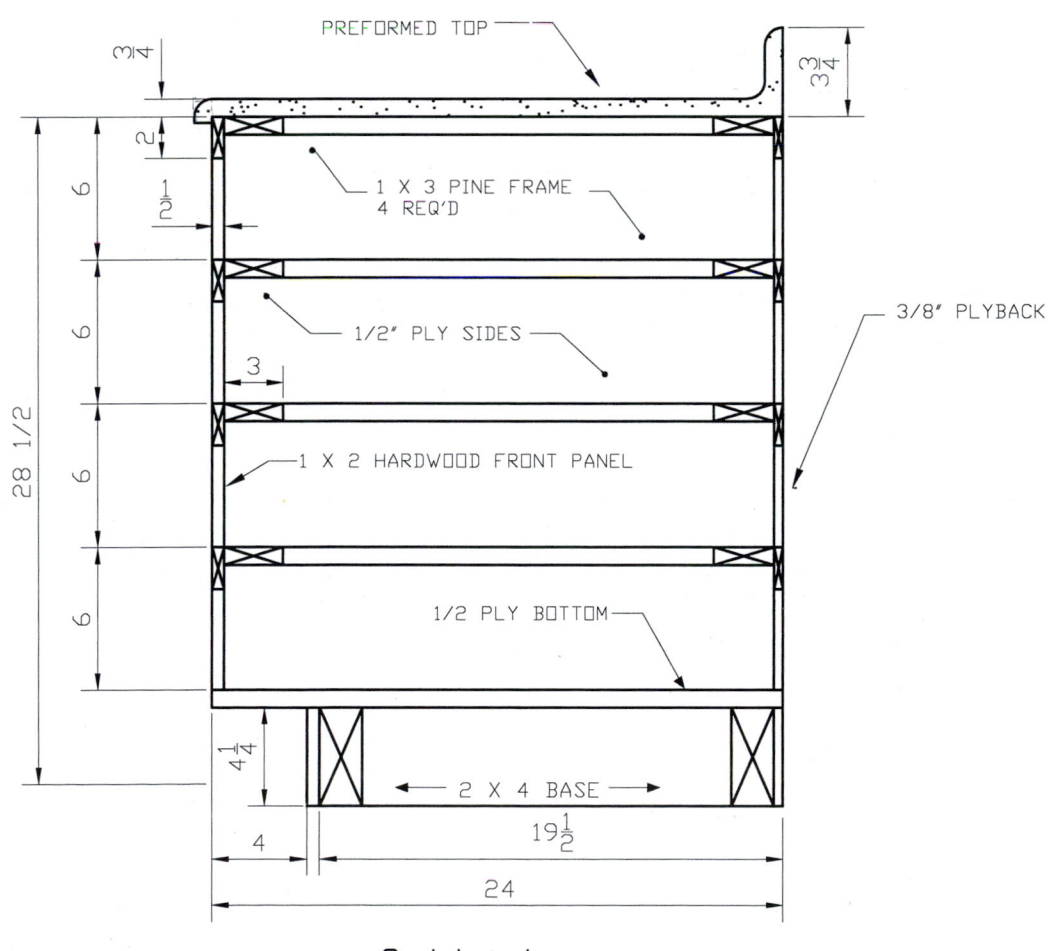

Cabinet

Exercise 2 Architectural

For the Lot Plan you will need to change your UNITS to Surveyors.
The Floor Plan is drawn with Architectural Units.
To get the area of the lot, set your OSNAP to ENDpoint, then use the AREA command.
You will be prompted for the corners of the lot. Press Enter for the area.

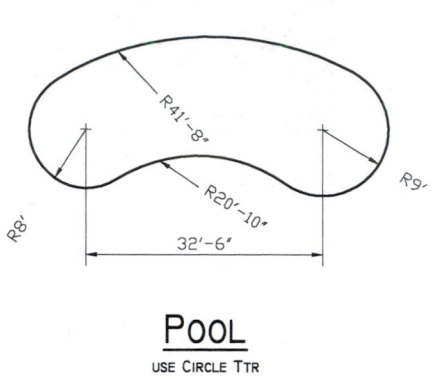

POOL
USE CIRCLE TTR

LOT (SURVEYORS' UNITS)

LOT PLAN

LOT PLANS SHOW THE PROPOSED STRUCTURE AND THE LOT
UPON WHICH IT IS TO BE BUILT. THE LOT IS SHOWN IN
SURVEYORS' UNITS, AND OFTEN INCLUDES EXISTING WATER,
GASS AND ELECTRICAL SERVICES. THE NORTH ARROW SHOWS
THE ORIENTATION OF THE BUILDING. THE EASEMENTS AND
GRADING OF THE LOT WOULD BE NEEDED TO COMPLY WITH
LOCAL BUILDING CODES IN ORDER TO GET A BUILDING
PERMIT.
IN THE NORTHERN HEMISPHERE, THE MAJORITY OF THE
WINDOWS SHOULD BE ON THE SOUTH SIDE OF THE BUILDING
TO TAKE ADVANTAGE OF PASSIVE SOLAR HEATING IN THE
WINTER. DECIDUOUS TREES SHOULD BE PLACED OUTSIDE
THESE WINDOWS TO SHADE THEM IN SUMMER.

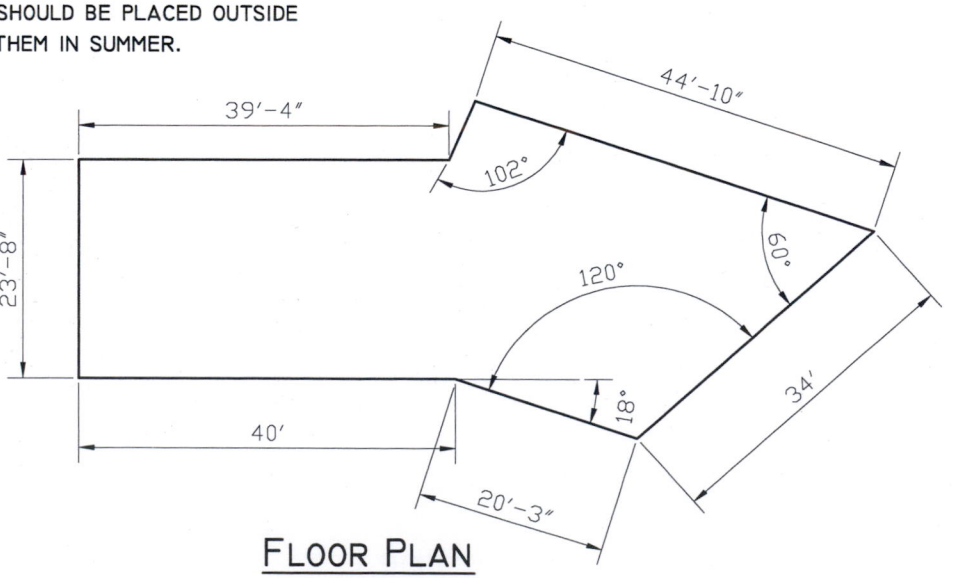

FLOOR PLAN

Exercise 2 Mechanical

TANgent OSNAP will be important for both of these.
Start with a circle at 0,0. ZOOM, then place your next circle relative to it.
These objects are not drawn to scale within the page.

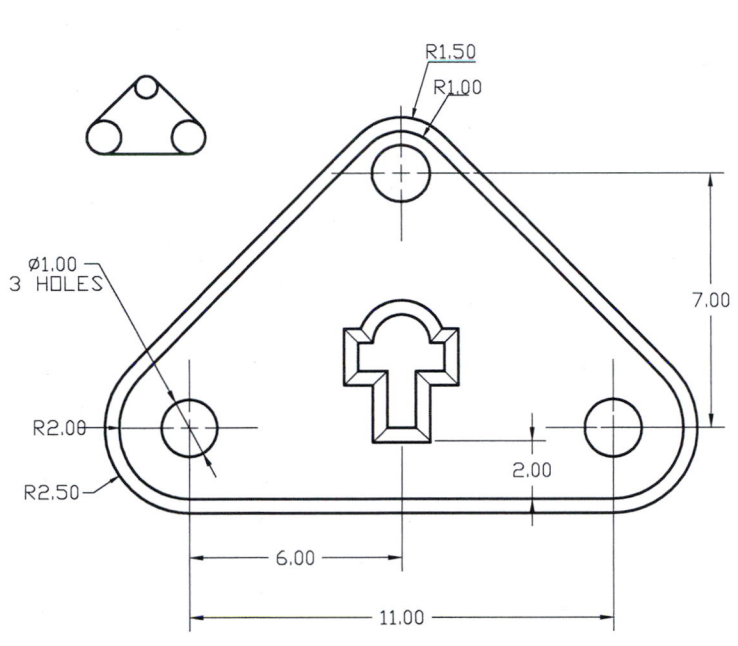

R1.50
R1.00
Ø1.00
3 HOLES
R2.00
R2.50
7.00
2.00
6.00
11.00

Escutcheon

Edges chamfered at 45°

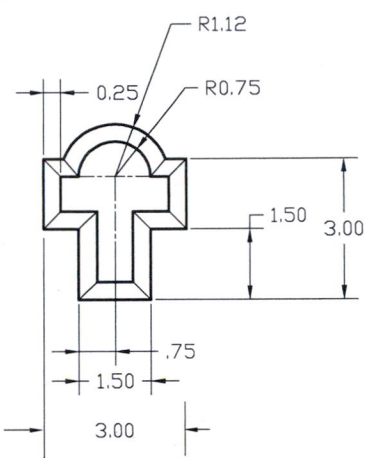

R1.12
0.25
R0.75
1.50
3.00
.75
1.50
3.00

Escutcheon Detail

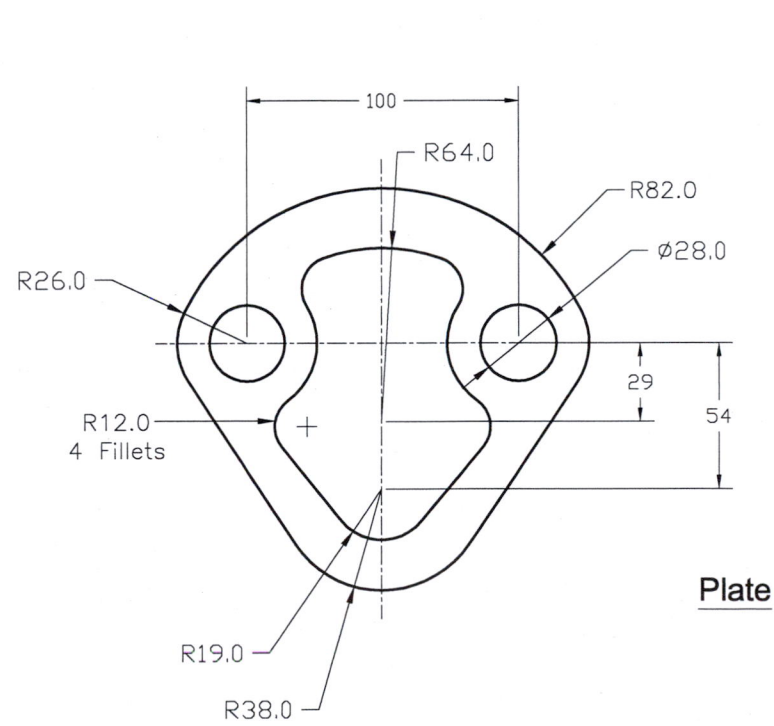

100
R64.0
R82.0
Ø28.0
R26.0
R12.0
4 Fillets
29
54
R19.0
R38.0

Plate

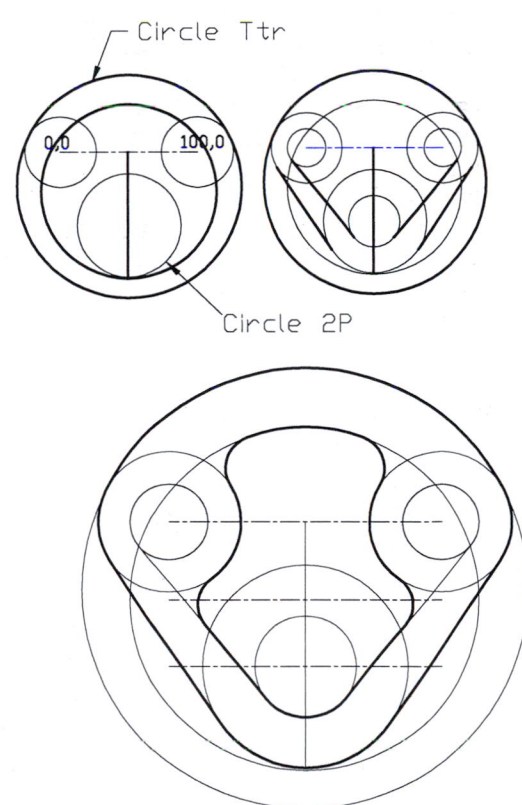

Circle Ttr
0,0
100,0
Circle 2P

Exercise 2 Challenger

Start with the origin at the center of the left circle.
Place the circles according to the coordinates.
Fillets and lines at angles are demonstrated in Chapter 1.

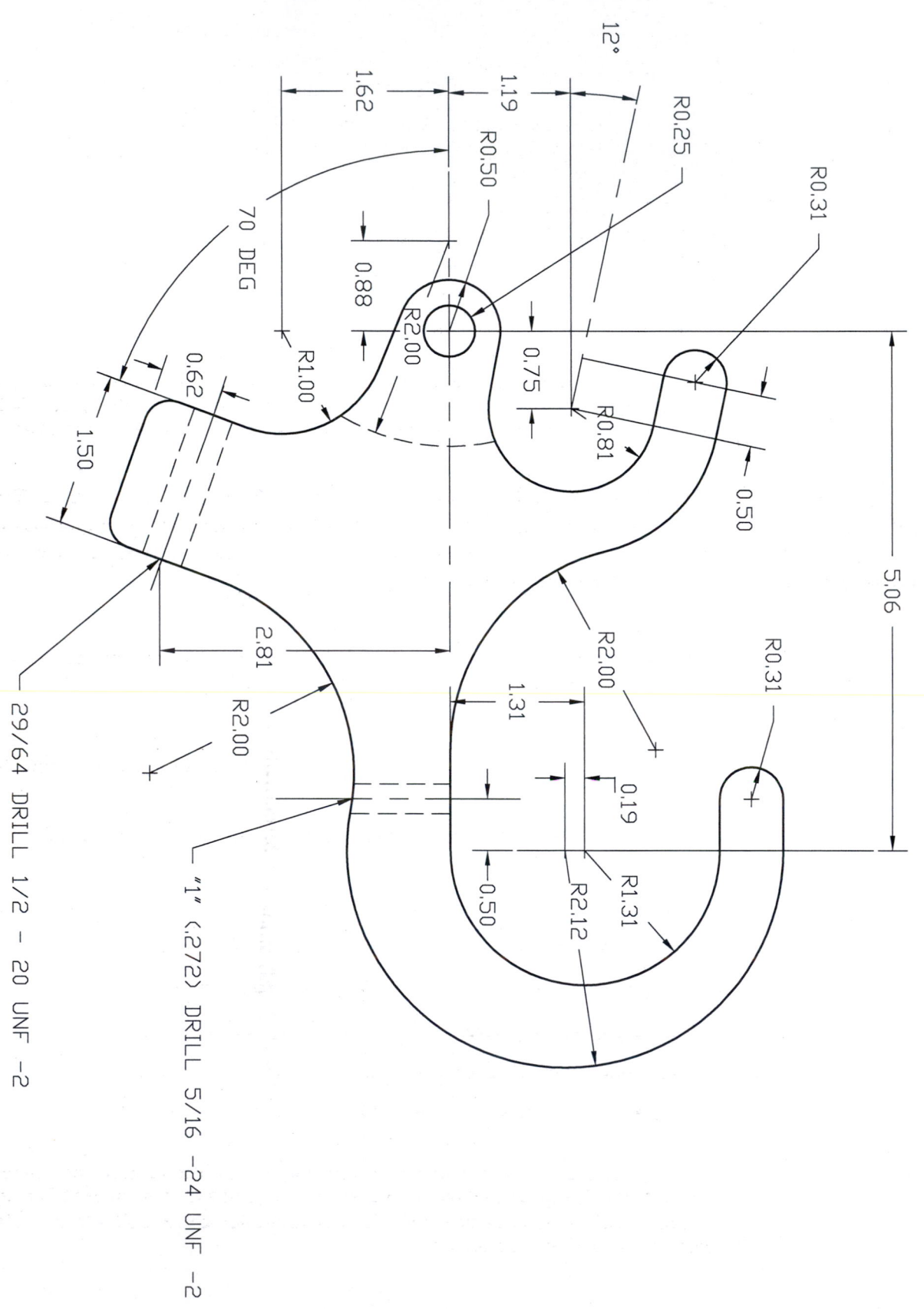

3 Object Selection and Modify Commands

On completion of this chapter, you should be able to:

1. Use the various options for selecting objects
2. Edit objects with the MOVE command
3. Edit objects with the COPY command
4. Edit objects with the MIRROR command
5. ROTATE objects
6. SCALE objects
7. Use grips to edit objects
8. Change the LINETYPE.

Once objects are drawn on the screen, editing commands under the Modify menu or toolbar are used to cut down on drawing time. When editing, OSNAP is particularly important to get the exact position of objects when they are moved, copied, or rotated.

Selecting Objects Within the Modify Commands

In virtually every editing command, you will be prompted to 'Select objects'. This makes sense because the editing command will change the position or the parameters of objects, and thus the system needs to know which objects you want to change. As you select items, they are highlighted (dotted lines).

Objects can be selected in a variety of ways. If you are selecting the object after you have invoked the command you can select the objects by:

- digitizing the desired item with the cursor pick box

- indicating a group of items with the **C**rossing option – type **C** or pick right to left

- indicating a group of items with the **W**indow option–type **W** or pick left to right

- indicating a series of items with **F**ence – type **F**

- indicating a group of items with **CP**olygon– type **CP**

- indicating a group of items with **WP**olygon – type **WP**

- indicating the last entered item using the **L**ast option – type **L**

- indicating the previous selection set with **P**revious – type **P**

- indicating all objects by typing in **ALL.**

Once you have identified your selection set or chosen the objects that are to be edited, AutoCAD will keep prompting you to select objects to be added or removed from the selection set. To continue with the Modify command after selecting objects, press ↵ to signal the end of object selection.

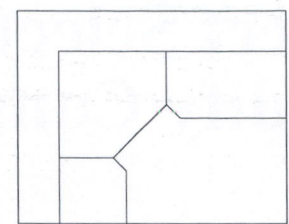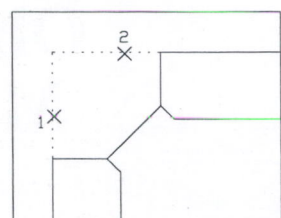

Figure 3.1

The object selection default is to select the items one by one, as shown in Figure 3.1. Only the two objects selected will be affected by the editing command.

```
Select objects:(pick 1, 2)
```

Selection Windows

You can also select objects by drawing a rectangular window or crossing area in response to the Select objects: prompts as shown in Figure 3.2.

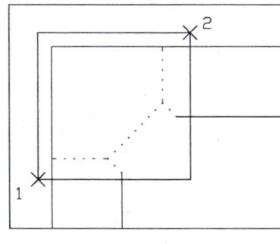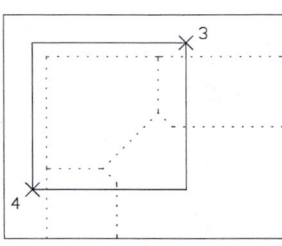

a   b

Figure 3.2

Window is Blue	***Crossing is Green***
Command: **ERASE**	Command: **ERASE**
Select objects:**(pick 1)**	Select objects:**(pick 3)**
Specify opposite corner:**(pick 2)**	Other corner:**(pick 4)**

Picking from the left to the right, as in Figure 3.2 a will create a Window (blue after Release 2007) that will highlight for modification all objects contained within it. Picking from right to left, as in Figure 3.2b will identify for modification all objects that touch the crossing line (green after Release 2007).

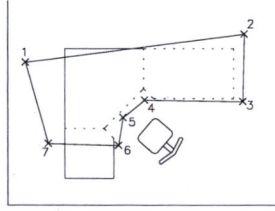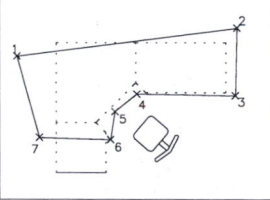

a   b

Figure 3.3

Selection Polygons

To select objects in an irregularly shaped area, use Window Polygon or Crossing Polygon. The Wpolygon (Figure 3.3a) and Cpolygon (Figure 3.3b), work like Window and Crossing windows. Type in WP or CP after the Select objects: prompt.

Selection Lines or Fence

With the Fence option, the objects touching the fence line will be picked up, as in Figure 3.4. Fence is useful for non-adjacent objects.

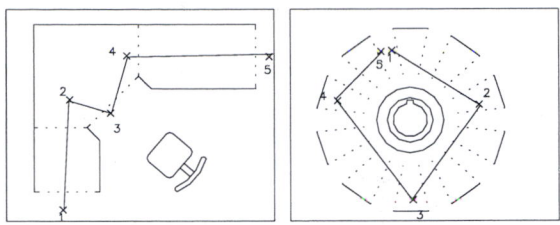

Figure 3.4

```
Select objects:F (Fence) (pick points)
```

You can use any of the above methods for object selection in any of the editing commands. The following examples show how to choose first the command, and then the selection set. Choosing the selection set first with GRIPS is covered on page 56.

Modify Commands

The COPY Command

The COPY command takes an item or group of items and places a copy at another location or at multiple locations. The COPY command assumes the creation of multiple copies.

> **Toolbar** From the Modify toolbar choose
>
> **Pull-down menu** From the Modify menu choose Copy.

The command line equivalent is COPY.

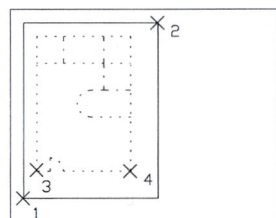 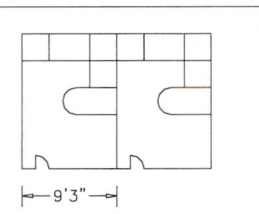

Figure 3.5

```
Command:COPY
Select objects:(pick 1, 2)
Select objects:↵
Specify base point or displacement[Multiple]:END of (pick 3)
Specify second point of displacement or <use first point as
    displacement>:END of (pick 4)
Specify second point of displacement or <>:↵
```

In Figure 3.5 the vector along which the objects will be copied is identified by point picks. You can also use an incremental value, as seen below.

```
Second point of displacement:@9'3",0 (uses the incremental
    value rather than picking points)
```

Notes

Since COPY assumes multiple copies, ↵ must be used to exit from the command.

Figure 3.6 illustrates the use of COPY with multiple entries.

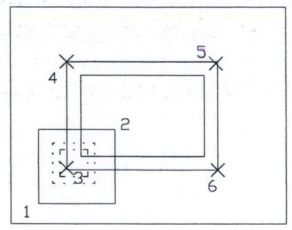 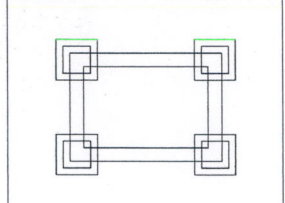

Figure 3.6

```
Command:COPY
Select objects:(pick 1, 2)
Select objects:↵
Specify base point or displacement[Multiple]:END of (pick 3)
Specify second point of displacement or <use first point as
   displacement>:END of (pick 4, 5, 6)
Specify second point of displacement or <>:↵
```

The MOVE Command

The MOVE command moves an object or series of objects from one point to another, relative to a defined point on the object or a base point.

Toolbar From the Modify toolbar choose

Pull-down menu From the Modify menu choose Move.

The command line equivalent is MOVE or M.

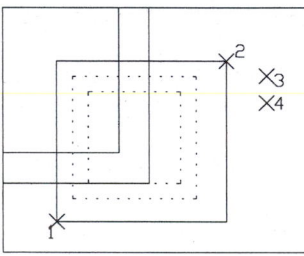 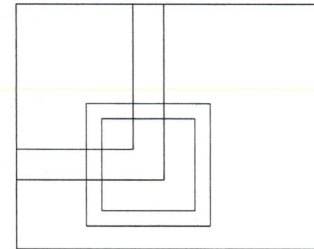

Figure 3.7

```
Command:MOVE
Select objects:(pick 1)
Specify opposite corner:(pick 2)
Select objects:↵
Specify base point or displacement:(pick 3) (from where)
Specify second point of displacement or <use first point as
   displacement>:(pick 4) (to where)
```

In Figure 3.7 the objects are moved according to point picks. As in the above example for copy, an increment or polar could also have been used:

```
Specify second point of displacement or <>:@-2,0
```

Moving Objects to 0,0

To move a selected group of objects from their current position to 0,0, first identify the selection set. Identify which point you would like to be at 0,0. Use 0,0 as the destination or displacement. In Figure 3.8, the two circles were created in the middle of the screen. They were then moved to 0,0 and then zoomed into the screen again.

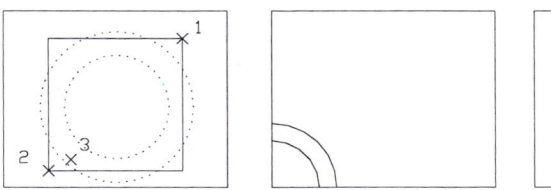

Figure 3.8

Notes

In Figure 3.8, pick 3 is where the objects are coming from, and 0,0 is where they are going to. The destination need not be ON the objects

```
Command:MOVE
Select objects:(pick 1)
Specify opposite corner:(pick 2)
Select objects:↵
Specify base point or displacement:CENter of (pick 3)
Specify second point of displacement or <use first point as
    displacement>:0,0   (to 0,0)
```

The MIRROR Command

The MIRROR command creates a mirror image of an item or group of items through a specified mirroring plane selected by a real or imaginary line.

Toolbar From the Modify toolbar choose

Pull-down menu From the Modify menu choose Mirror.

The command line equivalent is MIRROR.

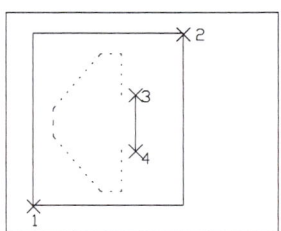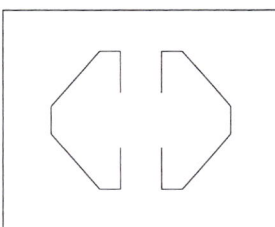

Figure 3.9

```
Command:MIRROR
Select objects:(pick 1)
Specify opposite corner:(pick 2)
Select objects:↵
Specify first point of mirroring plane:END of (pick 3)
Specify second point of mirroring plane:END of (pick 4)
Delete source objects? [Yes/No]<N>:↵
```

In Figure 3.9 the objects are mirrored through a plane created by picks 3 and 4.

Step 4

Use FILLET to create the arcs between the parts.

```
Command:FILLET
Current settings: Mode =
  Trim,Radius = <5.0000>
Select first object or[Undo/
  Polyline/Radius/Trim/Multiple]:r
Specify fillet radius <1.000>:300
Select first object or [Undo/Polyline/Radius/Trim/Multiple]:
  (pick 1)
Select second object or shift-select to apply corner:
(pick 2)

(pick 3 to 6) as shown
```

Your drawing should look like this.

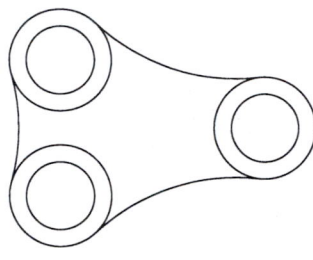

Step 5

Now use ROTATE with the copy option to make two rocker arms at a 240 degree angle.

```
Command:ROTATE
Current positive angle in UCS:
  ANGDIR=clockwise ANGBASE=0
Select objects:(pick 1, pick 2)
Select objects:⏎
Specify base point:CEN of (pick 3)
Specify rotation angle or [Copy/Reference]:c
Rotating a copy of the selected objects
Specify rotation angle or [Copy/Reference]:240
```

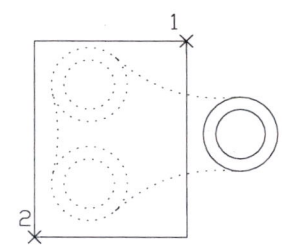

Your object should now look like this.

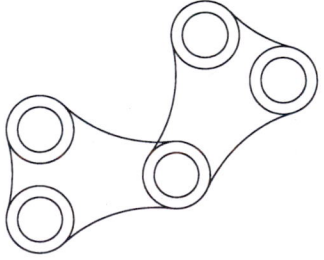

Step 6

Change the linetype of the new rocker arm.

Load a hidden linetype with the Linetype button.

Pick the objects using a crossing window and place them on this linetype.

Use LTSCALE if needed to make the new linetype appear as a hidden line.

In this example we will make a bay window using ROTATE, COPY, and MIRROR. Open a new file in imperial units and change your units to architectural – use UNITS.

Step 1

Use LINE to draw in the window itself, not the dimensions, just the window.

```
Command:LINE
Specify start point:0,0
Specify next point or (Undo):8,0
Specify next point or (Undo):8,60
Specify next point or (Undo):0,60
Specify next point or (Undo):c


Command:ZOOM   All
Command:ZOOM   .8X
```

Notes

Depending on how your system is set up, you may have to use STARTUP to access imperial units.

You can also use your mouse and roller ball to zoom the part onto the screen.

```
Command:LINE
Specify start point:0,6
Specify next point or (Undo):@8,0
Specify next point or (Undo):↵
Command:LINE
Specify start point:0,54
Specify next point or (Undo):@8,0
Specify next point or (Undo):↵
Command:LINE
Specify start point:MID of (pick 1)
Specify next point or (Undo):PER of (pick 2)
Specify next point or (Undo):↵
```

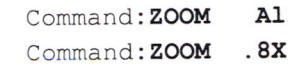

Step 2

Use COPY to create the glass section. Use POLAR if you like, if not type in the displacement value.

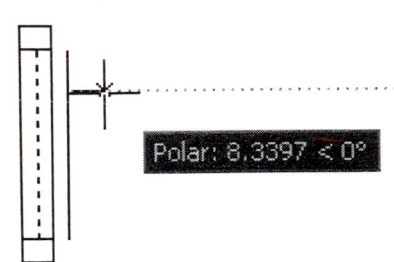

```
Command:COPY
Select objects:(pick line)
Select objects:↵
Specify base point or
   [Displacement] displacement: (pick anywhere)
Specify second point of displacement
or <use first point as displacement>:@1,0
```

Step 3

Make sure that the POLAR and OTRACK options are on to add the next 8 × 8 section.

```
Command:LINE
Specify start point:(pick the
   corner)
Specify next point or (Undo):8
Specify next point or (Undo):8
Specify next point or (Undo):8
Specify next point or (Undo):↵
```

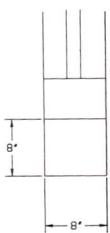

Exercise 4b Practice

OFFSET and ARRAY will be useful for drawing these objects.

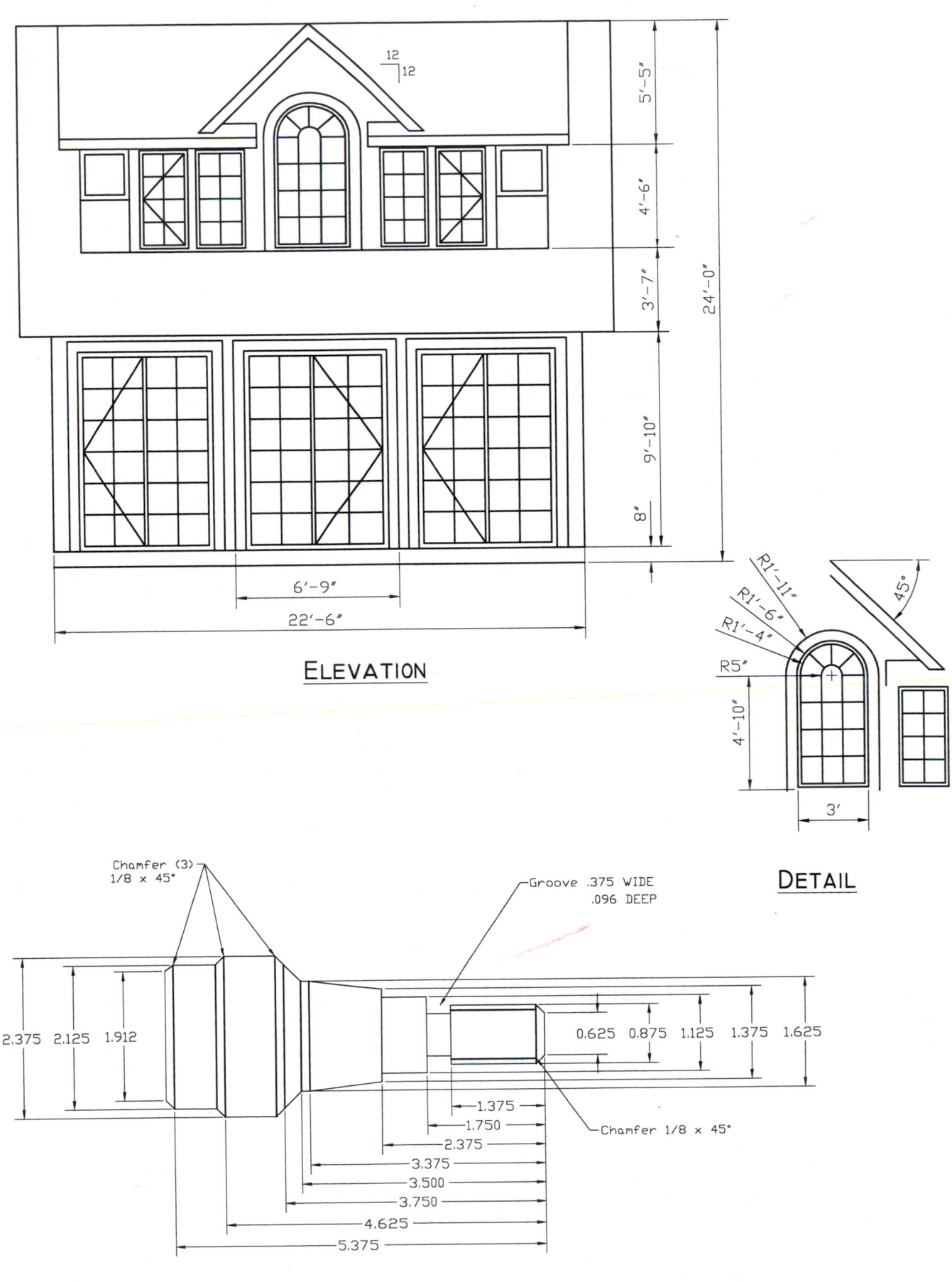

ELEVATION

DETAIL

NOZZLE

Exercise 4 Architectural

Use OFFSET to place the LINEs for the walls. Use FILLET radius 0 to clean them up.
The indow measurements are on the first floor plan on page 137.
Section A-A is on page 117.

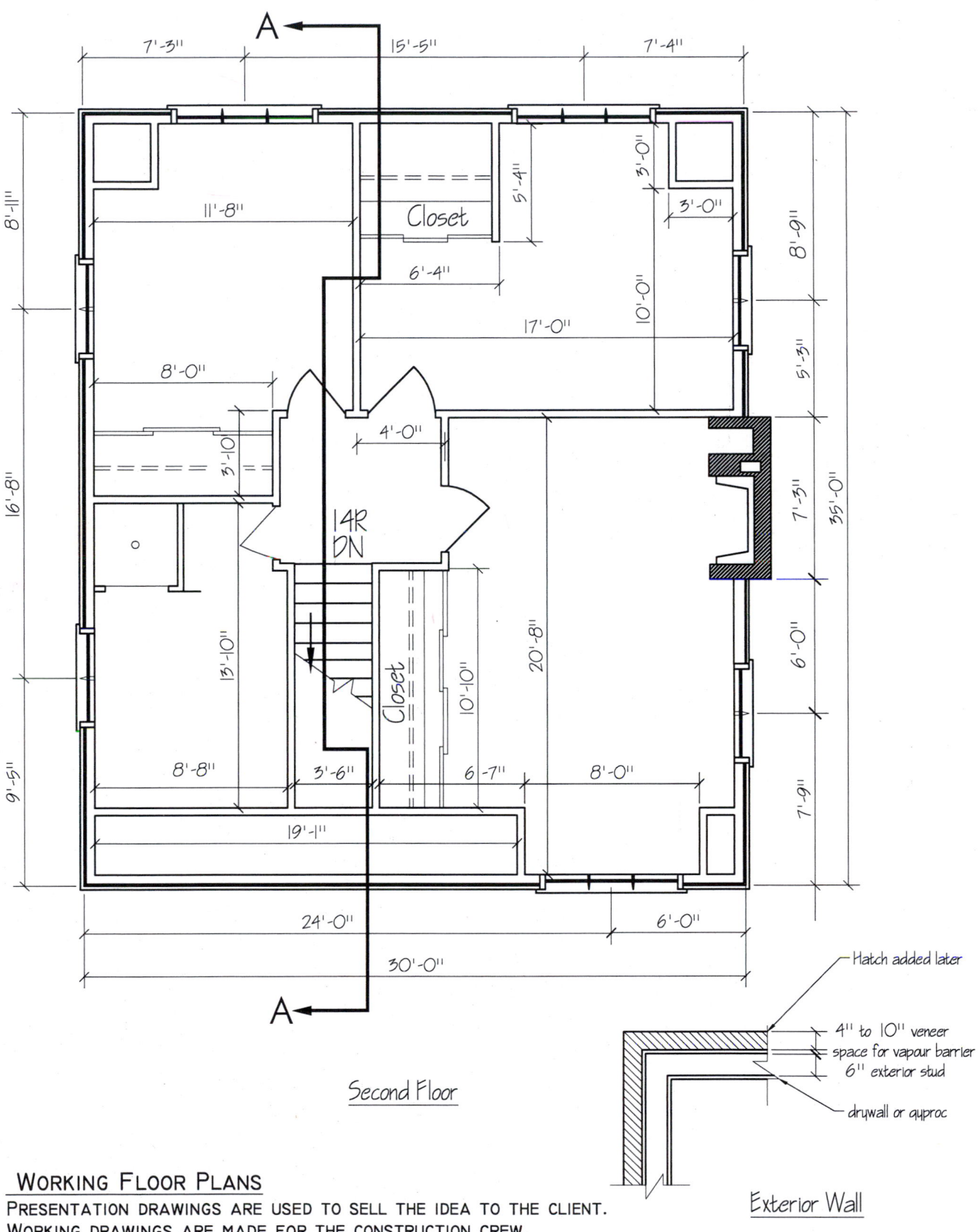

Second Floor

Exterior Wall

WORKING FLOOR PLANS

PRESENTATION DRAWINGS ARE USED TO SELL THE IDEA TO THE CLIENT.
WORKING DRAWINGS ARE MADE FOR THE CONSTRUCTION CREW.
MEASUREMENTS ARE FROM THE OUTSIDE OF THE FRAME BECAUSE
THE CARPENTER IS THE FIRST ONE TO USE THEM.
EXTERIOR WALLS ARE 6" STUD WALLS, WITH A 4 TO 10" VENEER.

Exercise 4a Mechanical

Use the commands on the facing page to create this gear.
Save your file so you can add the dimensions to it later.

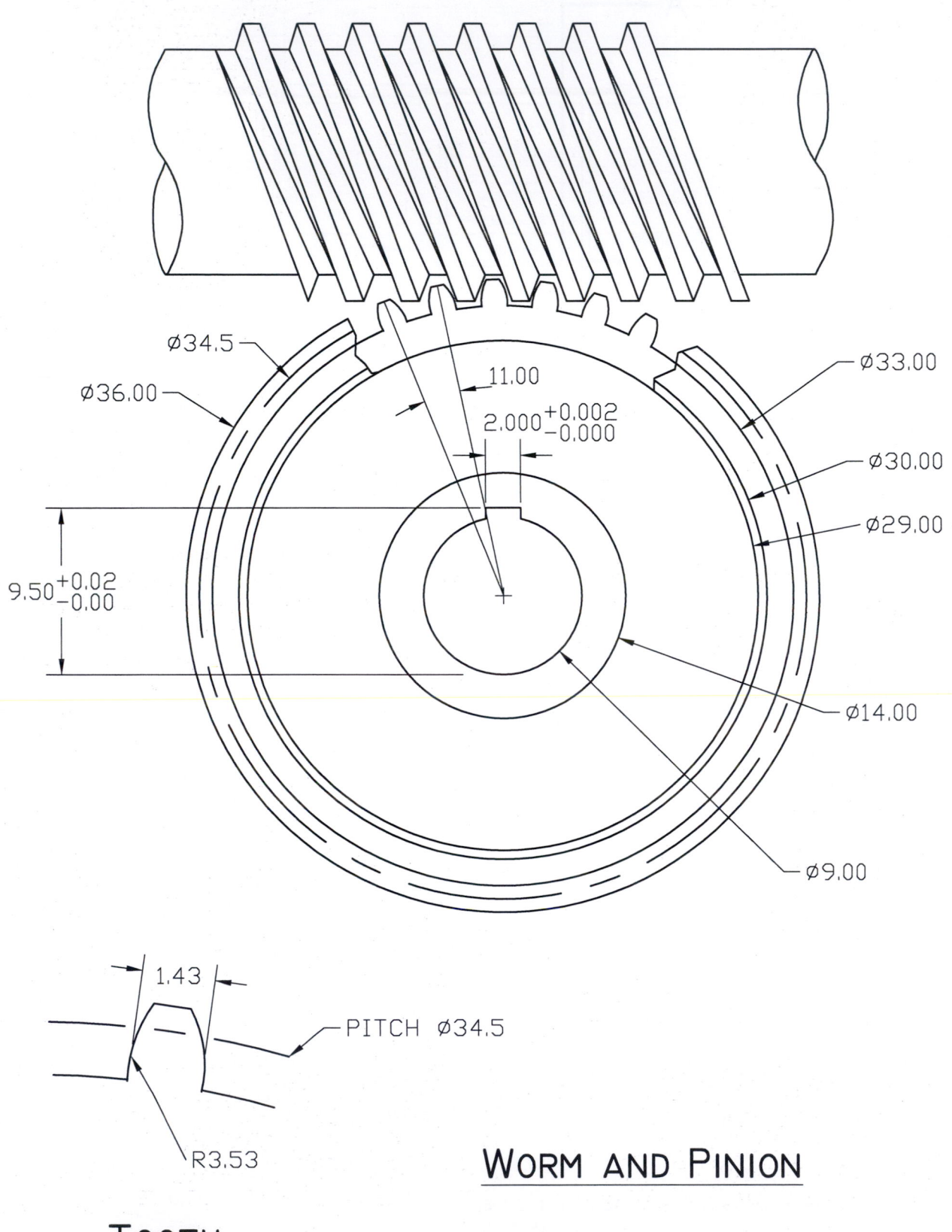

Ø34.5
Ø36.00
11.00
$2.000^{+0.002}_{-0.000}$
Ø33.00
Ø30.00
Ø29.00
$9.50^{+0.02}_{-0.00}$
Ø14.00
Ø9.00

1.43
PITCH Ø34.5
R3.53

WORM AND PINION

TOOTH

Exercise 4b Mechanical

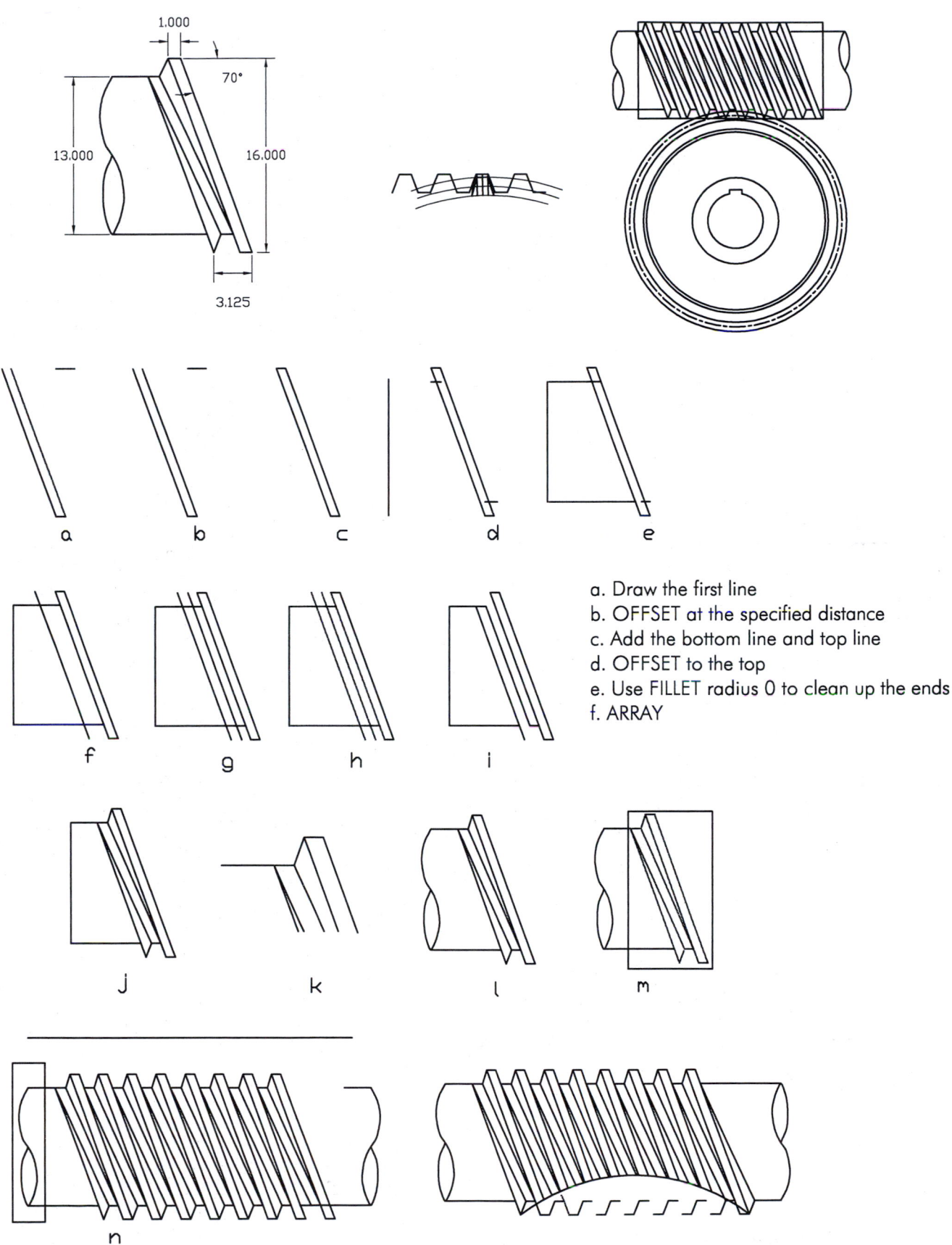

a. Draw the first line
b. OFFSET at the specified distance
c. Add the bottom line and top line
d. OFFSET to the top
e. Use FILLET radius 0 to clean up the ends
f. ARRAY

Exercise 4 Challenger

This is an interesting part that you can model in 3D later.

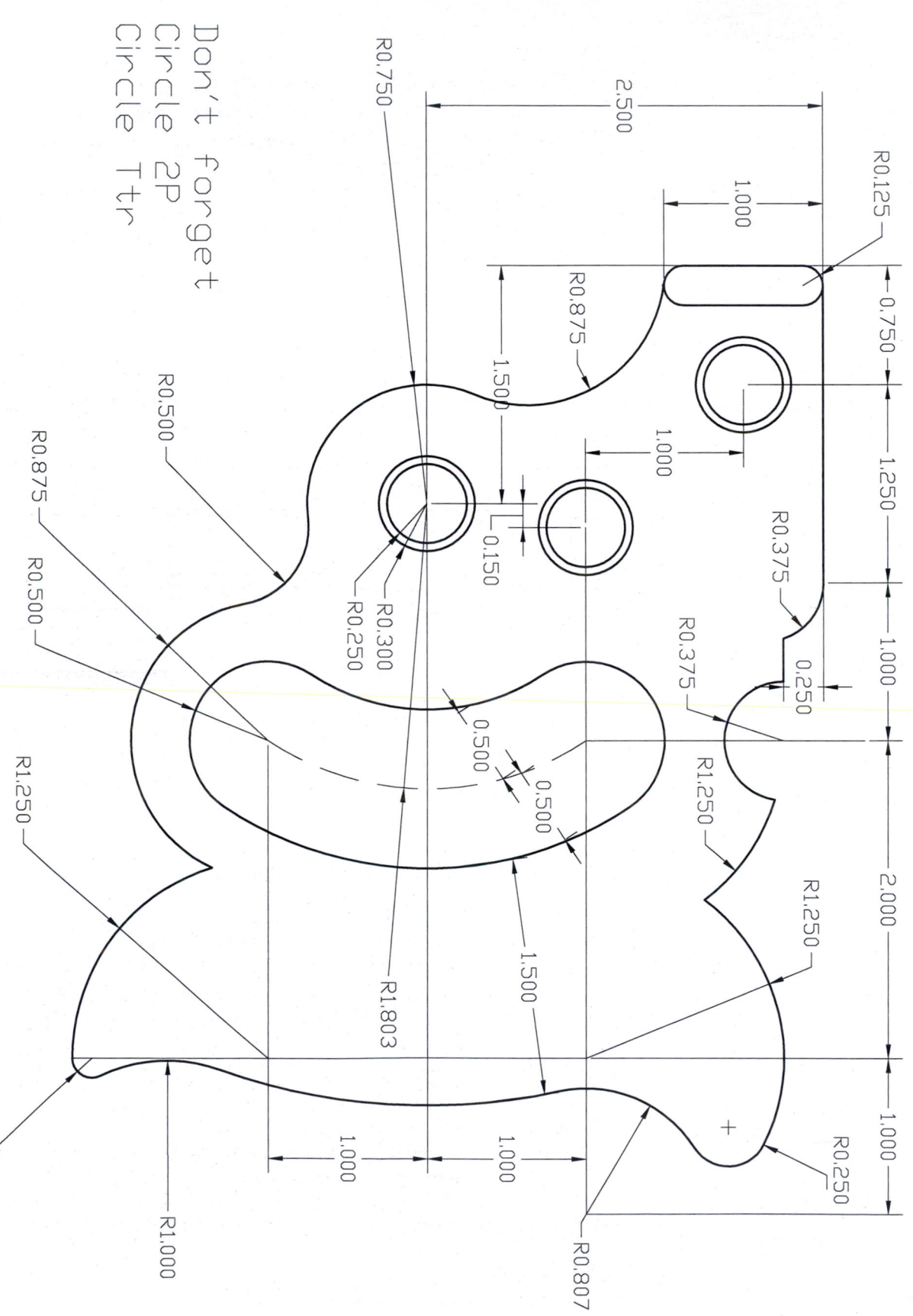

5 Entity Commands with Width

On completion of this chapter, you should be able to:
1. Create a polyline PLINE with acceptable corners and widths
2. Edit a pline using PEDIT to change the width and curve factors
3. Create a POLYGON
4. Create a SOLID
5. Create a DONUT with a specific width
6. Enter simple TEXT
7. Add Multilines with the MLINE command.

The PLINE Command

A PLINE or polyline is a single-drawing entity that includes line and curve sections that may vary in thickness, and may be edited using PEDIT (polyline edit) for the Spline and Fit curve options to create contour lines or airfoils. The individual segments are connected at vertices; the direction, tangency, and line width are stored at each vertex.

The PLINE command can create rectangles as single entities as well as curved segments of varying thickness.

Toolbar From the Draw toolbar choose

Pull-down menu From the Draw menu choose Polyline.

The command line equivalent is PLINE.

```
Command:PLINE
Specify start point:(pick a point)
Current line width is 0.0000
Specify next point or
 [Arc/Close/Halfwidth/Length/Undo/Width]:
```

Where:
Arc = a change from line entry to arc entry
Close = a closed pline, in which the first point will be joined to the last entered point in the pline to make a closed object; more than two points are needed to have a closed pline
Halfwidth = a specified halfwidth on either side of the pline vector
Length = the length of the pline
Undo = an undo of the last point entered
Width = a specified width of the line or arc segments on either side of the pline vector

The first PLINE command prompt asks for a point at which the polyline will start.

```
Command:PLINE
Specify start point:(pick a point)
```

You *must* enter the first point, after which you can choose one of the options (Width, Arc, etc.). If you want to change the width, note that you must change the start *and* the end width to create a wide line as in Figure 5.1.

Figure 5.1

```
Command:PLINE
Specify start point:(pick a point)
Specify next point or
   [Arc/Close/Halfwidth/Length/Undo/Width]:W
Specify starting width<0.0000>:.50
Specify ending width<0.500>:↵
Specify next point or
   [Arc/Close/Halfwidth/Length/Undo/Width]:11,0
```

The command default is to enter the second point. If you pick a second point, this assumes a straight segment or a line. If you continue picking points, the object created will look like a series of lines, but it will be a single object that can be edited using PEDIT or other editing commands.

PLINE is often used to create borders around views or drawings, as in Figure 5.2.

```
Command:PLINE
Specify start point:0,0
Current line width is 0.0000 units
Specify next point or
   [Arc/Close/Halfwidth/Length/Undo/
   Width]:W
Specify starting width<0.0000>:.25
Specify ending width<0.2500>:↵
Specify next point or
   [Arc/Close/Halfwidth/Length/Undo/
   Width]:11,0
```

Figure 5.2

You can continue drawing with this line at the current thickness or change it at any time. POLAR, OTRACK, and ORTHO work the same on PLINE as on LINE.

```
Specify next point or [Arc/Close/.../Undo/Width]:11,8.5
Specify next point or [Arc/Close/.../Undo/Width]:0,8.5
Specify next point or [Arc/Close/.../Undo/Width]:C
```

To achieve a perfect corner on a box or rectangle, use the Close option. This will attach the first point to the last entered point and create a clean, beveled corner.

When changing the width both start and end points must be entered.

```
Command:PLINE
Specify start point:(pick 1)
Current line width is 4.00 units
Specify next point or
   [Arc/Close/Halfwidth/Length/Undo/Width]:W
Specify starting width<4.00>:.25
Specify ending width<0.25>:↵ (don't pick!)
```

Do not pick! If you pick, you will create an object similar to that in Figure 5.3. This is a common mistake. AutoCAD will measure the distance from your first pick to your second and use that as the end width of your PLINE.

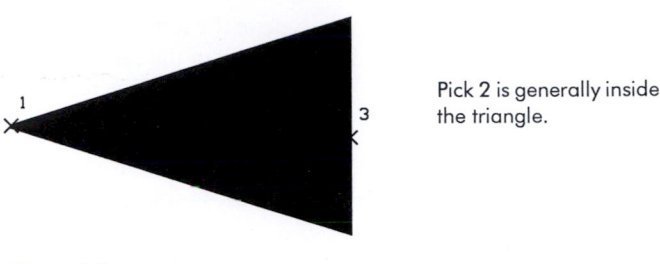

Pick 2 is generally inside the triangle.

Figure 5.3

Pline with Varying Width

You can create objects such as arrows using one PLINE with a series of different segments with varying widths. You can change the width at every vertex, as in Figure 5.4.

Figure 5.4

```
Command:PLINE
Specify start point:(pick 1)
Current line width is 4.00 units
Specify next point or
   [Arc/Close/Halfwidth/Length/Undo/Width]:W
Specify starting width<4.00>:0
Specify ending width<4.00>:.35
Specify next point or
[Arc/Close/Halfwidth/Length/Undo/Width]:(pick 2)
Specify next point or
   [Arc/Close/Halfwidth/Length/Undo/Width]:W
Specify starting width<0.35>:.10
Specify ending width<0.10>:.10
Specify next point or
   [Arc/Close/Halfwidth/Length/Undo/Width]:(pick 3)
```

PLINE Corners

The end of each pline is calculated relative to the points or vertices used to create it. If the pline has a width, the ends are perpendicular to the pline itself. Pline corners must be smooth, or a gap will result, as in Figure 5.5.

When making plines, use a continuous series of points as in Figure 5.5b, don't pick twice in the same spot as in Figure 5.5a.

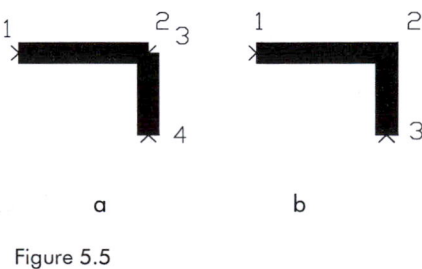

a   b

Figure 5.5

Polyarcs

The PLINE command can be used to create arcs within polyline segments or on their own, as shown in Figure 5.6.

Simply type in the letter A to toggle to the arc mode, then type in the letter L to get back to straight line segments. The resulting polyline will be one entity with arc and line segments.

Editing the polyline curve options with PEDIT will override these line and arc segments.

Figure 5.6

```
Command:PLINE
Specify start point:(pick a point)
Current line width is 0.1000
Specify next point or [Arc/Close/
 Halfwidth/Length/Undo/Width]:A
[Angle/CEnter/CLose/Direction/Halfwidth/
 Line/Radius/Second point/Undo/Width]:
```

The command default is to create a two-point arc. Once the first arc segment is in, the system assumes that you want to continue with a series of arcs until the L option is entered, which will return you to a line segment.

The PEDIT Command

One of the great advantages of PLINE is that, once the pline is entered, it can be modified using PEDIT.

In the introductory stages, this command sequence is used most often to change the width of a pline. In Figure 5.7, the border is edited from .25 units to .10 units.

The PEDIT command changes the width of all the segments of the identified pline.

Toolbar From the Modify II toolbar choose

Pull-down menu From the Modify menu choose Polyline.

The command line equivalent is PEDIT.

Figure 5.7

```
Command:PEDIT
Select polyline:(pick the polyline)
Enter an option
[Close/Join/Width/Edit Vertex/Fitcurve/Spline/Decurve/Ltype
  gen/Undo:W
Specify new width for all segments:.10
Enter an option
 [Close/Join/Width/Edit Vertex/Fitcurve/Spline/Decurve/Ltype
  gen/Undo]:↵
```

More segments can be added to the pline by using the Join option. This will add lines or arcs to a pline which can then be edited for width.

In Figure 5.8, the arc and the two lines identified by picks 2, 3, and 4 are added to the pline selected with the first pick before the command option Join. Only segments with a common end point can be joined.

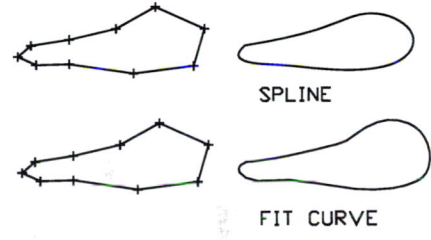

```
Command: PEDIT
Select polyline: (pick 1)
Enter an option
   [Close/Join/Width/Edit
  Vertex/Fit
   curve/Spline/Decurve/Ltype
  gen/Undo]: J
Select objects: (pick 2)
Select objects: (pick 3)
Select objects: (pick 4)
Select object: ⏎
Enter an option [Close/Join/Width/Edit
  Vertex/Fit curve/Spline/DecurveLtype gen/Undo]: ⏎
```

Figure 5.8

You can change polylines back to regular lines by using the command EXPLODE. This will also remove the width given.

If you apply PEDIT to an object that is not a polyline, you will be given the option of turning it into one. Then use Join.

PEDIT with Spline and Fit Curve

To create contour lines and other items used in surveying, among other fields, a polyline can be modified to become a spline or fit a curve through a series of points or vertices.

The Spline curve option of PEDIT will edit the pline according to the series of points used to create it. Make sure you use the Close option when entering the pline to get a continuous spline as seen in Figure 5.9.

SPLINE

FIT CURVE

Figure 5.9

PLINE with PEDIT can also be used to show wood grain as in reverse sections for woodworking or molding, as can be seen in Figure 5.10. First put in a PLINE. Then use PEDIT with either the Fit Curve or Spline options.

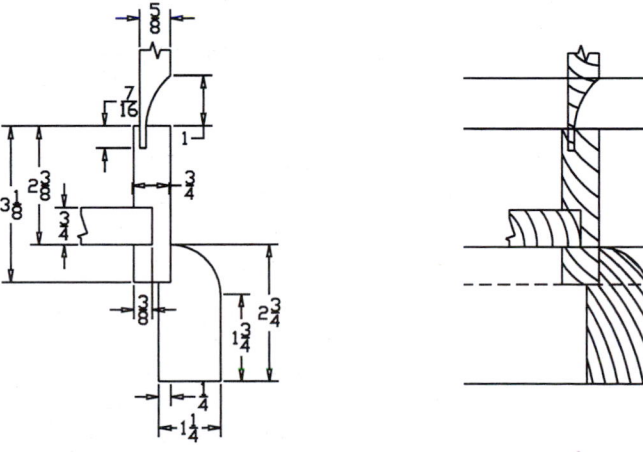

Figure 5.10

The POLYGON Command

The POLYGON command is like PLINE in that you are creating an object that has many vertices.

Toolbar From the Draw toolbar choose the Polygon flyout then

Pull-down menu From the Draw menu choose Polygon.

The command line equivalent is POLYGON.

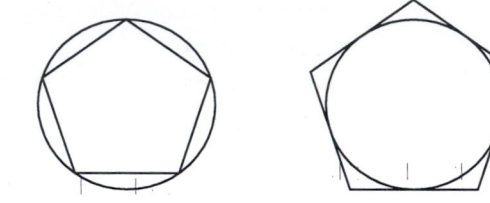

Figure 5.11

```
Command:POLYGON
Enter number of sides<4>:5
Specify center of polygon or [Edge]:(pick a center point)
Enter an option [Inscribed in circle/ Circumscribed about
   circle] <I>:I
Specify radius of circle:1.5
```

Figure 5.11 illustrates a five sided-polygon.

Hexagons

In mechanical drawing, hexagons (six-sided regular polygons) are quite common.

If you need to draw a hexagon that is measured by the distance across the flats choose the option Circumscribed about circle. The diameter across the circle will equal the distance across the flats, as shown in Figure 5.12.

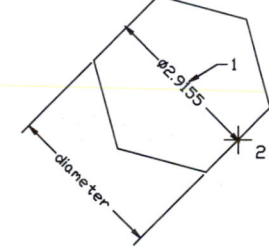

Figure 5.12

```
Command:POLYGON
Enter number of sides<4>:6
Specify center of polygon or [Edge]:(pick center)
Enter an option [Inscribed in circle/
   Circumscribed about circle] <I>:C
Radius of circle:(pick 2)
```

You can also use the POLYGON command to draw regular polygons by specifying the length of an edg,e as shown in Figure 5.13.

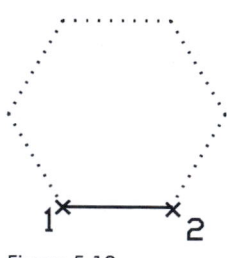

Figure 5.13

```
Command:POLYGON
Enter number of sides<4>:6
Specify center of polygon or [Edge]:E
Specify first endpoint of edge:(pick 1)
Specify second endpoint of edge:@4<0
```

The SOLID Command

SOLID and DONUT are commands that have a filled-in area or solid mass. This can also be achieved with the BHATCH command with the solid fill hatch. These commands are always available when you type them in, and can be found in some AutoCAD Releases (not Releases 2007 or 2008) on the pull-down menus and toolbars.

Toolbar From the Surfaces toolbar choose

Pull-down menu From the Draw pull-down menu choose Surfaces then 2D SOLID.

The command line equivalent is SOLID.

```
Command:SOLID
Specify first point:(pick 1)
Specify second point:(pick 2)
Specify third point:(pick 3)
Specify fourth point:(pick 4)
Third point:↵
```

Figure 5.14

Figure 5.14 illustrates the order in which points are entered. If you continue to digitize after the fourth entry, the system will add in a paired sequence until you terminate with ↵. To create a triangle, use ↵ after the third point prompt.

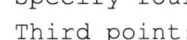

Figure 5.15

The order in which points are entered is very important. Figure 5.15 provides some examples.

Notes

Do not confuse SOLID with the SOLIDS option given under the Draw menus.

The DONUT Command

The DONUT command is used to create a thick or solid circle. The inside diameter is used to determine the hole of the doughnut. Use an inside diameter of zero to create a solid circle; use a larger diameter to create a ring. Once DONUT is active, a donut will be drawn every time you digitize until you press ↵.

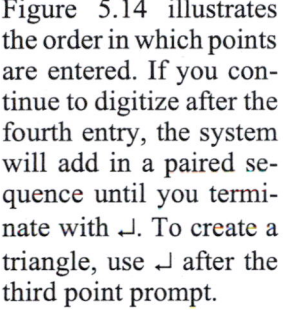

Figure 5.16

Figure 5.16 illustrates the inside and outside diameters of a DONUT .

DONUT is useful for surveying and civil engineering applications for property and position markers. It is also useful within dimensioning if the options available aren't giving you what you think you want.

```
Command:DONUT
Specify inside diameter of donut<.5>:0
Specify outside diameter<1.0>:2
Specify center of donut or <exit>:(pick 1)
Specify center of donut or <exit>:(pick 2)
Specify center of donut or <exit>:↵
```

Like the SOLID command, DONUT has been replaced by CIRCLE and BHATCH. For many applications, however, DONUT is a faster solution for a solid circle.

The TEXT Command

Chapter 8 deals with many aspects of text, including text style, editing, fonts, paragraph text, and a wide variety of text alignments. Here we introduce simple, one-line text entry. The MTEXT command will make paragraphs and large text formats easier. The TEXT command is useful for title blocks and view titles.

When entering TEXT, AutoCAD will prompt you to choose a height for each character, a rotation angle for your string, and a point at which to place the text string on the model or drawing.

> **Toolbar** From the Draw toolbar neither TEXT nor DTEXT are available.
>
> **Pull-down menu** From the Draw menu choose Text, then Single-Line Text.

The command line equivalent is TEXT.

Figure 5.17 illustrates the justification of the TEXT command. The default is left justification at the baseline of the text string. The other options will center, right justify, or fit to either top or bottom corners.

DEFAULT   CENTERED

ALIGNED   MIDDLE
FIT FIT RIGHT

Figure 5.17

```
Command:TEXT
Current text style: "Standard" Text height 0.200
Specify start point [Justify/Style]:J
Align/Fit/Center/Middle/Right/TL/TC/TR/ML/MC/MR/BL/BC/BR:
```

Once you have chosen a point at which to place your text, the command will prompt you for the height of the letters, the rotation angle, and the text or string of characters itself. Figure 5.18 illustrates the placement of single line text.

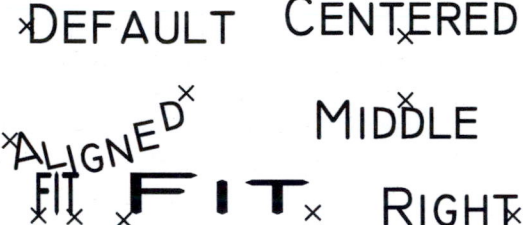

SCALE:
1/4"=1'0"

Figure 5.18

```
Command:TEXT
Current style: "Standard" Text height
  0.200
Specify start point [Justify/Style]:(pick 1)
Specify height <.2000>:.15
Specify rotation angle of text <0>:↵ (to accept the default)
Enter text:Scale:
Enter text:↵
Command:↵
Current text style: "Standard" Text height 0.150
Specify start point [Justify/Style]:(pick 2)
Specify height<.2000>:.25
Specify rotation angle of text <0>:↵
Enter text:1/4"=1'0"
Enter text:↵
```

Multilines

The MLINE or multiline command is used to create multiple, parallel lines.

The command line equivalent is MLINE.

MLINE is particularly useful for drawing walls and other architectural features. From a specified point, AutoCAD draws a multiline segment using the current multiline style, and continues to prompt for other points. Like LINE, using Undo undoes the last vertex point on the multiline. If you create a multiline with two or more segments, the Close option will be included in the command string as seen in Figure 5.19.

```
Command:MLINE
Current settings; Justification  = Top, Scale = 1.00,
  Style = Standard
Specify start point or [Justification

 /Scale/STyle]:(pick 1)
Specify next point:(pick 2)
Specify next point or
 [Close/Undo]:(pick 3)
Specify next point or
 [Close/Undo]:(pick 4)
Specify next point or
 [Close/Undo]:C
```

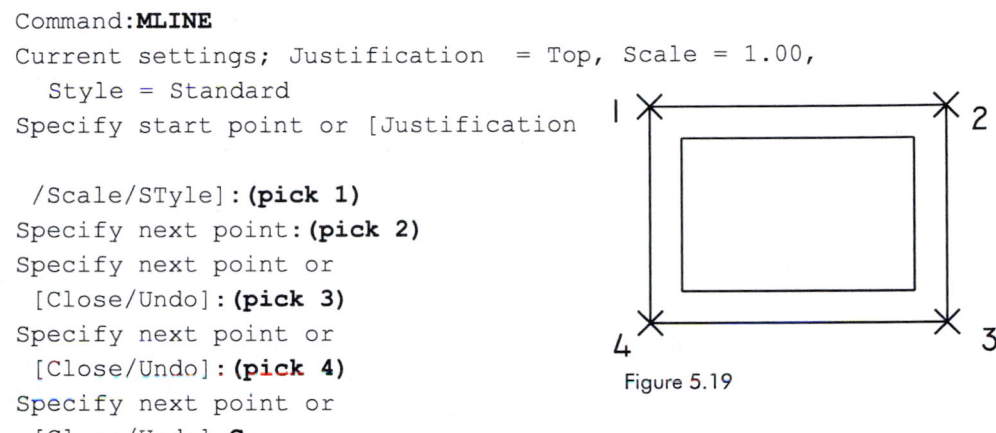

Figure 5.19

Justification

This option determines how the multiline is drawn between the points you specify.

Enter Justification type Top, Zero, or Bottom, as shown in Figure 5.20.

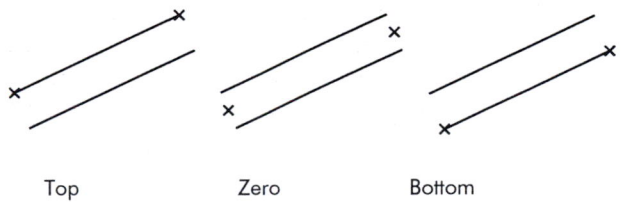

Top   Zero   Bottom

Figure 5.20

Scale

This option determines the distance between the two lines of the multiline.

The scale is based on the width established in the multiline style, as shown in Figure 5.21. All lines in the MLINE will be scaled to the entered scale factor.

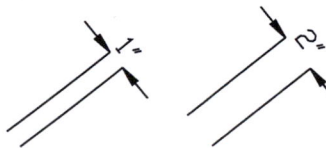

Figure 5.21

Tutorial 5 Using PLINE and SOLID

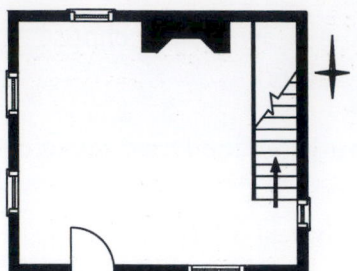

By using PLINE and SOLID effectively, we can create a presentation view of a small cabin with a staircase and a fireplace. Like many architectural applications, this is more easily drawn with SNAP on because it is in regular intervals.

Step 1

First change the units to Architectural. At the command prompt type UNITS.

```
Command:UNITS
Architectural
Precision: 0'-0"
```

Step 2

Make your LIMITS -1',-1' and 24',18'. Set SNAP to 4".

```
Command:LIMITS
Reset model space limits
Specify lower left corner or [ON/OFF] <0'-0",0'-0">:-1',-1'
Specify upper right corner <12.0000,9.0000>:24',18'
```

Step 3

Add the PLINEs as shown on the illustration.

Notes

When entering co-ordinates, DYNamic gets in the way. Using GRID and SNAP is the fastest way to do this.

This is a presentation drawing not a construction drawing, the dimensions are from the middle of the plines on the corners. Note how to start the PLINE so that your corner is correct in the command string below.

Set your SNAP and GRID to 4 to make it easier.

The polyline width is 8", the size of a regular exterior wall.

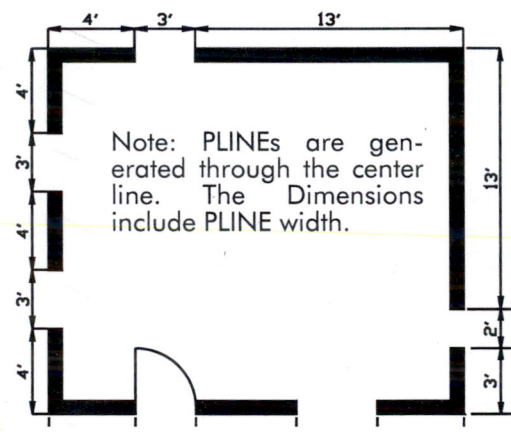

Note: PLINEs are generated through the center line. The Dimensions include PLINE width.

```
Command:PLINE
Specify start point:4',0
Current line width is 0.00 units
Specify next point or
   [Arc/Close/Halfwidth/Length/Undo/Width]:W
Specify starting width<0.00>:8
Specify ending width<8.00>:↵
Specify next point or
   [Arc/Close/Halfwidth/Length/Undo/Width]:0,0
Specify next point or
   [Arc/Close/Halfwidth/Length/Undo/Width]:0,4'
Specify next point or
   [Arc/Close/Halfwidth/Length/Undo/Width]:↵
```

Repeat PLINE for the exterior walls. Use the spacebar to bring back the PLINE command to make it faster.

Step 4

Now use ZOOM Window to get a closer look at the northeast corner of the building. Add the lines as shown, with the pline width at 2″.

Once sufficiently zoomed, enter the staircase with PLINE.

```
Command:PLINE
Specify start point:NEAR to (pick 1)
Current line width is 8.00 units
Specify next point or [Arc/Close/
 Halfwidth/Length/Undo/Width]:W
Specify starting width<8.00>:2
Specify ending width<2.00>:↵
Specify next point or [Arc/Close/
 Halfwidth/Length/Undo/Width]:(pick 2)
```

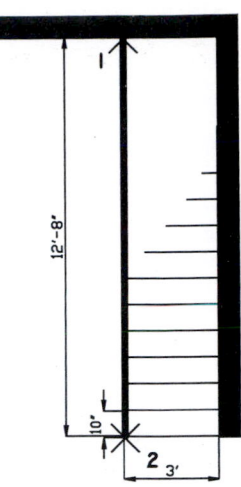

Step 5

Now use PLINE to place an arrow pointing up the stairs. The dimensions in the command string below are simply a guide. You can use your co-ordinate readout to place these rather than typing them in. Just make the arrow look like the one shown.

```
Command:PLINE
Specify start point:18'2",4'6"
Current line width is 2.00 units
Specify next point or [Arc/Close/Halfwidth
 /Length/Undo/Width]:18'2",7'
Specify next point or [Arc/Close/Halfwidth
 /Length/Undo/Width]:W
Specify starting width <2.00>:7
Specify ending width<7.00>:0
Specify next point or [Arc/Close/Halfwidth
 /Length/Undo/Width]:18'2",8'
Specify next point or [Arc/Close/Halfwidth
 /Length/Undo/Width]:↵
```

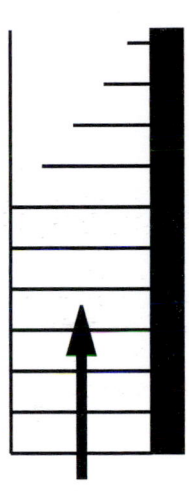

Step 6

Now use PLINE to create a break line on the staircase. A break line is used on floor plans to indicate that the stairs extend up or down to another level.

```
Command:PLINE
Specify start point:(pick 1)
Current line width is 0.00 units
Specify next point or [Arc/Close/
Halfwidth/Length/Undo/Width]:(pick 2,
  3, 4, 5, 6, 7 in sequence)
```

> **Notes**
>
> Since PLINEs are created along a center line with width added to *both* sides, be sure to deduct half the wall thickness to obtain the correct wall length.

A break line in a staircase is always necessary on a floor plan. The plan view shows only one floor or level in the building. The stairs extend up to another level. The break line is the visual image that tells the viewer that the stairs continue.

Step 7 Now use PAN to move the drawing to the right, at the same scale factor, and then use SOLID to create a fireplace. You should have SNAP on and OSNAP off for this command.

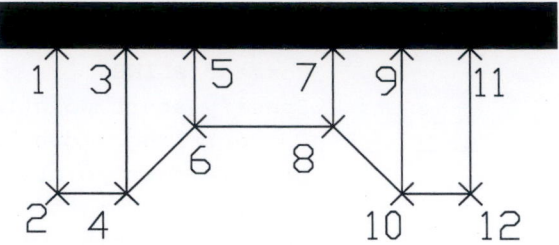

Notes

PLINEs are entered from their MIDDLE point, so calculations will need to be adjusted.

```
Command:SOLID
Specify first point:(pick 1)
Specify second point:(pick 2)
Specify third point:(pick 3)
Specify fourth point or [Exit]:(pick 4)
Specify third point:(pick 5)
Specify fourth point or [Exit]:(pick 6)
Specify third point:(pick 7)
Specify fourth point or [Exit]:(pick 8)
Specify third point:(pick 9)
Specify fourth point or [Exit]:(pick 10)
Specify third point:(pick 11)
Specify fourth point or [Exit]:(pick 12)
Specify third point:⏎
```

Step 8 Use PAN again to move the drawing over to the left so that you can create a North arrow outside the building. Draw in a vertical line at 3 ′, and a horizontal line at 1.5′ across the lower section as shown. Then draw in two smaller lines to make an x. Use the ends of these lines and OSNAP ENDpoint to draw in a North arrow.

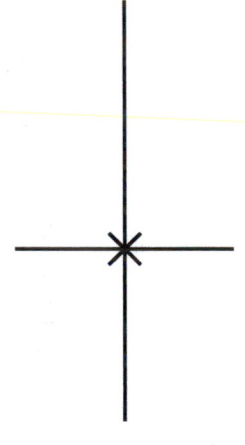

North arrows are used on floor plans to show the direction that the house is facing. In the northern hemisphere, the north side of the building will be cooler than the south, so the fireplace is often placed on that wall.

Step 9 Then, with your SNAP still set to 2, draw in two diagonal lines through the intersection. Now use SOLID to fill in the sides of the North arrow.

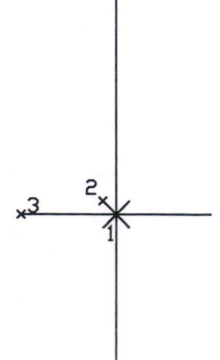

```
Command:SOLID
Specify first point:(pick 1)
Specify second point:(pick 2)
Specify third point:(pick 3)
Specify fourth point or [Exit]:⏎
Specify third point:⏎
```

Continue adding solids until you have a North arrow.

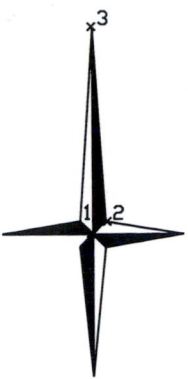

Step 10   Finally, add lines for windows and doors, and your simple floor plan is complete.

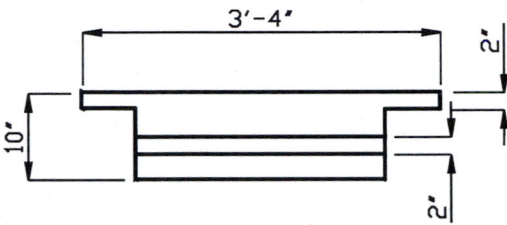

Once the first window is drawn in, use COPY, ROTATE, and STRETCH to make the others. Note that you can adjust the size of the window openings with STRETCH to make the window on the south side longer, taking advantage of the sun and solar energy.

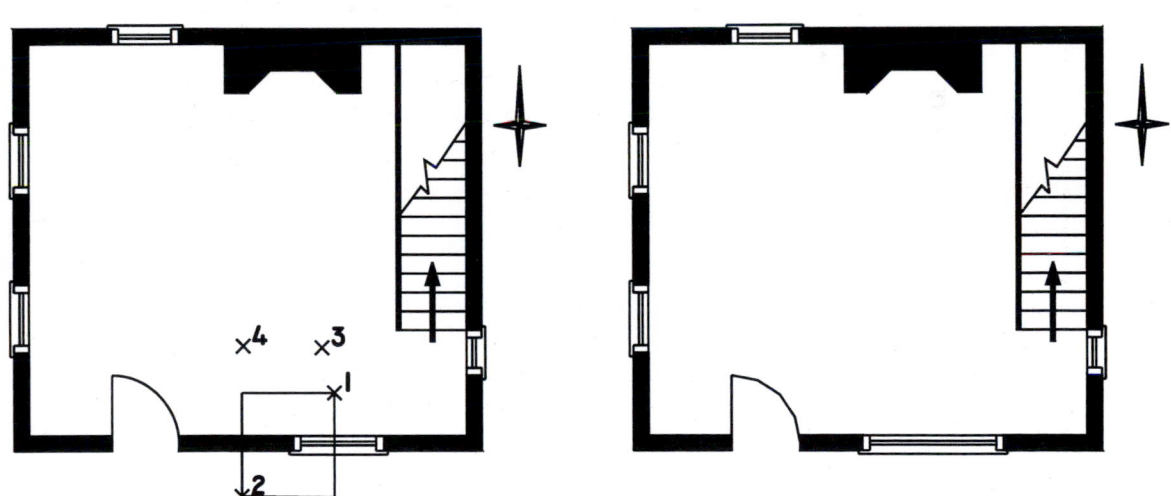

<u>PRESENTATION DRAWINGS</u>
PRESENTATION DRAWINGS, GENERALLY DRAWN IN PLINES, ARE USED TO SHOW A CLIENT THE GENERAL LAYOUT OF A BUILDING. THE POSITIONS OF WINDOWS, DOORS, STAIRCASES, AND FIREPLACES ARE SHOWN, BUT THERE ARE NO DEFINITE DIMENSIONS, NO NOTATIONS OF MATERIALS, AND NO STRUCTURAL INFORMATION. THESE WOULD BE ON THE CONSTRUCTION DRAWINGS (SEE PAGE 85).

Exercise 5 Practice

Draw this intersection using the line widths and types shown.
Use PLINE and width for the direction arrows.

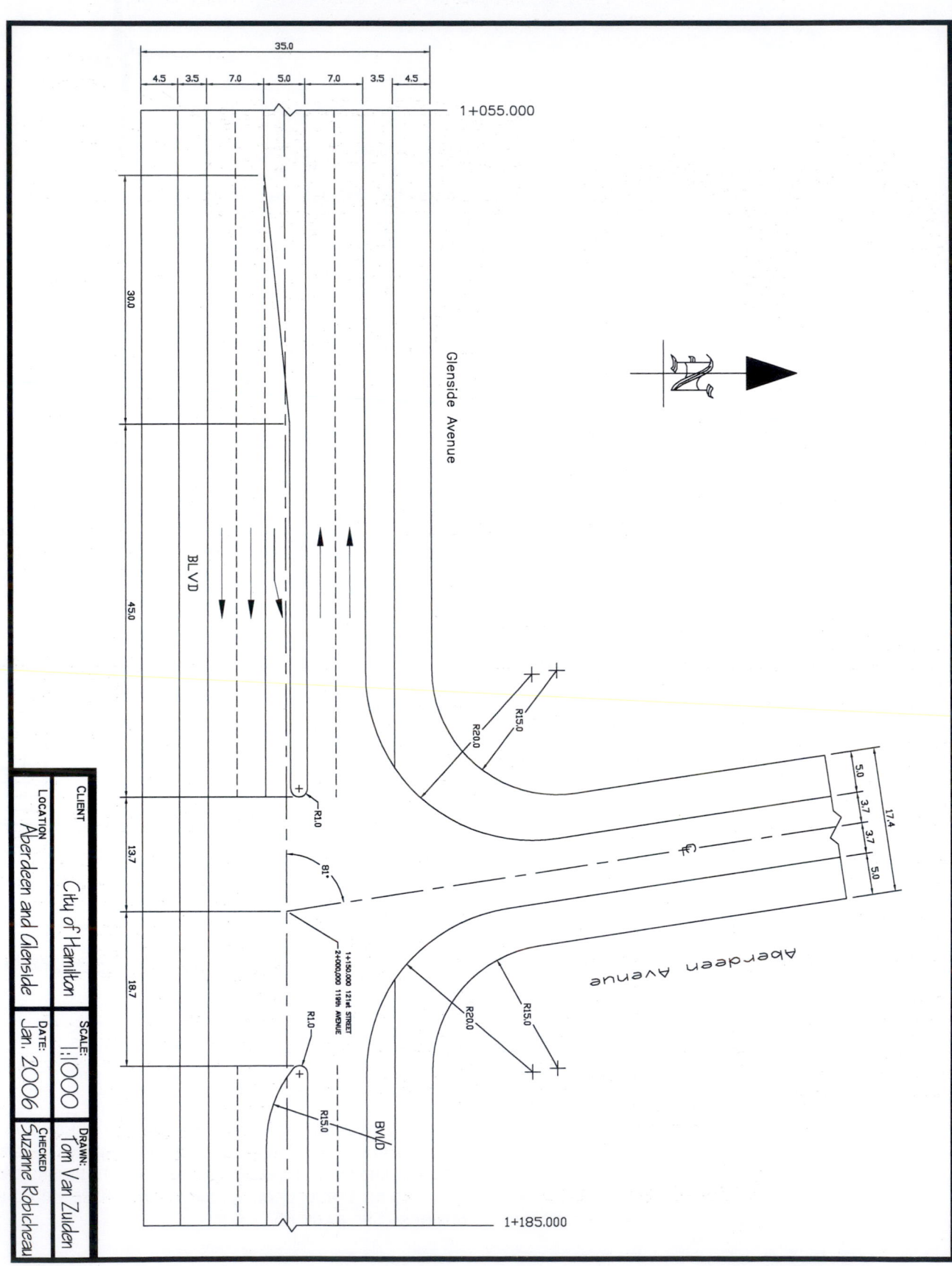

Exercise 5 Architectural

To appreciate architecture you must have some knowledge of architectural styles.
Here are two examples. Use PLINE for the acanthus leaves.
Set PDMODE to 34, then use DIVIDE to place the bricks along the arc.
MIRROR, COPY, and SCALE will also be useful.

TRIGLYPHS

EGG AND DART
ECHINUS

VOLUTE

ACANTHUS LEAVES

TORUS

BYZANTINE AND ROMANESQUE

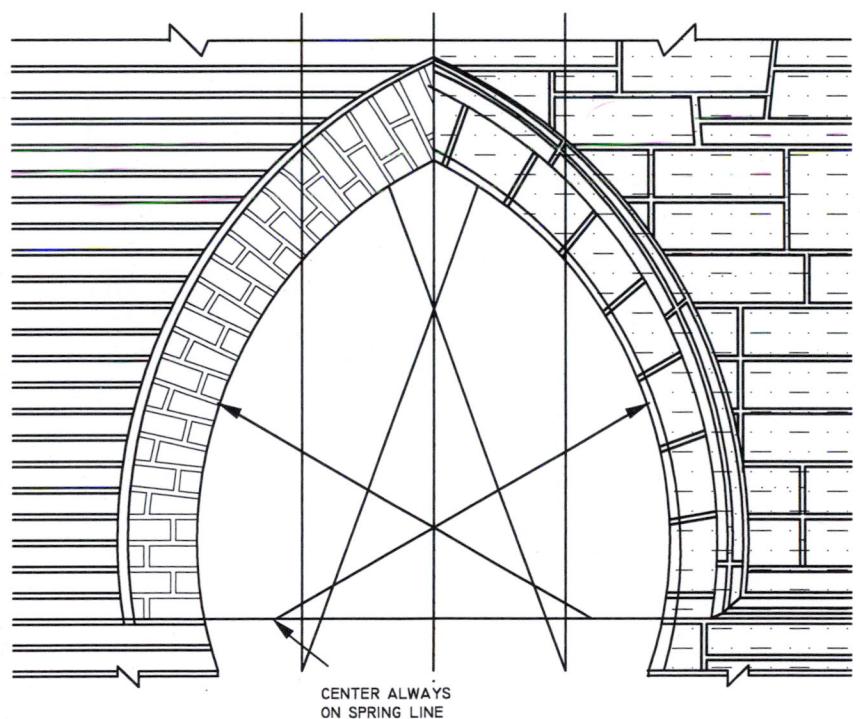

CENTER ALWAYS
ON SPRING LINE

Pointed Horseshoe

ARCHITECTURAL DETAILS

THE ARCHITECTURAL DETAILS
AROUND DOORS AND WINDOWS ARE
USED TO DETERMINE THE
BUILDING'S STYLE AND DATE. THIS
IS WHERE CRAFTSMANSHIP AND A
SENSE OF PROPORTION ARE
IMPORTANT.
WHERE STRUCTURAL DETAILS
SHOW THE STRUCTURE OF A
BUILDING, ARCHITECTURAL DETAILS
DETERMINE THE STYLE.
SEE
WWW.ONTARIOARCHITECTURE.COM
FOR MORE INFORMATION ABOUT
THESE STYLES.

NOTE : POINTED HORSESHOE ARCHES ARE FOUND PRIMARILY IN VENETIAN GOTHIC ARCHITECTURE.
STONE JOINTS MAY BE HANDLED IN A VARIETY OF WAYS

Exercise 5 Mechanical

The PLINE showing the break in the shaft will be the hardest part to draw in this example.

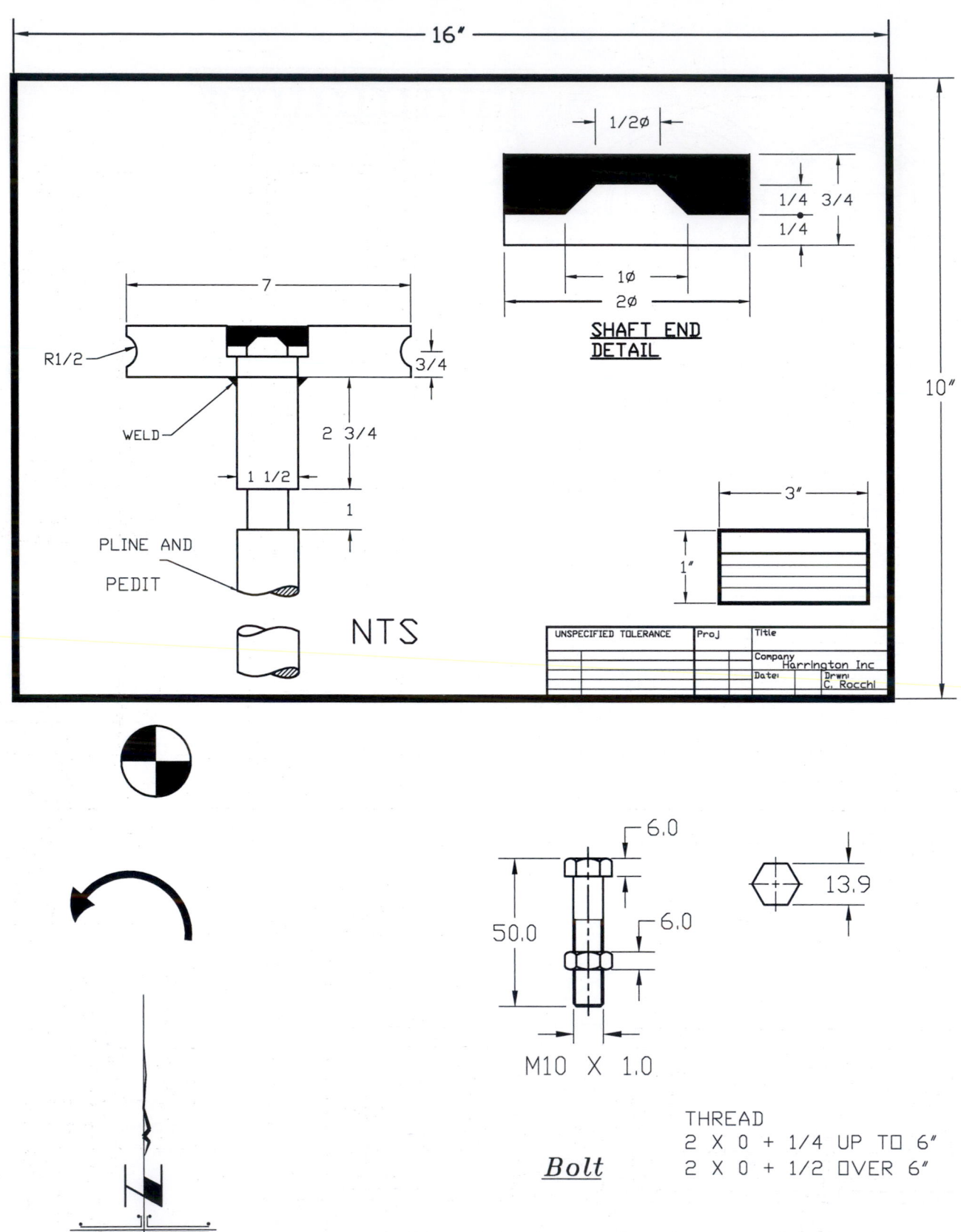

SHAFT END
DETAIL

R1/2

WELD

PLINE AND
PEDIT

NTS

UNSPECIFIED TOLERANCE	Proj	Title	
		Company	Harrington Inc
		Date:	Drwn: C. Rocchi

M10 X 1.0

Bolt

THREAD
2 X 0 + 1/4 UP TO 6"
2 X 0 + 1/2 OVER 6"

Exercise 5 Wood

These clock pieces have reverse sections to show how the clock is constructed.
The break lines show that the body of the clock is longer than is shown.
Draw the clock at a scale of 1=1.

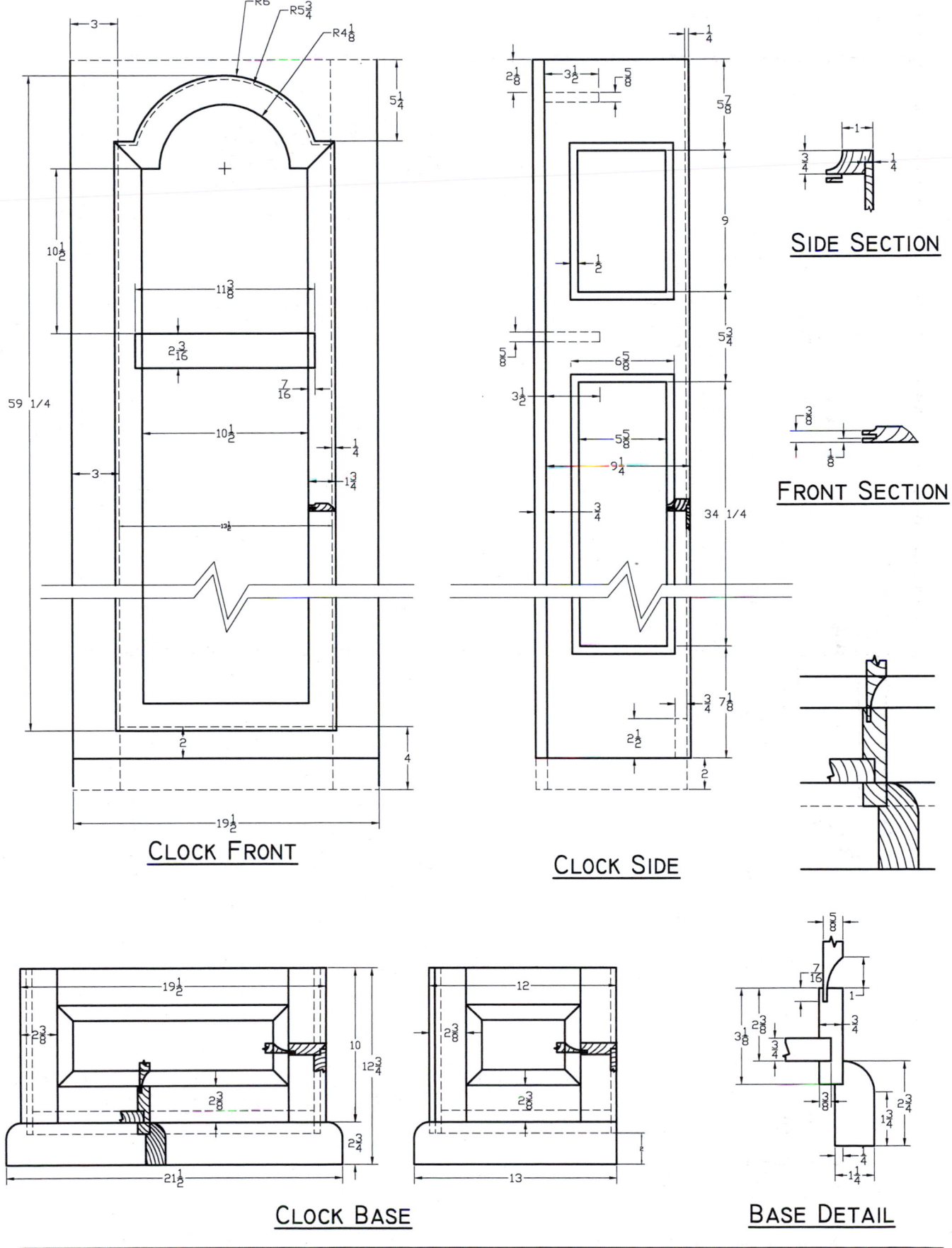

CLOCK FRONT

CLOCK SIDE

SIDE SECTION

FRONT SECTION

CLOCK BASE

BASE DETAIL

Exercise 5 Challenger

Profiles of wood and stone molding are often shown with section lines as in this 13th century church doorway in Italy.

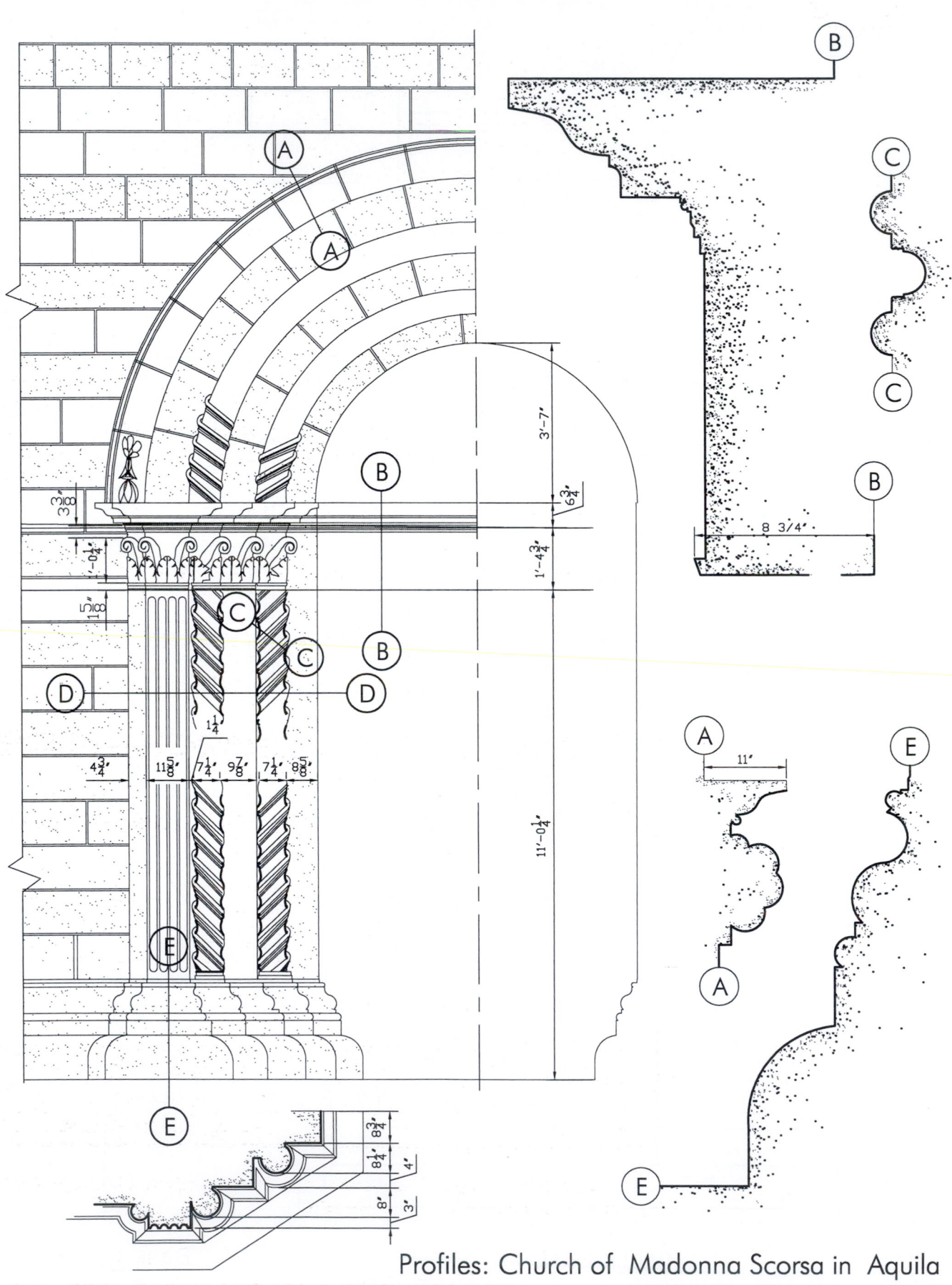

Profiles: Church of Madonna Scorsa in Aquila

Entity Properties: Layers, Colors, and Linetypes

On completion of this chapter, you should be able to:

1. Set up LAYERs and create geometry on them
2. Set colors to layers and use the COLOR command
3. Set LINETYPEs to layers and use the LINETYPE command
4. Change the properties of objects
5. List the properties of objects
6. Freeze/thaw and turn on/off the LAYER
7. Lock and unlock layers.

About LAYERs

The LAYER command allows you to control the drawing by means of visible entities, much like transparencies, as shown in Figure 6.1. Different colors and linetypes can be associated with different layers. To help complete the drawing, these layers can be either displayed or undisplayed, active or inactive, accessible or inaccessible. The number of layers you can use is unlimited.

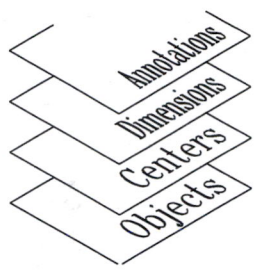

Figure 6.1

Toolbar From the Object Properties toolbar choose

Pull-down menu Under the Format menu choose Layer...

The command line equivalent is LAYER.

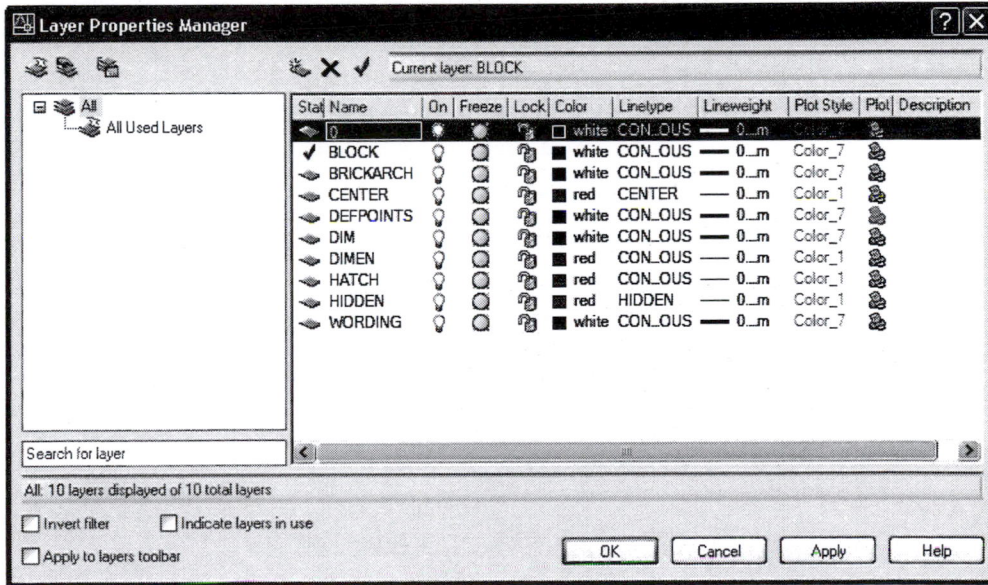

Layers can only be set up and modified with the Layer Properties Manager, but the layer can be can be set with the Layer toolbar.

The Layer Properties Manager dialog box will help you to view the layers, change the colors, turn layers off, freeze and thaw them, and lock or unlock them.

Current	Current	sets the highlighted layer to current – the one you are working on
	Name	the name of the layer; no spaces or dots allowed
	Off or On	sets the selected layers off or on, which makes the layer invisible or visible
	Freeze	freezes selected layers, making them invisible and not regenerated, or thaws (unfreezes) them. Freeze is for model space layers
	Lock	locks selected layers, making them visible but not accessible, or unlocks them
	Color	the color set to that layer. All of the choices along the top so far have been toggle switches, on or off; this allows you to choose a color
	Linetype	the linetype of the layer
	Lineweight	the lineweight of the layer
	New	allows you to enter a new layer starting with the layer name
	Delete	allows you to delete a layer if there are no objects in it
	OK	exits from the layer dialog box while saving changes
	Show Details	offers more information on the status of the highlighted layer
	Cancel	cancels the layer changes or additions you have made
	Help	provides help files on the layer functions

Creating a New Layer

Pick the button for New on the top left side of the layer dialog box, highlighted in Figure 6.2, and you will see a new layer appear in the list. Type in the name for the layer to override Layer1, the default name.

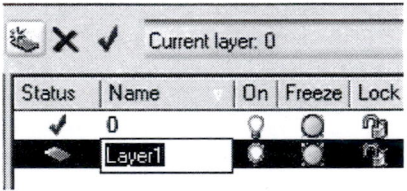

Figure 6.2

Once the name is entered the layer will be added to the layer list. The next time you enter the dialog box the layers will be listed alphabetically.

When typing layer names, add no spaces, slashes or periods.

Activating Layers for Changes

In order to change the name, state, color, linetype, or lock value of the layer, it must be identified or selected. Move the cursor to the layer name and pick it. The layer's line should be highlighted.

When selecting more than one layer for changes, use the regular Windows method for identifying multiple objects: use the Shift key to highlight all layers between the top and bottom ones that you choose, and the Ctrl key to highlight multiple layers individually. The selected layers will be highlighted and modified.

Color and Linetype options will affect all highlighted objects, as shown in Figure 6.3.

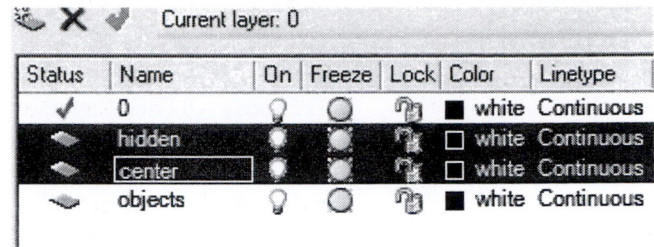

Figure 6.3

Changing a Layer Name

If you should type in the name incorrectly, double-click on the name. The area now activated will turn blue. Move your cursor away from the line, and the name itself should be highlighted. Type in a new name or revise the current name. The new name will replace the old.

Making a LAYER Current

In order to draw on a layer you must make it *current*. The easiest way to do this is through the Layer Properties Manager shown in Figure 6.4. Highlight the layer that you would like to have current, and pick the word Current.

Figure 6.4

The Layers toolbar shown on Figure 6.5 is possibly quicker, if that toolbar is displayed. Pick the down arrow and select the name of the layer that you would like to be current. To place objects on a different layer, highlight them, then using the down arrow, place them in the desired layer.

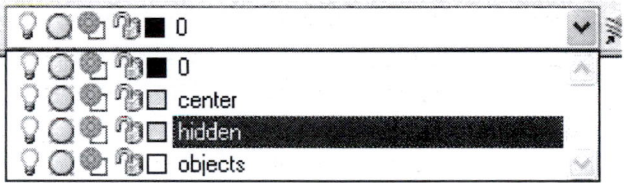

Figure 6.5

The Layer toolbar will show which layer is current. If this toolbar is not displayed, right-click the end of another toolbar, then pick layers from the menu.

Changing Layer Color

In the Layer Properties Manager select the the colored box under the letter C that corresponds to the layer you would like to change, as shown in Figure 6.6.

This will invoke the Select Color dialog box. Pick the color that you would like, then click OK.

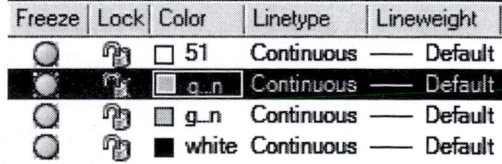

Figure 6.6

Problem Shooting

If your layers suddenly disappear, you have chosen the New Group Filter by mistake. In the Layer Properties Manager choose All Used Layers as shown in Figure 6.7 and your layers will reappear.

Figure 6.7

Setting Color Independent of Layer

You can set a color using the COLOR command. This will override your layer color setting. To do this, type in the word COLOR at the command prompt, or pick the color button on the Object Properties toolbar shown in Figure 6.8, which is generally just to the right of the Layer toolbar.

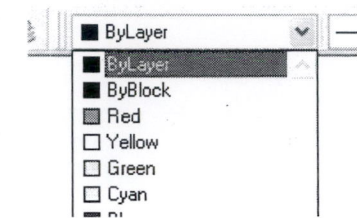

Figure 6.8

```
Command:COLOR
```

All subsequent objects will be drawn in this color. This may cause some confusion if your colors have been set to a particular layer because the color setting overrides the layer color. Keep the COLOR command set to Bylayer and you should have no problem.

Loading LINETYPEs

In addition to colors, you can also have linetypes associated with each layer. Load linetypes within the Layer Property Manager dialog box in the same way that you would load them in the Graphics area, see page 56.

Once you have set a color or a linetype outside the Layer dialog box, all future geometry added to this layer will use this color or linetype. **Until you are sure of what you are doing, both the Color setting and the Linetype setting should be ByLayer.**

Changing LTSCALE

Depending on the size of the object, the hidden lines may not show up as hidden. If this is the case, either the objects that you thought were on that layer are not, or the scale of the drawing is too large or too small for the linetype to show. You will need to change the linetype scale relative to the current drawing.

```
Command:LTSCALE
Enter new linetype scale factor<1.00>:12
```

Match Properties and CHPROP

The Match Properties command will take the linetype, color, and linetype scale of a selected object and apply these properties to any other objects that you choose.

```
Command: '_matchprop
Select source object:(Pick an object)
Current active settings:  Color Layer Ltype Ltscale
   Lineweight Thickness
PlotStyle Dim Text Hatch Polyline Viewport Table Material
   Shadow display
Select destination object(s) or [Settings]:(Pick other
   objects)
```

The last objects will have the properties i.e. color, linetype, dimension style, etc . of the first object chosen.

The CHPROP command cycles through the properties. With the dialog box you are prompted to pick the objects you would like modified; then you can choose Layer, Color, Linetype, Thickness, etc. and type in the new value. If you use the command line, here is the sequence:

```
Command:CHPROP
Select objects:(pick an object)
Select objects:⏎
Enter property to
   change[Color/LAyer/LType/LtScale/Thickness]:C
Enter new color <bylayer>:Red
Enter property to change
   [Color/LAyer/LType/LtScale/Thickness]:LT
Enter new linetype <hidden>:Center
Enter property to change
   [Color/LAyer/LType/LtScale/Thickness]:⏎
```

Layer Filtering

Sometimes you may want only certain layers to be listed in the Layer Control dialog box. The Filter option allows you to limit which layers are listed. You can filter on the basis of:

- layer names, colors, and linetypes

- whether layers are on or off

- whether layers are frozen or thawed

- whether layers are locked or unlocked

- whether layers have plot or don't plot status

- whether layers contain objects or not

- whether or not layers are part of externally referenced (xref) drawings.

Layer filtering is used on very large files, or files that belong to someone else or where you don't need to use many of the layers.

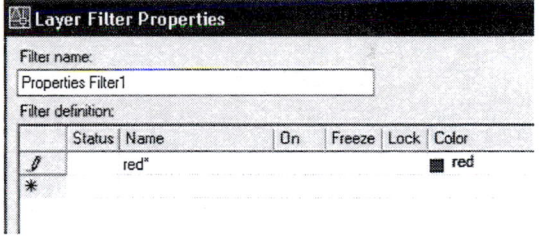

Figure 6.9

To Filter a Layer

In the Layer Properties Manager, pick the New Property Filter button. Create a new filter group as shown. In Figure 6.9, all layers that are red have been filtered.

Layer Management

The default layer is 0. If you enter any dimensions, you will also create a layer called Defpoints. Neither of these layers is renamable. The Defpoints layer does not plot.

While creating a model in AutoCAD, layers should be named in a logical, straightforward manner. Also, keep your LINETYPE and COLOR commands set to ByLayer.

Many industries have developed layering standards so that there is no question about where objects will be located. If you are starting work with a company, make sure that you know what the layering standards for the company are, and find out whether there are any standards outside of the company that you should be aware of.

Freezing layers when you are not using them will save a great deal of time with larger models. Until they are regenerated, layers that have been frozen will not display when they are thawed.

In this example we will make an adjustable bearing with hidden lines and center lines.

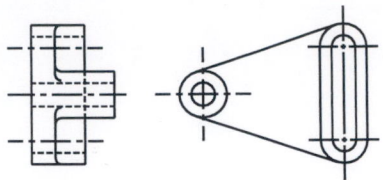

Step 1

First set up your layers.

Pick the New Layer button to add new layers. Add Centre, Hidden, and Object.

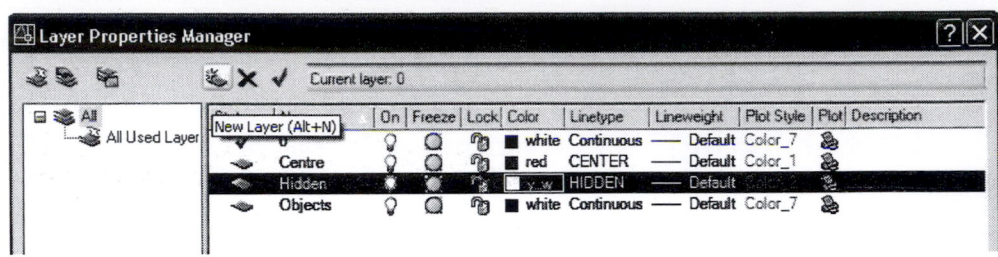

Step 2

Once the names are loaded, change the color of each layer by picking the layer name and then picking the Set color button. An overlay menu will offer you a selection of colors.

Choose a color for that layer. The nine base colors are the best to choose.

- **objects**   White
- **hidden**   Yellow
- **center**   Red

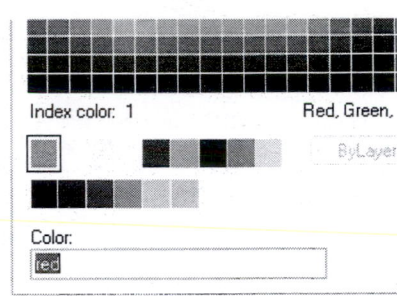

Step 3

Now load and set the linetypes. Pick the word Continuous under the linetype column across from the CENTER layer to load the Linetype manager.

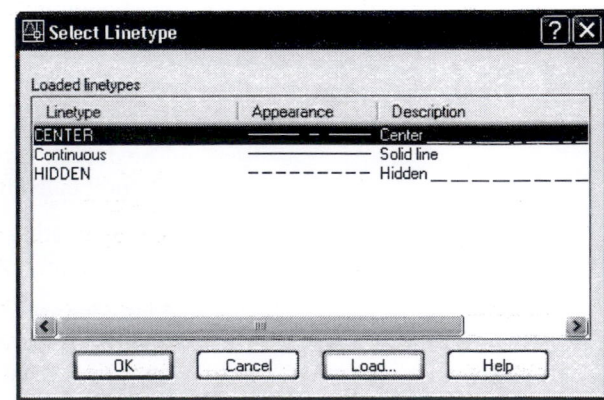

Choose Load to load new linetypes. Choose CENTER, hold the Ctrl key down on your keyboard and then pick HIDDEN from the list of linetypes to load them both.

Pick OK. Then choose CENTER under the Select Linetype menu shown to make your center lines 'center' lines.

Step 4   Now that the linetypes have been loaded, return to the Layer Properties Manager and load the hidden line type to the Hidden layer.

Activate the Hidden layer, choose the word Continuous under the Linetype column, then choose the hidden linetype from the list. Press OK. Use OK to exit from the Set Linetype dialog box, then OK again to exit from the Layer Control dialog box.

Step 5   Now that the layers are ready, use the same menu to make the Objects layer current.

Your screen should look like this:

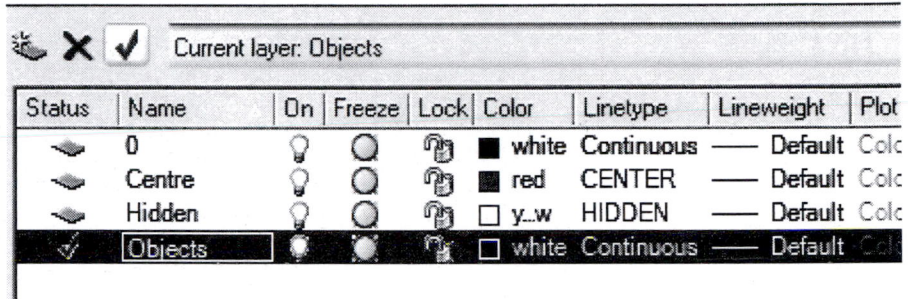

Step 6   Now with the Objects layer current, draw in the front view of the adjustable bearing as shown. Do not draw in the dimensions.

Draw a circle with a center of 0,0. ZOOM to the Extents of your geometry, and then zoom down.

Draw two more circles on at a center of 6,2. The radii will be .5 and 1.

Mirror these through the center line, and use LINE and TRIM to finish the part.

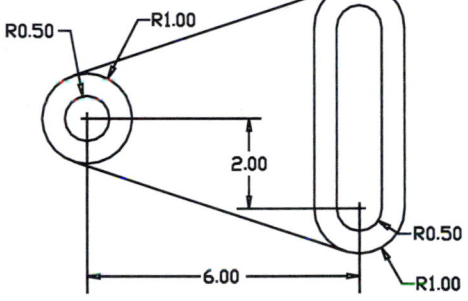

Danger

If you set a color or LINETYPE under the COLOR or LINETYPE commands, they will override your layer setting.

Step 7   Now return to the pull-down menu and make the Centre layer current.

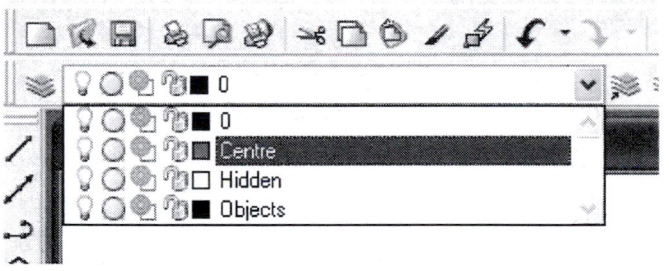

Draw in the lines as shown. They should be a different color as well as a different linetype.

Try changing the LTSCALE to see if there is any noticeable difference in the center lines.

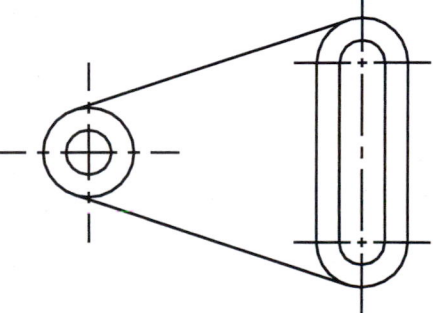

```
Command:LTSCALE
New scale factor<1.00>:1.5
```

Step 8   Return to the pull-down menu and make the Objects layer current to draw in the side view as shown by picking the word Objects.

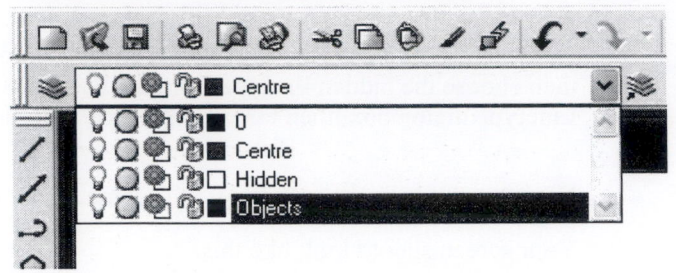

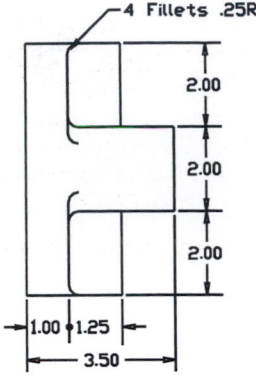

Again, do not put in the dimensions, simply put in the lines. Using OFFSET will probably prove the easiest way of doing this.

Step 9   Now return to your pull-down menu and make the Hidden layer current to place the hidden lines as shown.

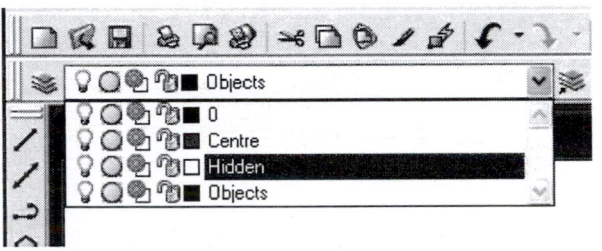

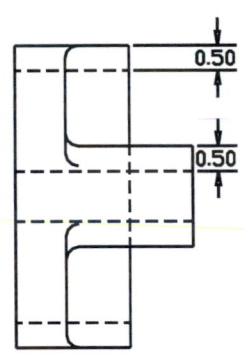

Make sure POLAR, OSNAP and OTRACK modes are on to line the hidden lines up with the front view of the object.

Step 10   Now make the Center layer current and add the center lines as shown. Again, SNAP set to .25 would be of use in this view.

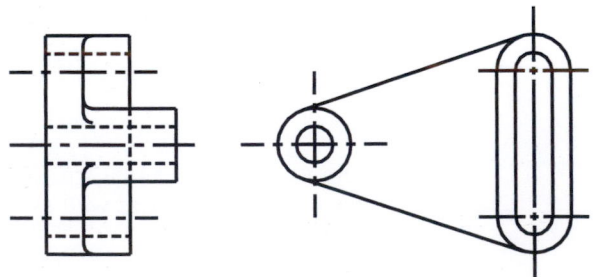

Step 11   Return to the Layer pull-down menu and change the colors of the layers to see if you can make it look nicer. If the colors remain the same as you change the layer color, it is because the current color of an object is overriding the layer color. Set color back to ByLayer, then use CHPROP to change all of the objects to be colored by layer.

Step 12   We are now going to play with the Lock/Unlock facility to see how it works. It may be time to save the file just in case.

```
Command:SAVE
(enter a name in the dialog box)
```

In the pull-down menu lock the Centre and Hidden layers. Once the layers are locked, you will be able to see them, but you will no longer be able to edit them. Make sure the Objects layer is current before you start.

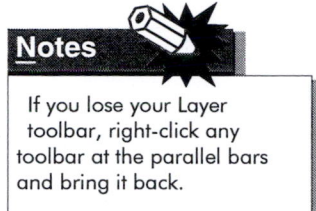
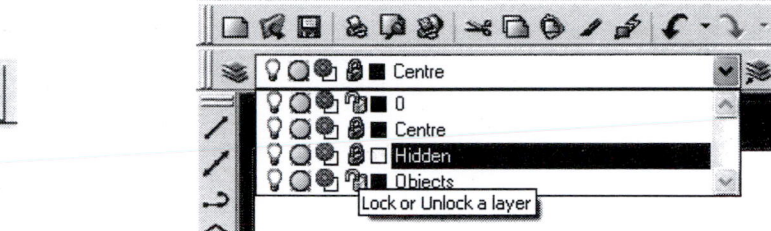

Step 13   Now back in the drawing or model, use a Window pick to highlight all of the objects on screen. Then from the Modify menu pick Properties and change the color to blue.

This can also be done on the command line as follows.

```
Command:CHPROP
  Select objects:ALL
  40 objects on locked layers
  Select objects:↵
  Enter property to change [Color/LAyer
    /LType/ltScale/Thickness]:C
  Enter new color <varies>:blue
  Enter property to change [Color/LAyer/
    LType/LtScale/Thickness]:↵
```

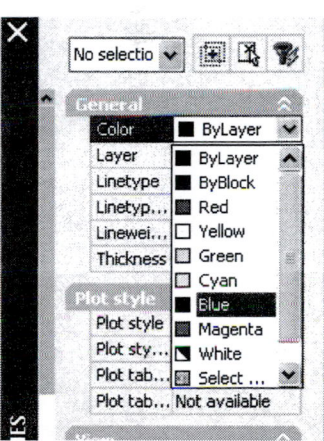

Notice that only the objects on the Objects layer were changed because the other layers were locked.

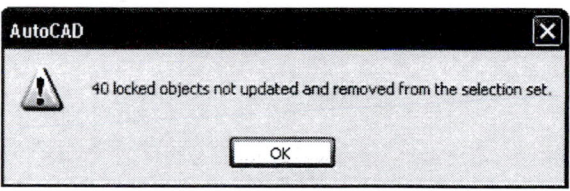

Step 14   Try turning the Layers Off and On, and Freezing and Thawing them. The data for the layers are still on file, but the information is not displayed.

Freeze/Thaw   On/Off

Exercise 6 Practice

Use different layers for the Hidden lines and Centre lines.
These are practice parts, not drawn to any scale. Draw at 1=1.

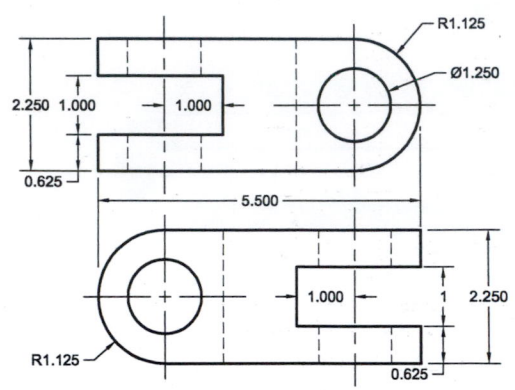

COUPLING

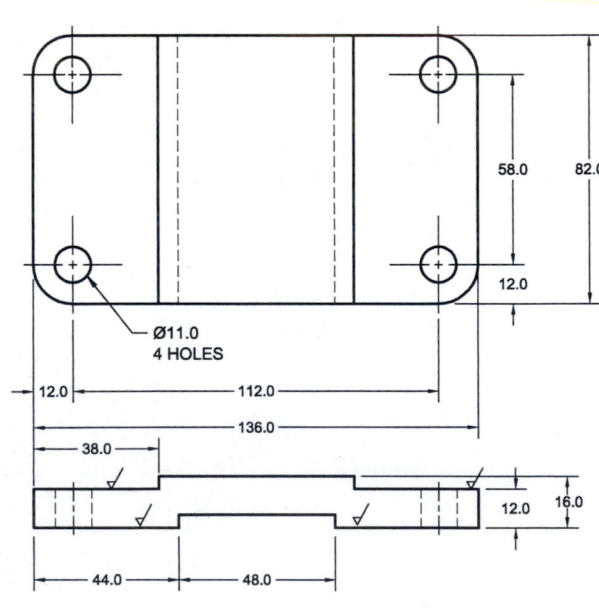

ROCKER ARM

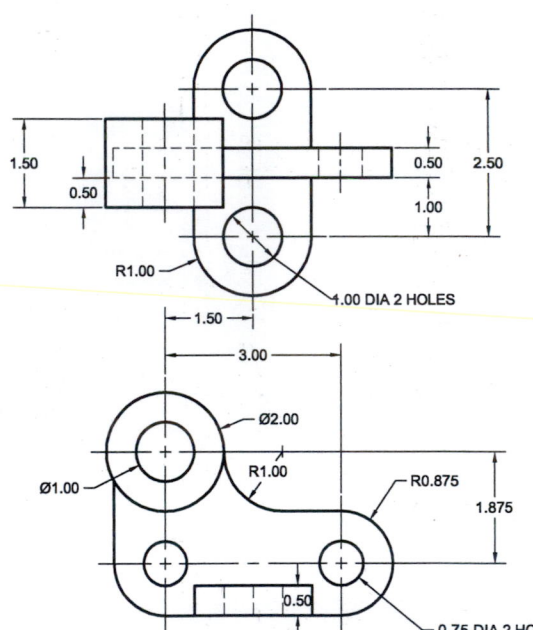

SWIVEL HAMMER

Top Plate

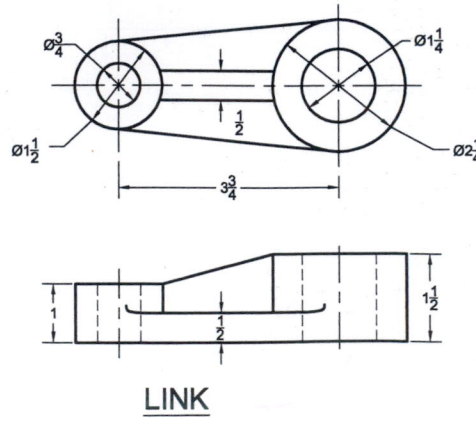

LINK

Exercise 6 Architectural

This is the section for the residential building found on pages 85 and 137.
The horizontal distances are found on page 85 and 137.

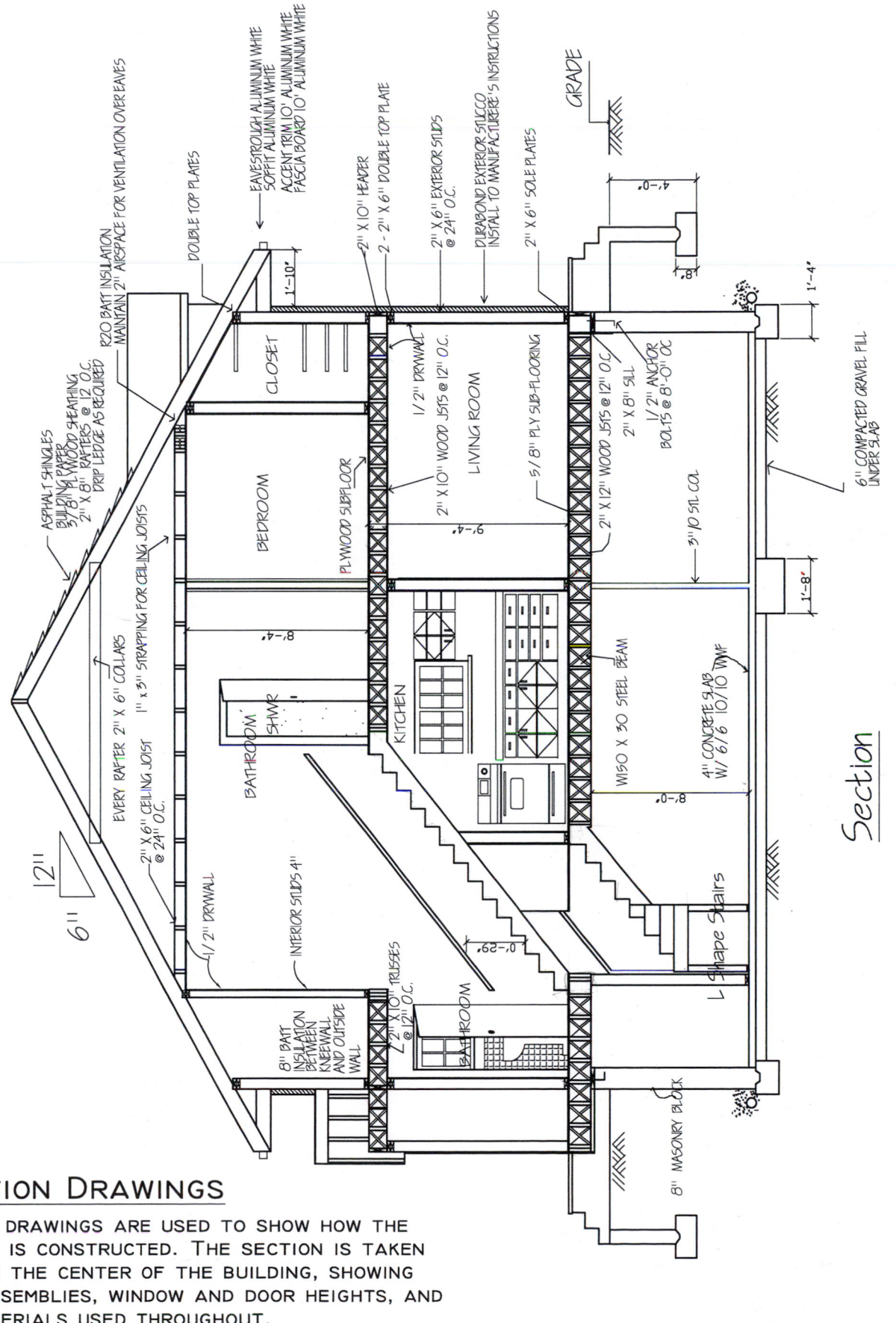

SECTION DRAWINGS

SECTION DRAWINGS ARE USED TO SHOW HOW THE
BUILDING IS CONSTRUCTED. THE SECTION IS TAKEN
THROUGH THE CENTER OF THE BUILDING, SHOWING
WALL ASSEMBLIES, WINDOW AND DOOR HEIGHTS, AND
THE MATERIALS USED THROUGHOUT.

Exercise 6 Mechanical

Draw 6A using different layers for hidden and center lines.
Draw the same two views - Front and Top - for 6B, 6C, and 6D.

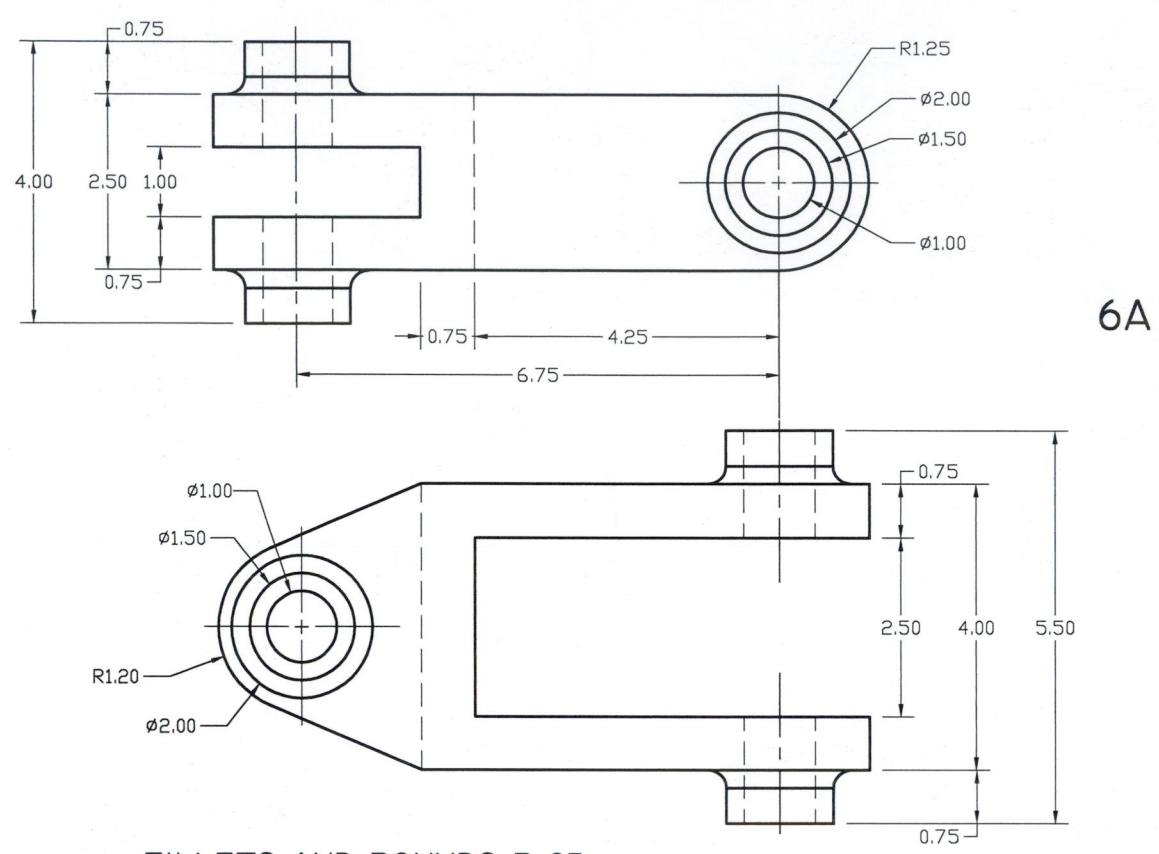

6A

FILLETS AND ROUNDS R.25

PICTORIAL VIEWS

PICTORIAL VIEWS, AS SHOWN BELOW, ARE 3 DIMENSIONAL
VIEWS USED TO ILLUSTRATE A PART.
CREATE WORKING DRAWINGS, AS SHOWN ABOVE, FROM THE
PICTORIAL VIEWS BELOW.

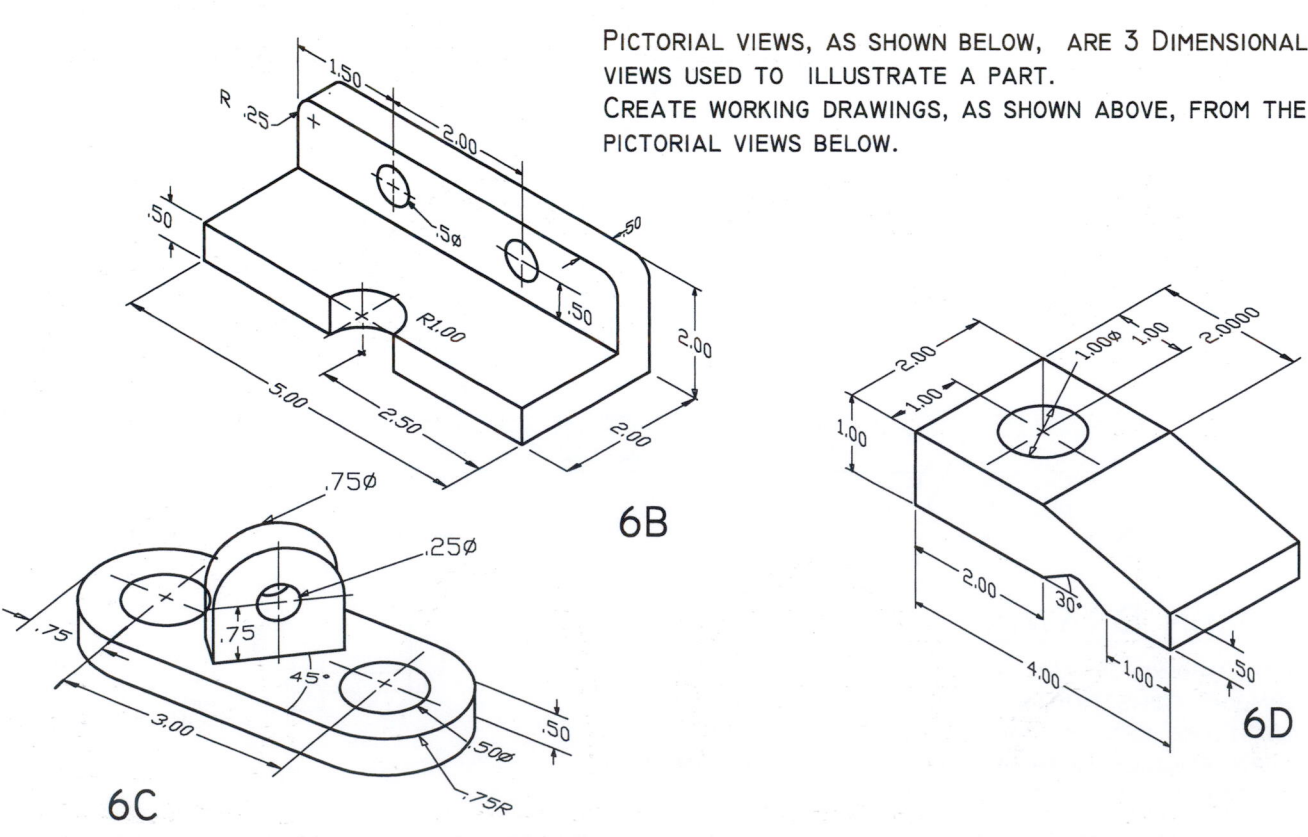

6B

6C

6D

Exercise 6 Challenger

Here are 40 different arches found in architecture.
Set PDMODE to 34, then use DIVIDE to section plines and arcs.

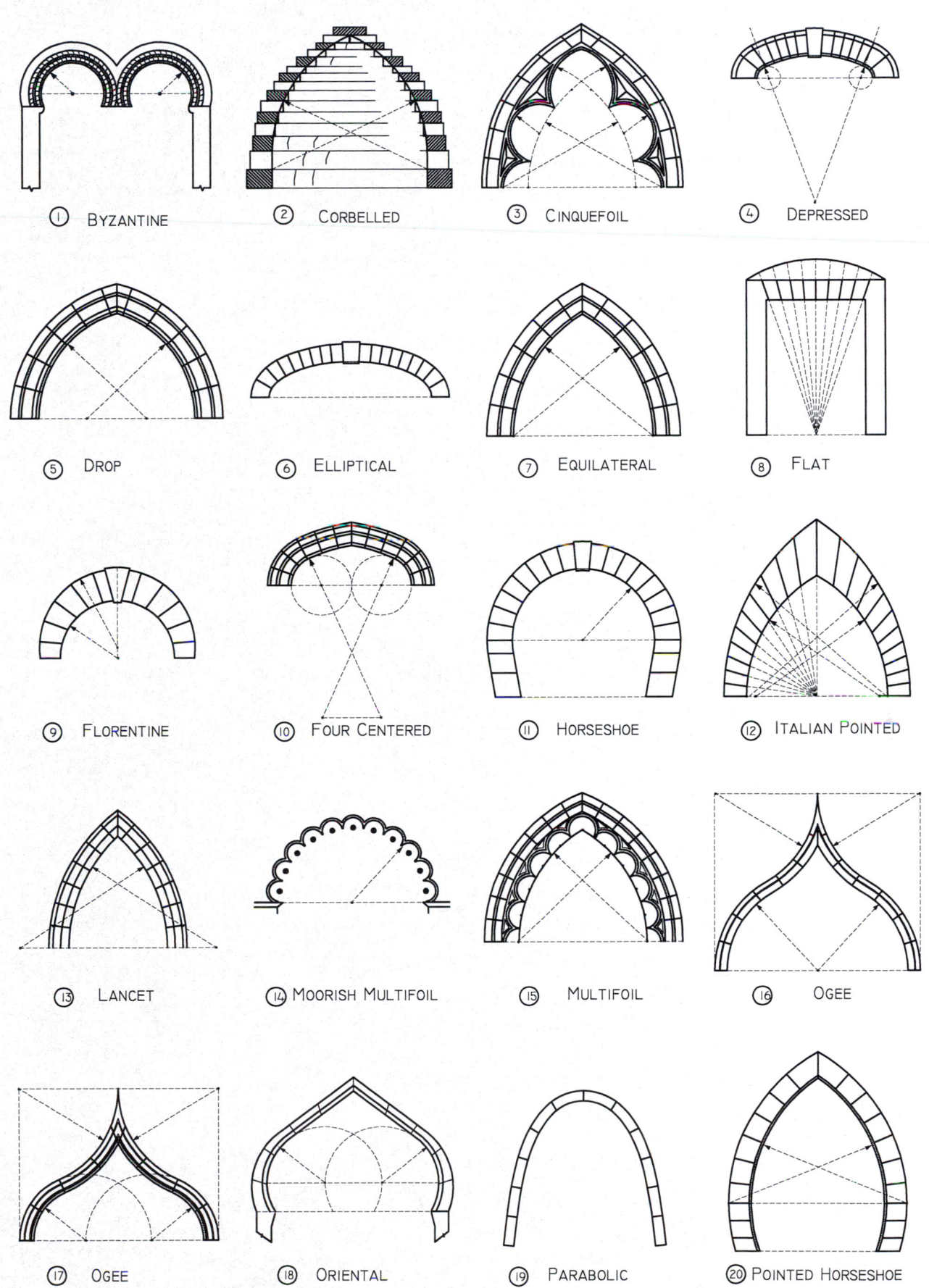

① BYZANTINE ② CORBELLED ③ CINQUEFOIL ④ DEPRESSED

⑤ DROP ⑥ ELLIPTICAL ⑦ EQUILATERAL ⑧ FLAT

⑨ FLORENTINE ⑩ FOUR CENTERED ⑪ HORSESHOE ⑫ ITALIAN POINTED

⑬ LANCET ⑭ MOORISH MULTIFOIL ⑮ MULTIFOIL ⑯ OGEE

⑰ OGEE ⑱ ORIENTAL ⑲ PARABOLIC ⑳ POINTED HORSESHOE

Exercise 6 Challenger (continued)

You can never get too much practice.

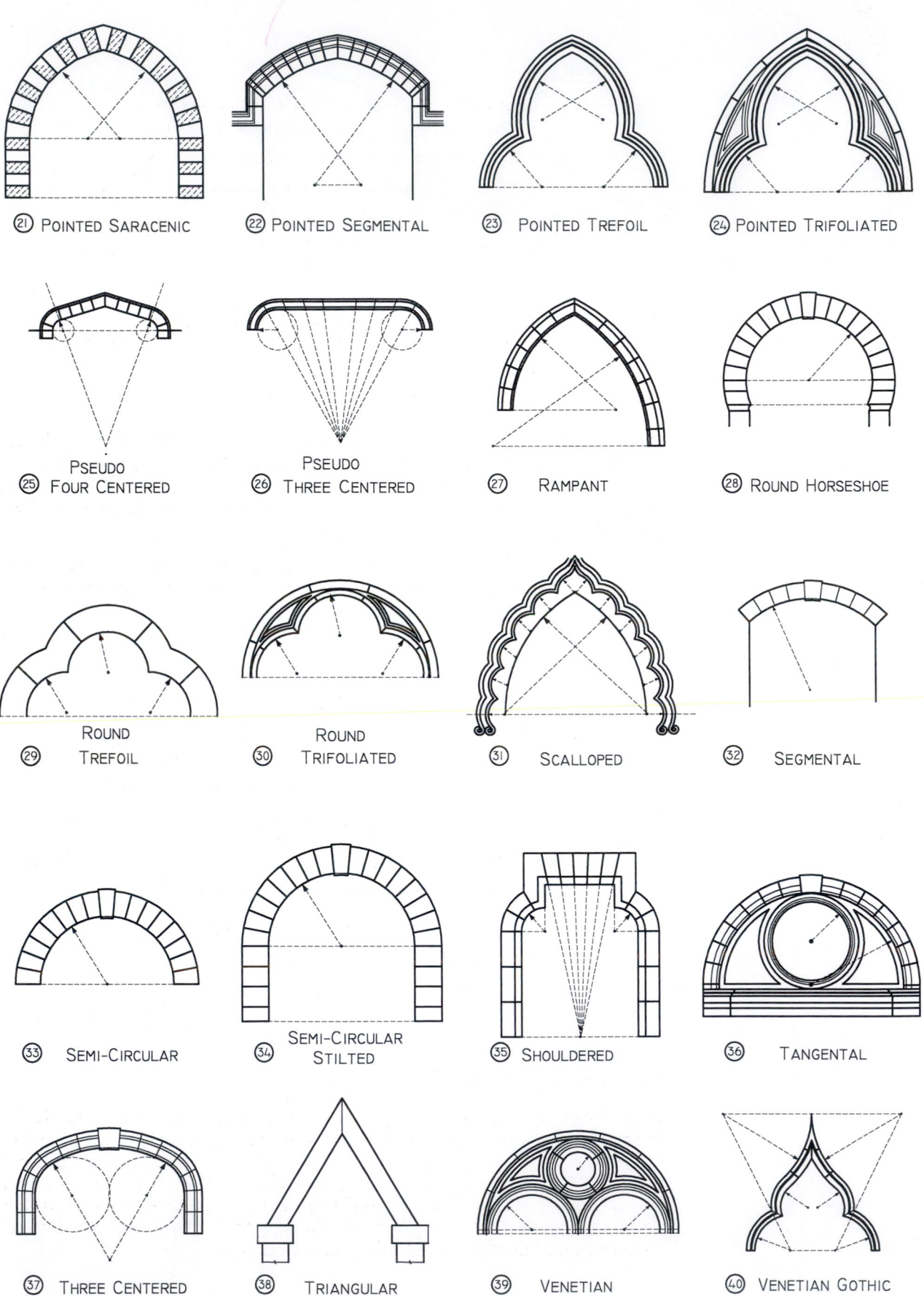

21 Pointed Saracenic

22 Pointed Segmental

23 Pointed Trefoil

24 Pointed Trifoliated

25 Pseudo Four Centered

26 Pseudo Three Centered

27 Rampant

28 Round Horseshoe

29 Round Trefoil

30 Round Trifoliated

31 Scalloped

32 Segmental

33 Semi-Circular

34 Semi-Circular Stilted

35 Shouldered

36 Tangental

37 Three Centered

38 Triangular

39 Venetian

40 Venetian Gothic

7 Dimensioning

On completion of this chapter, you should be able to:
1. Set up a dimensioning style, using the dialog box
2. Add vertical, horizontal, and aligned dimensions
3. Add baseline and continuous dimensions
4. Add diameter and radius dimensions
5. Alter the dimensions once they are in.

About Dimensioning

While the objects are being created, the size of the object is being programmed with each part's geometry. Lines, circles, arcs, etc. should be created perfectly every time. If you get into the habit of creating data in a lazy or slapdash manner, it will catch up with you later when dimensioning the drawing.

Dimensioning shows the measurements, the locations, and the angles of objects. The dimensioning commands are designed to extract the sizes that are already programmed with the part, and display them in accepted formats. Every discipline has a different set of drawing protocols. The dimensioning variables and dimension styles are used to set the dimensions to the required parameters.

AutoCAD offers a wide variety of ways to produce linear, baseline, radial, diameter, and angular dimensions. Figure 7.1 shows some of the basic dimension types.

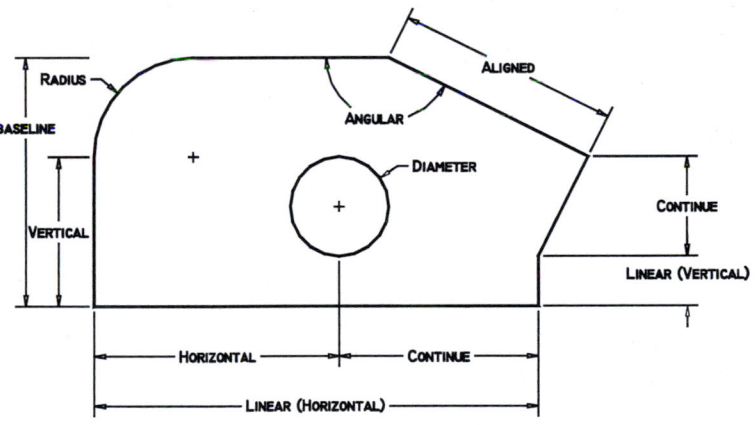

Figure 7.1

Dimensioning Components

Every dimension has several components. The ***dimension text*** states how big the object is. The ***dimension line*** holds the dimension text. The dimension text should be off-set from the dimension line by .625 mm or 1/16 inch. The ***extension line*** extends from the object to the dimension line. The extension line is offset by .625mm or 1/16 inch from the part itself and should extend 1.25mm or 1/8 inch past the dimension line. These measurements are universal drafting standards and must be maintained in computer-aided drafting as well as in manual drafting.

Figure 7.2 shows the components of a dimension.

The point at which you start your dimension is the reference point or ***definition point***. AutoCAD automatically puts a gap between this point and the start of the extension line. AutoCAD also creates a layer for this point called *Defpoints* that does not plot.

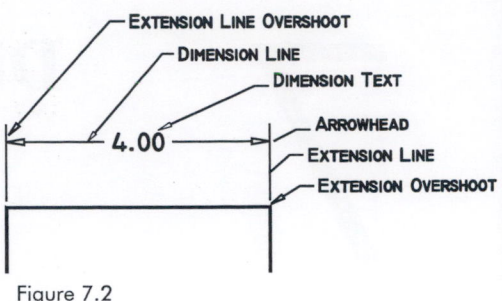

Figure 7.2

Accessing Dimensions

The Dimension toolbar, shown in Figure 7.3, can be accessed by right-clicking another toolbar at the parallel lines. Once in, move the toolbar until it is docked.

Figure 7.3

The pull-down menu shown in Figure 7.4 is also useful .

Typing in the dimensioning commands is not recommended.

You may want to lock the objects layer before dimensioning. This way the objects cannot be altered when using STRETCH or another Modify command.

Dimensions should be placed on the dimensioning layer and this should be a different color. The lineweight of the dimensions when plotted should be significantly smaller than that of the object lines. Having the layers in different colors helps to make sure that the dimensions will have the same lineweight.

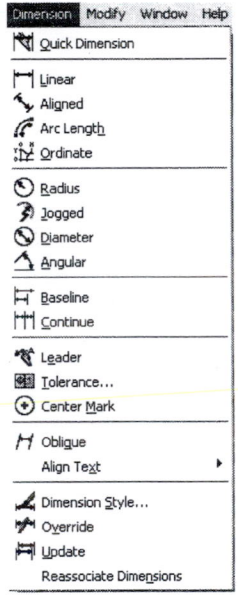

Figure 7.4

Entering Dimensions

Horizontal and Vertical Dimensions

Figure 7.5 illustrates linear dimensions. Use OSNAPs.

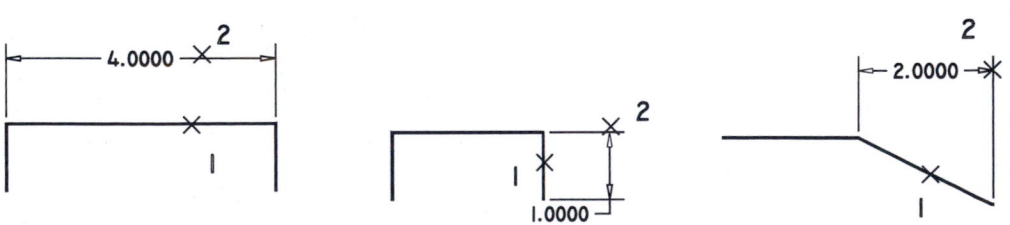

Figure 7.5

```
Command: (Dimension pull-down menu, Linear)
Specify first extension line origin or <select object>:↵
Select object to dimension: (pick 1)
Specify dimension line location
   [Mtext/Text/Angle/Horizontal/Vertical/Rotated]: (pick 2)
Dimension text  = 4.00
```

In the first example in Figure 7.5 the dimension is based on the line or object chosen. In the next two examples, the first extension line is picked, then the second, and finally the dimension line is placed. Use SNAP and OSNAP for accuracy.

Before specifying the dimension line location, you are prompted to change the dimension text (Text), the text angle (Angle), the dimension direction (Horizontal or Vertical), or the dimension line angle (Rotated).

In the third illustration in Figure 7.6, either the horizontal or the vertical length could be taken. Move your cursor to the position the dimension should be in, then pick the spot.

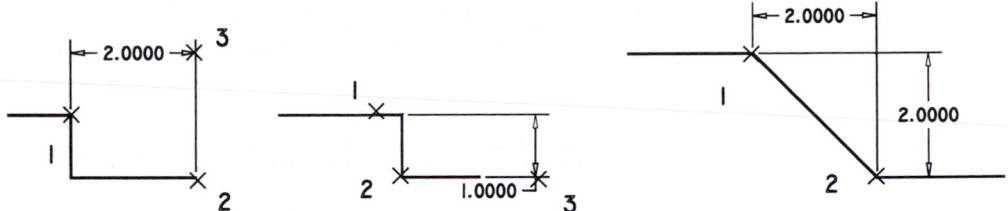

Figure 7.6

Aligned Dimensions

Linear dimensions give vertical or horizontal lengths. Aligned dimensions, as shown in Figure 7.7, give the dimension aligned to an angled surface.

```
Command: (Dimension pull-down menu,
    Aligned)
Specify first extension line origin
    or <select object>: (pick 1)
Specify second extension line
    origin: (pick 2)
Specify dimension line location
    or [Mtext/Text/Angle]: (pick 3)
Dimension text = 2.8284
```

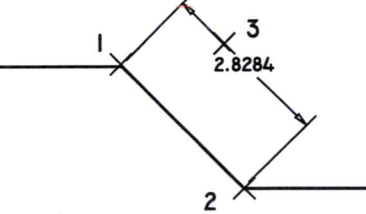

Figure 7.7

Like the vertical and horizontal, you can also simply choose the line itself to be dimensioned as in Figure 7.8.

```
Command: _DIMALIGNED
Specify first extension line origin or RETURN to select:↵
Select object to dimension: (pick 1)
Specify dimension line location or [Mtext/Text/Angle]: (pick
    2)
Dimension text = 2.8284
```

The distance between circle centers can also be dimensioned with either a linear dimension, if a vertical or horizontal is required, or with an aligned dimension.

To change the unit readout, change the dimension style (see page 126).

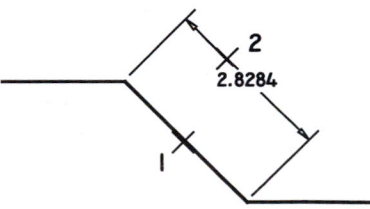

Figure 7.8

Continued Dimensions

Once you have a horizontal, a vertical, or an aligned dimension,
you can create dimensions that continue at regular intervals along the same edge as
shown in Figure 7.9.

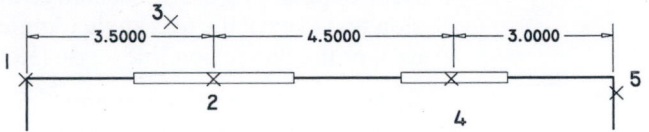

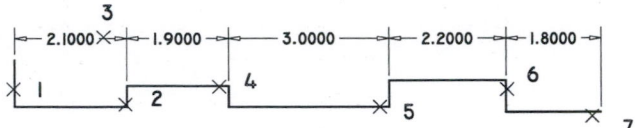

Figure 7.9

```
Command:_DIMLINEAR
Specify first extension line origin or <select object>:(pick
   1)
Specify second extension line origin:(pick 2)
Specify dimension line location [M/T /A/H/V/R]:(pick 3)
Command:_DIMCONTINUE (pull-down menu)
Specify a second extension line origin or [Undo/Select]
   <Select>:(pick 4)
Dimension text = 4.5000
Specify a second extension line origin or [Undo/Select]
   <Select>:(pick 5)
```

Baseline Dimensions

Similarly, for baseline dimensions, create the first dimension and continue from there.

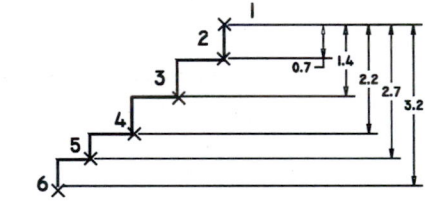

Figure 7.10

```
Command:_DIMLINEAR
Specify first extension line
origin or <select object>:(pick 1)
Specify second extension line origin:(pick 2)
Specify dimension line location [Mtext/Text
   /Angle/Horizontal/Vertical/Rotated]:(pick 3)
Dimension text  = 1
Command:_DIMBASELINE
Specify a second extension line origin or
   [Undo/Select]<select>:(pick 4)
Specify a second extension line origin or
   [Undo/Select]<select>:(pick 5)
Specify a second extension line origin or
   [Undo/Select]<select>:↵
```

Radial Dimensions

A *radial dimension* measures the radius of an arc or circle. The dimension appearance is determined by the Fit tab options, as shown in Figure 7.11.

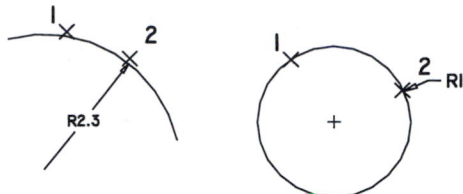

Figure 7.11

```
Command: (Dimension menu, Radius)
Select arc or circle:(pick 1 — the arc)
Specify dimension line location or [Mtext/
  Text/Angle]:(pick 2)
```

Diameter Dimensions

Diameter dimensions are affected by the same dimension variables as the radius dimensions as seen in Figure 7.12.

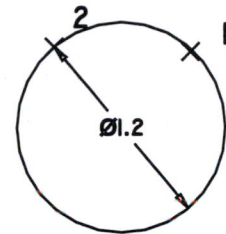
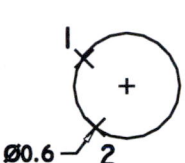

Figure 7.12

```
Command:_DIMDIAMETER
Select arc or circle:(pick 1)
Specify dimension line location
  [Mtext/(Text/Angle]:(pick 2)
```

Aswith the linear commands, the text and the text angle can both be changed. These are under the Text tab of the Dimension Styles dialog box.

Angular Dimensions

The angular dimension command measures the angle between two non-parallel lines or three points. It can also measure the angle around a portion of a circle or the angle subtended by an arc, as shown in Figure 7.13.

The dimension line for the angular measurement is an arc that spans the measured angle and passes through the measured point.

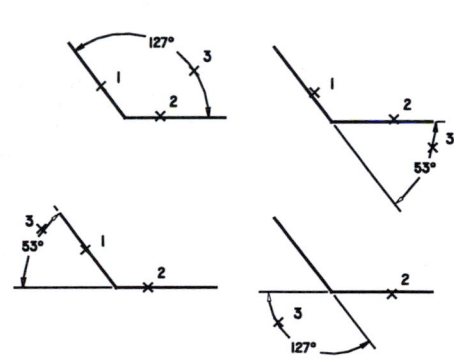

Figure 7.13

```
Command:_DIMANGULAR
Select arc, circle, line, or <specify vertex>:(pick 1)
Select second line:(pick 2)
Specify dimension arc line location or [Mtext/Text/
  Angle]:(pick 3)
```

Dimension Styles

Controlling Dimension Style

A ***dimension style*** is a named group of settings that determines the appearance of the dimension. Every dimension has an associated dimension style. If no style is applied before dimensioning, the Standard (imperial) or ISO-25 (metric) default style is used. The style controls the unit readout, the text style, the color, the linetype scale, and many other factors.

If you remember nothing else about dimensioning, remember NEVER CHANGE THE STANDARD DIMENSION STYLE.

Always make a new one as shown in Figure 7.14. You will regret it if you don't.

To create a new style, enter a style name in the Name box and pick Save. Changes that you make with the Modify Dimension Style menu will be filed with the saved name.

There are many variables in the Lines, Symbols and Arrows menu. Pick the style you want, but don't change any of the default sizes because that will affect your dimensions unless you really know what you are doing. Change the sizes in the Fit menu. See Tutorial 7a and 7b for how to do this.

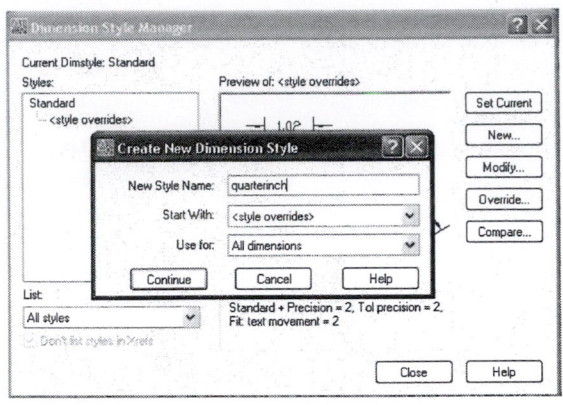

Figure 7.14

The Text Tab

The Text menu allows you to set the size and position of text on your dimension lines.

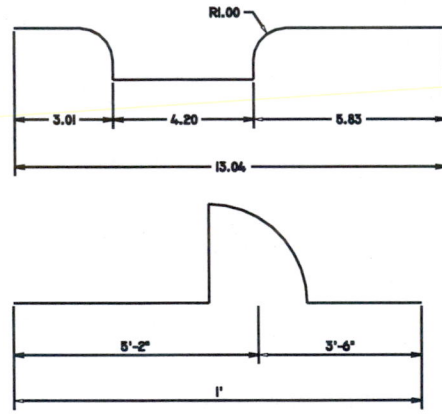

Figure 7.15

Mechanical drawings generally have the dimension text horizontally within the dimension line. Architectural drawings generally have the text above the dimension line and aligned with the dimension line. These alignments are changed using the vertical and horizontal text placement, as shown in Figure 7.15.

Again, do not change the text height here, change it in the Fit dialog box. Along with the text height and the arrowhead size there are gap distances, overshoot distances, extension line distances, and distances between the text and the dimension line itself. You must change ALL of these to make the dimensions work properly. It is easier to change them in the Fit dialog box.

A different text style can also be chosen for the printing of text. Keep the text style fairly simple for best results. In mechanical drawings Arial is generally the preferred text style. In architectural drawings, City Blueprint is often used. In either case none of the serif or script text fonts are appropriate for dimension text.

The Fit Tab

The Fit menu allows you to set the size of your dimensions and have everything printed at the correct scale.

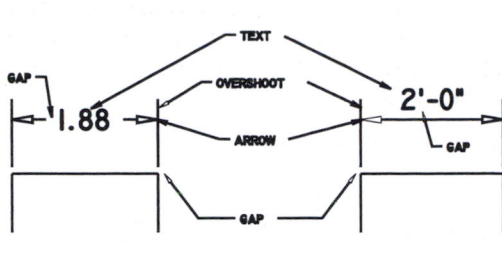

Figure 7.16

The overall size of the dimensions is determined in the Fit tab of the Modify Dimension Style dialog box. If you change the overall scale, you will change the arrowhead size, the extension line gap, the extension line overshoot, and the gap between the dimension and the dimension line. It is a much better idea to change the overall scale than to change each size individually. Figure 7.16 illustrates all the different preset elements.

If you change the overall scale, all of these factors will change. If you change just the text size or just the arrowhead size, all of the other parameters will remain the same.

Paper space dimensions are dealt with in Chapter 11.

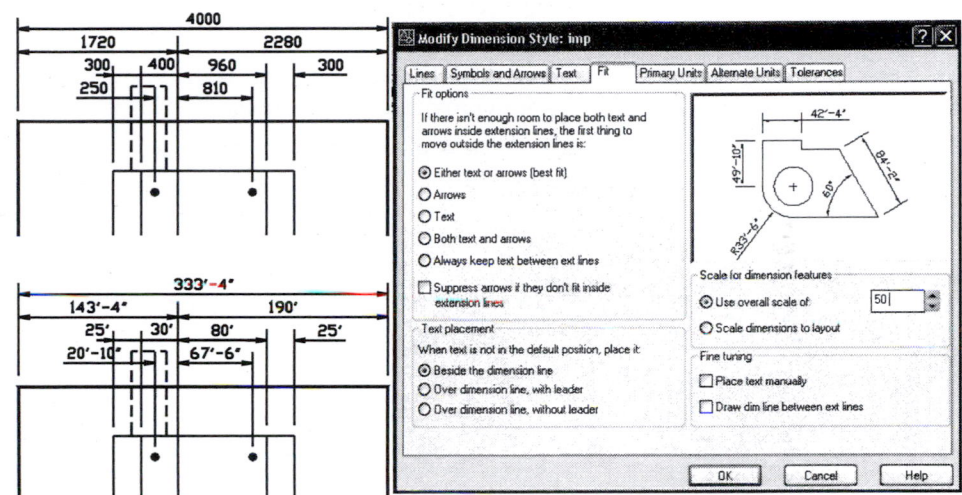

Figure 7.17

When you dimension a drawing and plot it, the final drawing is a fraction of the original size of the object.

Metric		*Imperial*	
1:25 = 1/25 of the original size		1/2″=1′-0″ = 1/24 of the original size	
1:50 = 1/50 of the original size		1/4″=1′-0″ = 1/48 of the original size	
1:100 = 1/100 of the original size		1/8″=1′-0″ = 1/96 of the original size	

Changing the overall scale under the Fit menu shown in Figure 7.17 will expand the dimensions to fit the drawing so that when printed the dimensions are the correct size. The dimensions are plotted at the same fraction as the drawing itself.

Metric		*Imperial*	
1:25 = overall scale 25		1/2″ = 1′-0″ = overall scale 24	
1:50 = overall scale 50		1/4″ = 1′-0″ = overall scale 48	
1:100 = overall scale 100		1/8″ = 1′-0″ = overall scale 96	

The Primary Units Tab

The Primary Units menu allows you to choose the units print-out. This does not change the size of the object, just the text units on the dimension line.

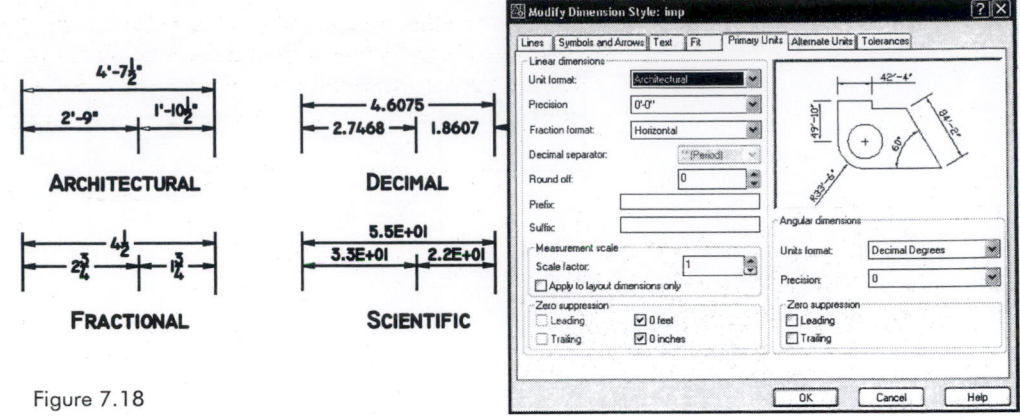

Figure 7.18

Choose the primary units that you would like the objects dimensioned in, then choose the precision that you would like to have. Angle measurements can also be programmed. The Measurement scale (known as DIMLFAC in previous releases) allows you to provide detailed drawings. This is an alternative to the paper space layouts.

Note: In Figure 7.18 the text placement has also been changed to reflect discipline standards.

The Alternate Units Tab

On rare occassions you may need to show both metric and imperial measurements on a drawing. These can both be displayed on the dimension line if they are set up under the Alternate Units tab.

The Symbols and Arrows Tab

The major criticism of the teaching of AutoCAD over the past 10 years is that the instructors concentrate on teaching AutoCAD, and ignore drafting conventions. Your final AutoCAD plots must be recognizable as drawings and employ the basic drafting protocols of your discipline. Here are some of the protocols.

Center Marks

The centers of circles must have center marks showing that the circle has a constant radius. The three styles of center mark are shown in Figure 7.19.

Figure 7.19

Also under the Symbols and Arrows tab you can choose which type of arrowhead you would like.

Mechanical operations use dots, architectural use ticks.

Make sure you have chosen an appropriate style.

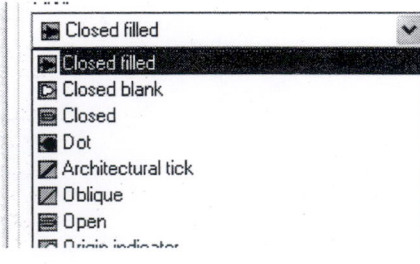

Saving Dimension Styles

Once you have created a dimension style that has all the correct settings for your application, it is automatically saved with your file and can be transferred to other files.

Variations on that dimension style can also be saved. For example, you can create a dimension style in metric with two decimal places of accuracy. On some drawings you may want to have the diameter lines forced inside, on others not. You can create different dimension styles to accomodate the different styles illustrated in Figure 7.20.

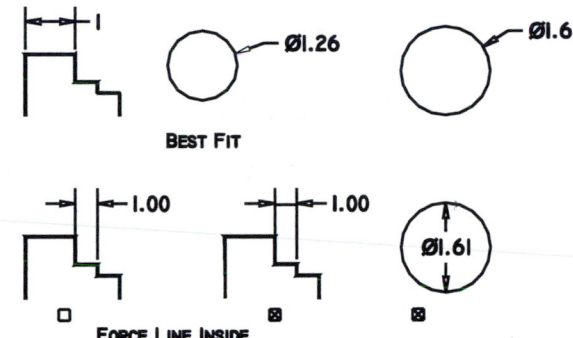

Figure 7.20

Once you have opened the Dimension Style dialog box, pick the New button and create a new dimension style. Add the name for your new dimension style, use 'Copy of Standard' to start, then pick Continue to set the parameters for your new style for your dimensions. Use OK to save. Figure 7.21 illustrates some of the variables you may want to set.

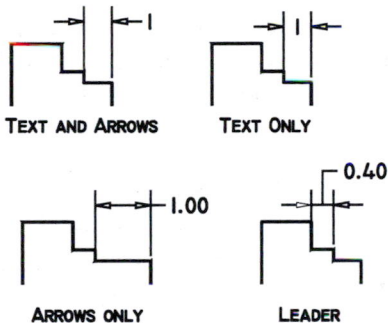

Figure 7.21

You will notice that once you have started dimensioning, a new layer called 'defpoints' will be part of your layer listing. This layer will not print, and is used to define the point to which your dimension is pointing.

Editing Dimensions

If you pick Properties from the Modify pull-down menu and select a dimension, you are given the Modify Dimension dialog box. Pick the box that offers what you would like to change. If you would like to change just the text, select DDEDIT, just as in modifying text.

Adding Prefixes

Under the Primary Units dialog box is the heading Prefix. You can add the prefix **%%c** for the diameter sign to add diameter signs to all of the dimensions shown.

Stretch

You can also stretch the object and have the dimensions automatically updated. The STRETCH command can be used in the same way that DIMEDIT is used to move the text across the dimension line.

EXTEND and TRIM can also be used to ensure dimensions show up *exactly* correctly.

Step 1

Quickly draw the part shown in metric units by selecting acadiso.dwt, or use STARTUP to make sure that you are starting with a metric file. The dimensioning defaults are set up differently in imperial and metric. It is VERY important to start with the correct units or this tutorial will not work.

The dimensions are shown for your convenience in this illustration. Do not try to add them yet. Starting with the lower left arc at 0,0 and a SNAP set to .25 is the easiest way to do this.

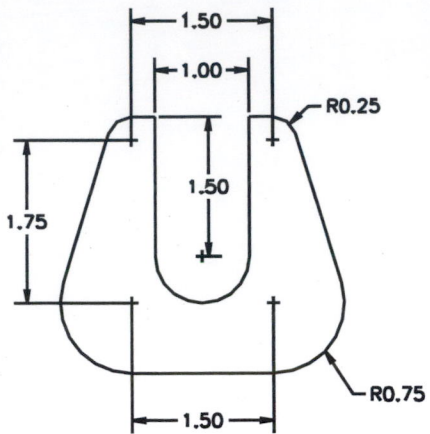

Step 2

Once the part is drawn in, let's see what a dimension would look like using the standard dimensioning format.

Before starting dimensioning, create a layer called Dims, make it red, and make it current.

Now set your object snap to ENDpoint and CENter. The object snap dialog box is hidden in the Drafting Settings box under Tools.

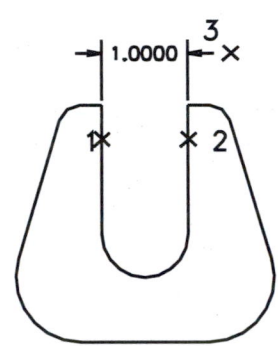

```
Command:LINEAR
Specify first extension line origin or <select object>:END
   of (pick 1)
Specify second extension line origin:END of (pick 2)
Specify dimension line location or
   [Mtext/Text/Angle/Horizontal/Vertical/Rotated]:(pick 3)
Dimension text = 1.0000
```

The dimension line should be 1/2 inch away from the object line. With SNAP set at .25, this should be easy to place.

Step 3

Four decimal points of accuracy are not needed, and the scale of the dimensions is larger than needed. The next step is to make a dimension style that will incorporate the changes.

First, access the Dimension Style dialog box, create a style called Mechanical, and click Continue to change the precision.

Note that the default style here is an ISO-25.

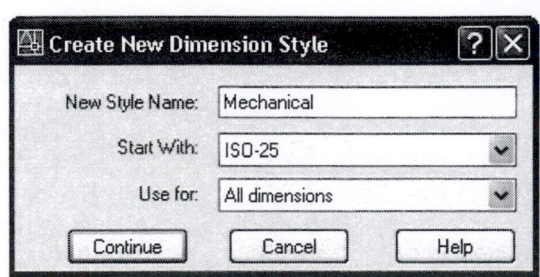

If you were using imperial measurements with your STARTUP command in imperial, the default would be called Standard. If you have started this drawing in the wrong measurements, close this file and start again in metric.

Having set up your mechanical style, now set the precision to two decimal places under the Primary Units tab.

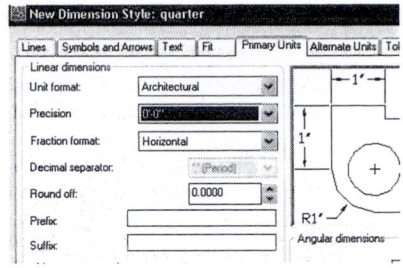

Now pick the Fit tab, and change the overall scale to .75 as shown on the right below.

Pick OK to return to the Dimension Style dialog box.

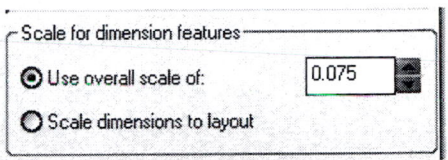

Make sure that your new dimension style is current by choosing the Set Current button on the Dimension Style Manager menu, then close the dialog box.

Now erase the first dimension and add the new one made with the right parameters as shown on the right. Make sure that your dimensions are going on to the Dims layer.

```
Command:LINEAR
Specify first extension line
   origin ...>:END of (pick 1)
Specify second extension line
   origin:CEN of (pick 2)
Specify dimension line location
   or ........]:(pick 3)
```

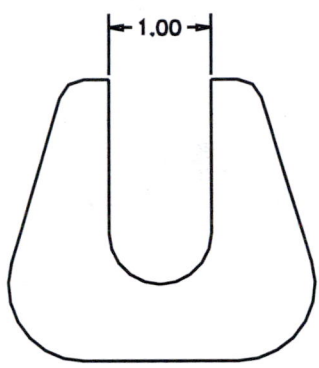

Step 4   Now that the style has been set, add the linear dimensions. First create the horizontal dimension between the centers of the two arcs on the top, then create the vertical dimension on the right.

```
Command:LINEAR
Specify first extension line
   origin or ...>:CEN of (pick 1)
Specify second extension line
   origin:CEN of (pick 2)
Specify dimension line location
   or [Mtext.....]:(pick 3)
Command:↵
Specify first extension line origin or <select object>:CEN
   of (pick 4)
Specify second extension line origin:CEN of (pick 5)
Specify dimension line location or
   [Mtext/Text/Angle/Horizontal/Vertical/Rotated]:(pick 6)
```

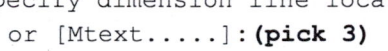
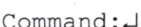

Notes

If you are not getting the two zeros, make sure that there are no zeros suppressed in Primary

Step 5   Now add the other vertical and horizontal dimensions in the same manner. Don't worry if the dimensioning defaults are different on your system.

```
Command:LINEAR
Specify first extension line
   origin ....>:END of (pick 1)
Specify second extension line
   origin:CEN of (pick 2)
Specify dimension line location or
   [....Rotated]:(pick 3)
Command:↵
Specify first extension line
   origin or ...>:CEN of (pick 4)
Specify second extension line
   origin:CEN of (pick 5)
Specify dimension line location or
   [Mtext/Text/Angle/Horizontal/Vertical/
   Rotated]: (pick 6)
```

Step 6   Now add the radial dimensions. When you pick radial for the dimensions, you can move the leader line either inside or outside the radius. Place the radii outside the object as shown.

```
Command:RADIUS
Select arc or circle:(pick 1)
Specify dimension line location or
   [Mtext/Text/Angle]:(pick 2)
Command:RADIUS
Select arc or circle:(pick 3)
Specify dimension line location or
   [Mtext/(Text/Angle]:(pick 4)
```

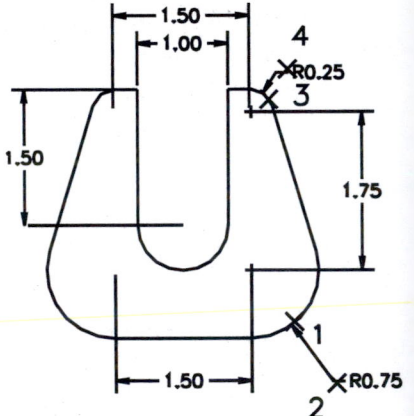

If the radii do not respond in the way that you want, return to the Dimension Style Manager and make sure you are working in the correct dimension style, then use Modify to change the parameters. The lines of the radius dimension are under the Fit tab.

Step 7   Your dimensions are fine, so now add center marks. From either the pull-down menu or the toolbar, choose Center Mark.

```
Command:CENTER
Select arc or circle:(pick the radii)
```

The lines associated with the center mark are determined under the Lines tab.

Save your drawing now if you have not done so already.

Step 8   The dimensions added are associative dimensions. This means that the dimensions are entered relative to the points that you have identified. These are called defpoints, and there is a special layer created for them under your Layer menu. This layer does not plot, so be careful not to make it current.

This means that you can STRETCH the part and the dimensions will automatically update.

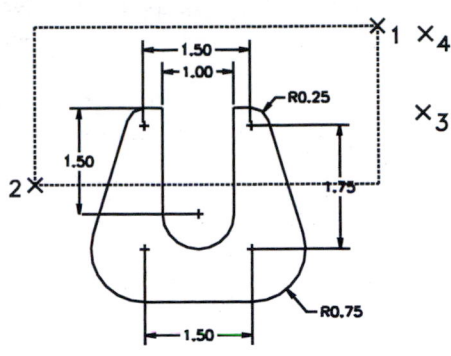

```
Command:STRETCH
Select objects to stretch by
   window ...
Select objects:C
First corner: (pick 1)
Other corner: (pick 2)
Select objects:↵
Base point: (pick 3)
New point:@1<90 (or pick 4)
```

The resulting object should look like this.

Your objects should stretch and the vertical dimension should change. If only the dimensions stretch and not the part itself, it is because the Objects layer, or whatever layer you have placed the objects on, is locked. Under the Layer toolbar, unlock the Objects layer.

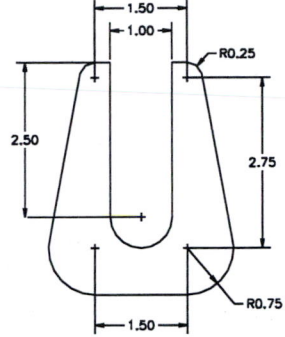

Step 9

Finally, if you have entered all the dimensions and then find that they are the wrong scale or the wrong units, that the arrowheads should be ticks, or some other problem, you can edit the dimensions singly or in groups by changing the options in the Dimension Style dialog box. What if the dimensions should really be in fractional rather than decimal? You can change the dimension style and apply it to the current drawing.

First, save the drawing under a different name by using the 'Save as' function under the File pull-down menu. Now you have two copies, one with fractions and one with decimals.

The next step is to make a new dimension style that incorporates the necessary changes. Access the Dimension Style dialog box .

From the main dialog box, pick New and create a style called Fractional using Mechanical as the base.

Once this is entered, pick Continue and go to the Primary Units tab to change the units to fractions. Pick OK and you will return to the Dimension Style Manager.

Use the Set Current button to set the current style to Fractional and close the dialog box.

To update all of your dimensions to fractional, use Matchproperties or the UPDate command.

Make a dimension anywhere on the screen. Pick the Matchproperties icon. Pick the new dimension, then all the other dimensions. They will update.

To use update, change the mode to dimensioning mode, then use UPD.

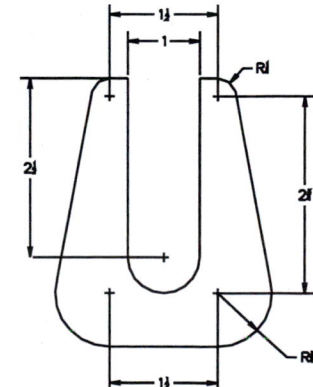

```
Command:DIM
Dim:UPD
Select objects: (pick all the
   objects with crossing or window)
Select objects:↵
```

Step 1

Open a floor plan drawn in architectural units.

If you don't have one, open a new file in imperial units using acad.dwt, change your units to architectural, and draw in a few lines for a small house, 25′ by 35′ with a simple window and door.

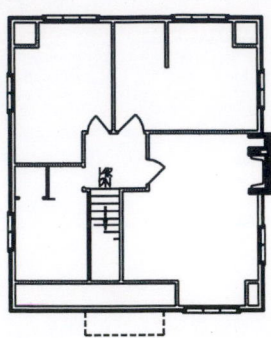

Step 2

The Standard dimensioning is set up to be printed at 1 - 1. For a floor plan, you will be printing at 1/4″= 1′-0″ or 1/2″ = 1′-0″. You will need to set up a dimension style right away.

DO NOT CHANGE THE STANDARD DEFAULTS!

Create a new dimension style and call it *Quarter* for 1/4″= 1′ -0″.

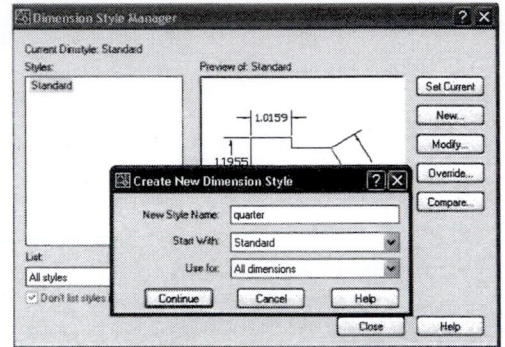

Step 3

You will need to change three things right away.

First, set your primary units to architectural with a precision of 1″. Having a 1/8″ or even a 1/2″ precision on a floor plan is unrealistic. These are for details.

The rounding-off factor is also important. In this case do not change it. By changing this you can sometimes have drawings where the smaller dimensions add up to more than the overall dimension.

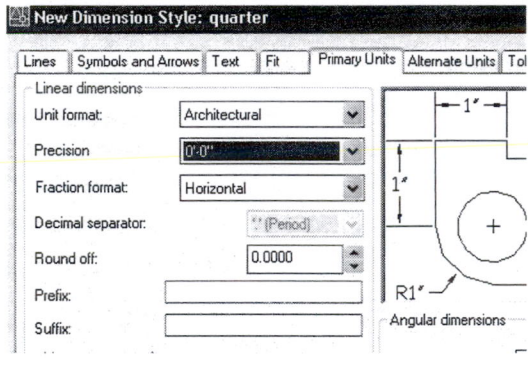

Now set your text position to be above the dimension line and aligned with it. This is under the Text tab.

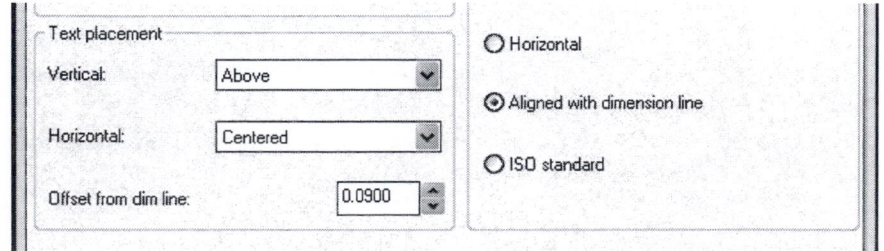

Architectural and construction applications generally have the text placed above the dimension line. This is a standard discipline protocol. In a set of drawings, it is not unusual to find that those prepared by the designers and architects have different protocols to the structural and mechanical drawings in the same set.

Now pick the Fit tab, and change the overall scale to 48 as shown on the right below.

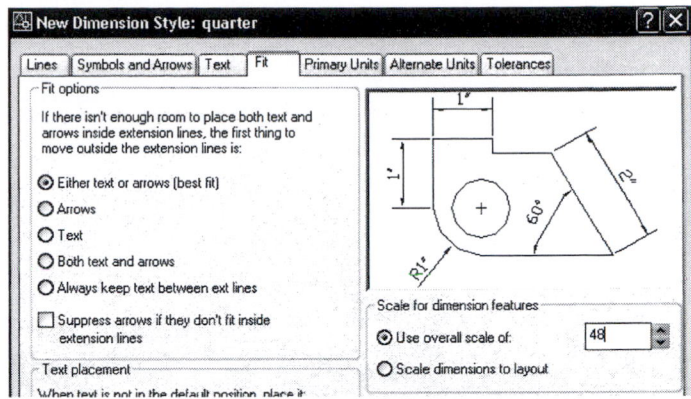

Often architectural drawings have ticks rather than arrowheads. Under the Symbols and Arrows tab, change the arrows to architectural ticks.

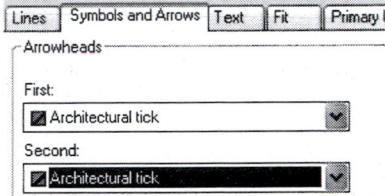

Your dimension style should be set and ready to use. Pick OK at the bottom of the dialog box, then choose Set Current from the Dimension Styles dialog box.

 Set up a new layer called Dims, and start dimensioning.

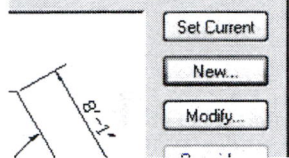

Step 4   You will want to set your OSNAP to ENDpoint and MIDpoint.

The dimensions on a frame building are set up for the carpenter, so the outside of the frame is used as the definition point, not the outside of the veneer. Since the window position is known, but the exact size of the window is not always known until it is shipped to the site (due to changes in orders, etc.) the windows and doors are generally dimensioned to the center point.

The interior walls are dimensioned to the outer edge of the wall to give the interior dimension, or sometimes to the wall center.

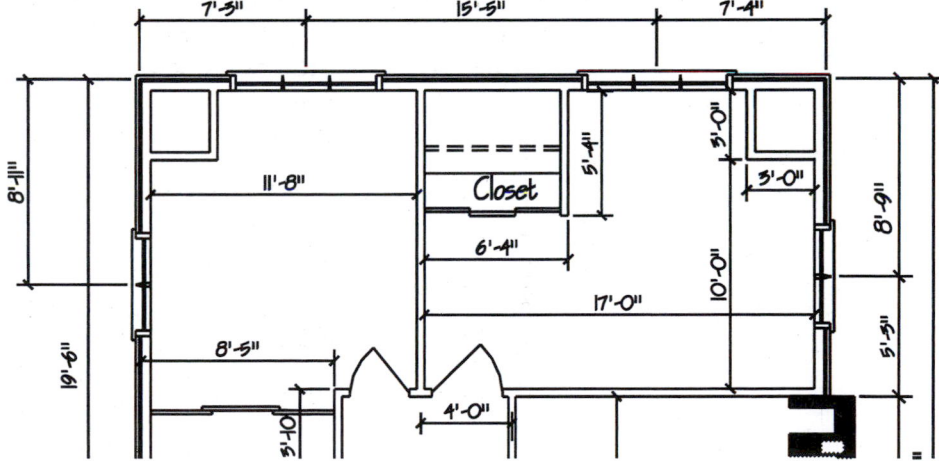

If the building is insulated with concrete or other masonry, these dimensions are usually measured from the outer edges.

Save your file when you are done. Don't forget to use STRETCH to place the text where you want it.

Exercise 7 Practice

This reinforced concrete footing is used in Tutorial 14c page 276.

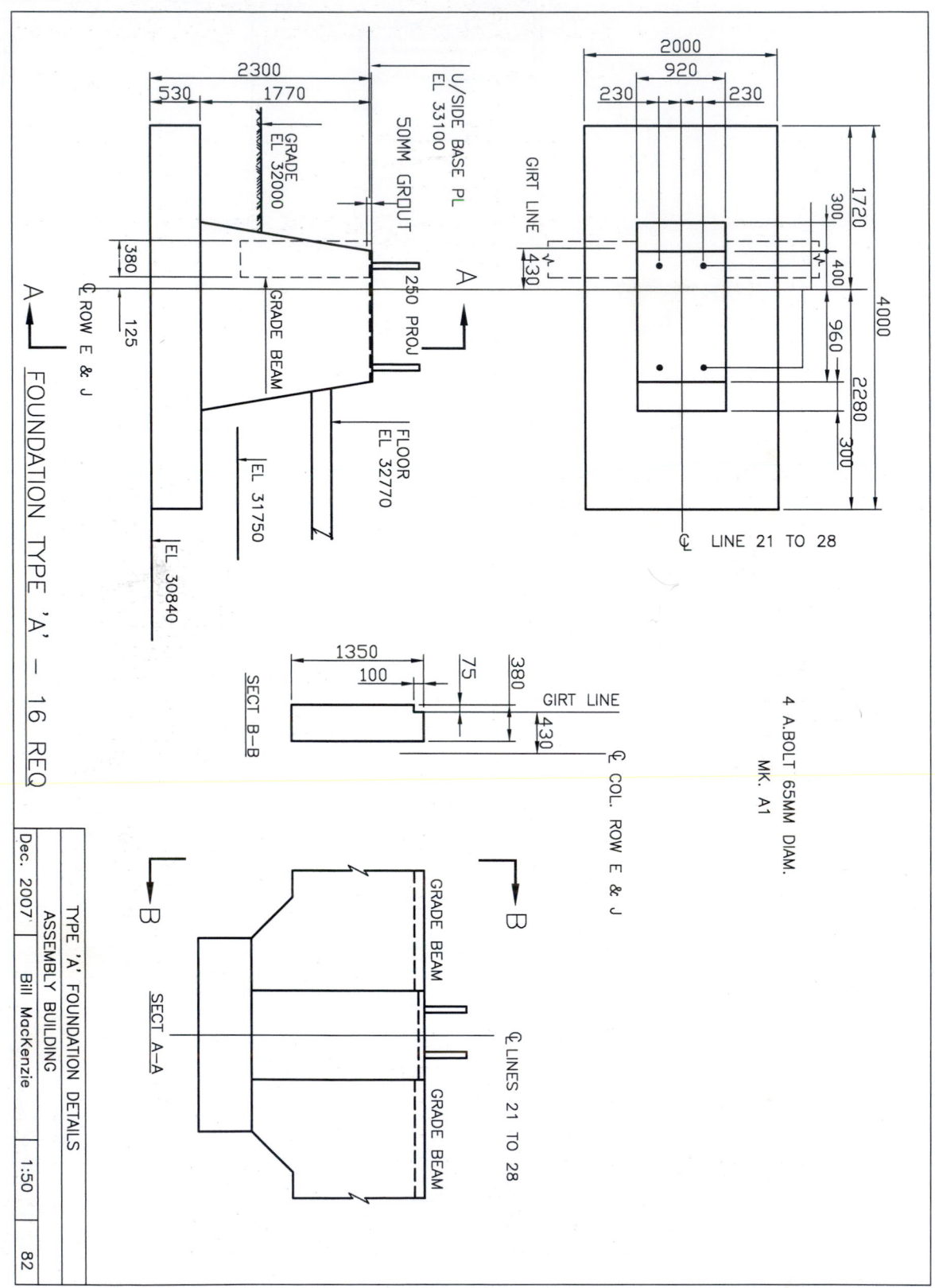

FOOTING AND FOUNDATION DRAWINGS

FOOTING AND FOUNDATION DRAWINGS ILLUSTRATE THE SIZE OF THE FOUNDATION AND WHAT FASTENERS ARE USED TO ATTACH IT TO THE STRUCTURAL MEMBERS. THESE DRAWINGS ARE GENERALLY PREPARED BY CIVIL OR STRUCTURAL ENGINEERS AS OPPOSED TO ARCHITECTS.

Exercise 7 Architectural

This is the first floor of the house shown on page 85.
Open the second floor plan shown on page 85. Use Save as and call it First Floor.
Erase what you don't need, and use STRETCH to adjust the windows.
Add the interior walls and fireplace, then dimension. Add the section arrow last.

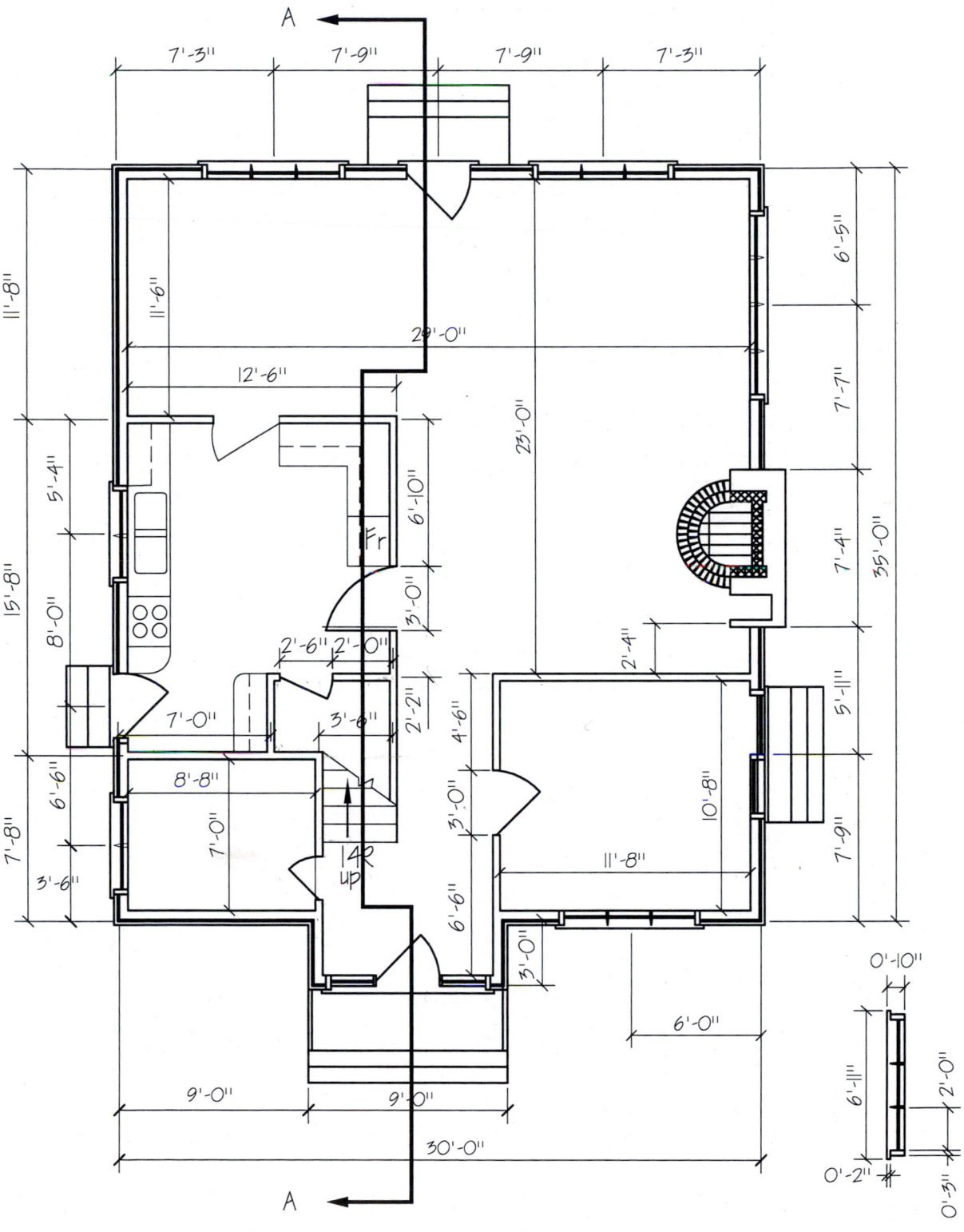

Exercise 7 Mechanical

With multiple views you must make sure all the views line up.

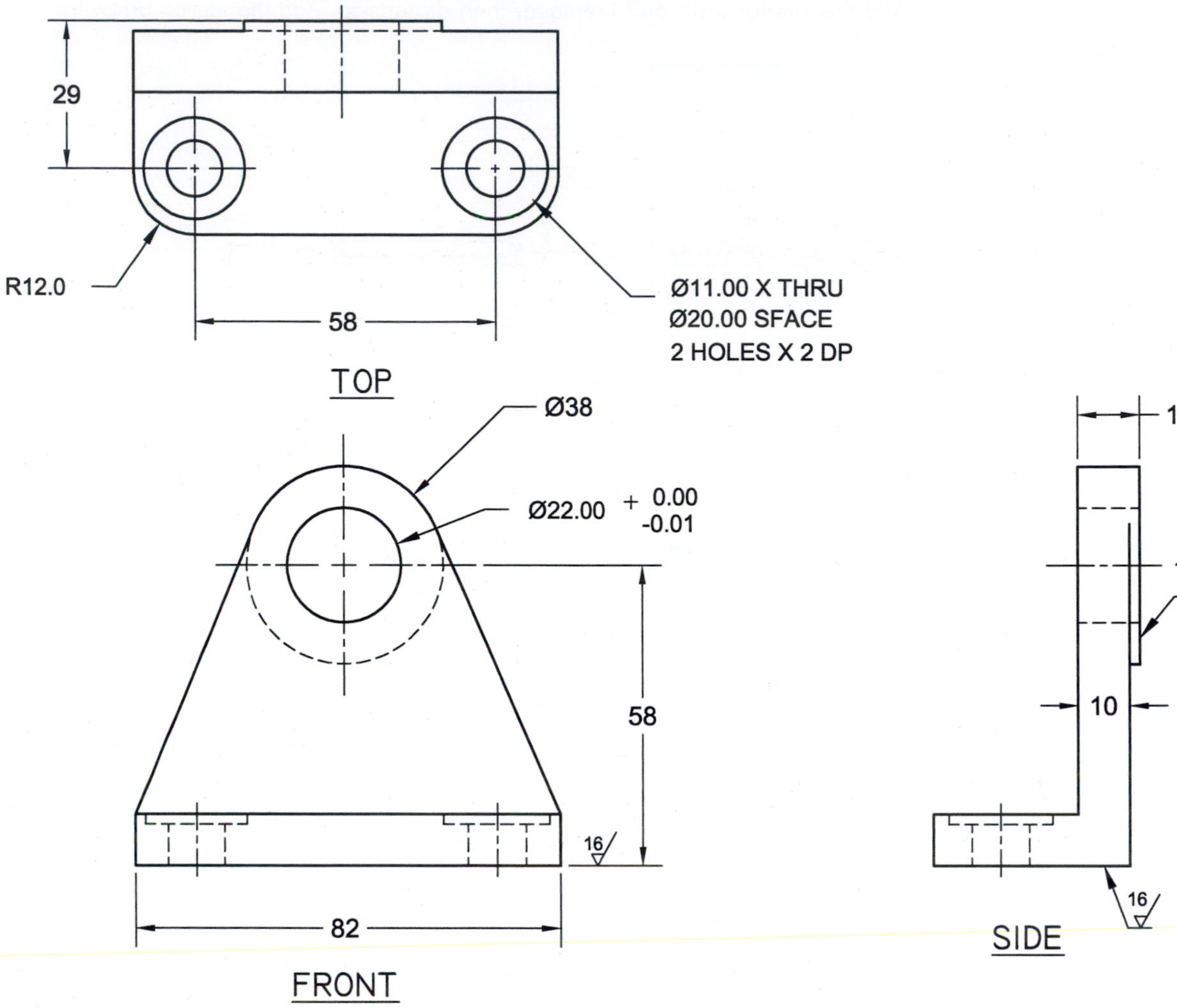

29

R12.0

58

TOP

Ø11.00 X THRU
Ø20.00 SFACE
2 HOLES X 2 DP

Ø38

Ø22.00 $+ \begin{array}{c} 0.00 \\ -0.01 \end{array}$

58

16

82

FRONT

12

16

10

16

SIDE

Axle Support

ORTHOGRAPHIC PROJECTION

THE PURPOSE OF SIMPLE ORTHOGRAPHIC PROJECTION IS TO SHOW THREE OR MORE VIEWS
OF AN OBJECT IN 2D IN ORDER TO COMPLETELY DESCRIBE IT. THE VIEWS ARE SHOWN
AT RIGHT ANGLES TO ONE ANOTHER, THE VIEWER IS LOOKING FROM A POINT
PERPENDICULAR TO THE MAIN FACE.
COMMON DIMENSIONS ARE FOUND ON ONE VIEW ONLY.

16

THIS SYMBOL MEANS THE SURFACE IS CUT RATHER THAN CASTE.

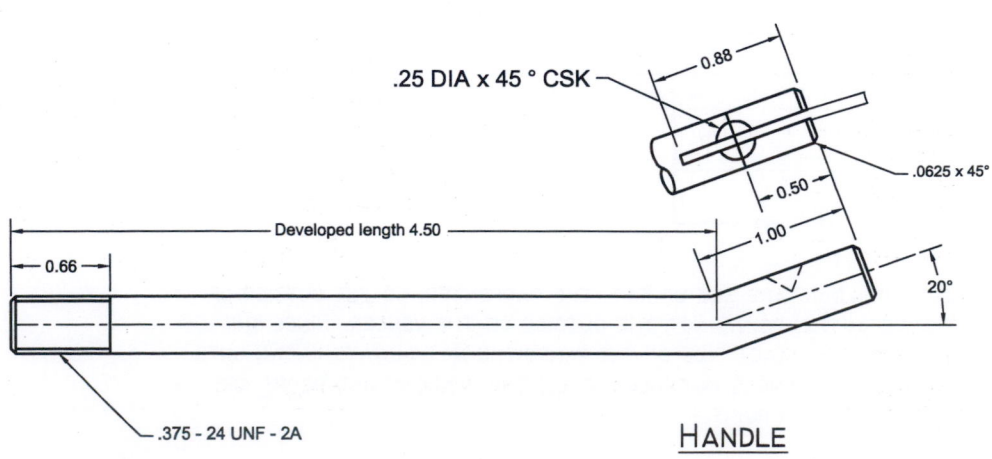

.25 DIA x 45 ° CSK

0.88

.0625 x 45°

0.50

1.00

Developed length 4.50

0.66

20°

.375 - 24 UNF - 2A

HANDLE

Exercise 7 Wood

This is the top or bonnet for the clock base found on page 105.
To create the numbers use ARRAY then DDEDIT or ED.
The design for the top is created with PLINE.

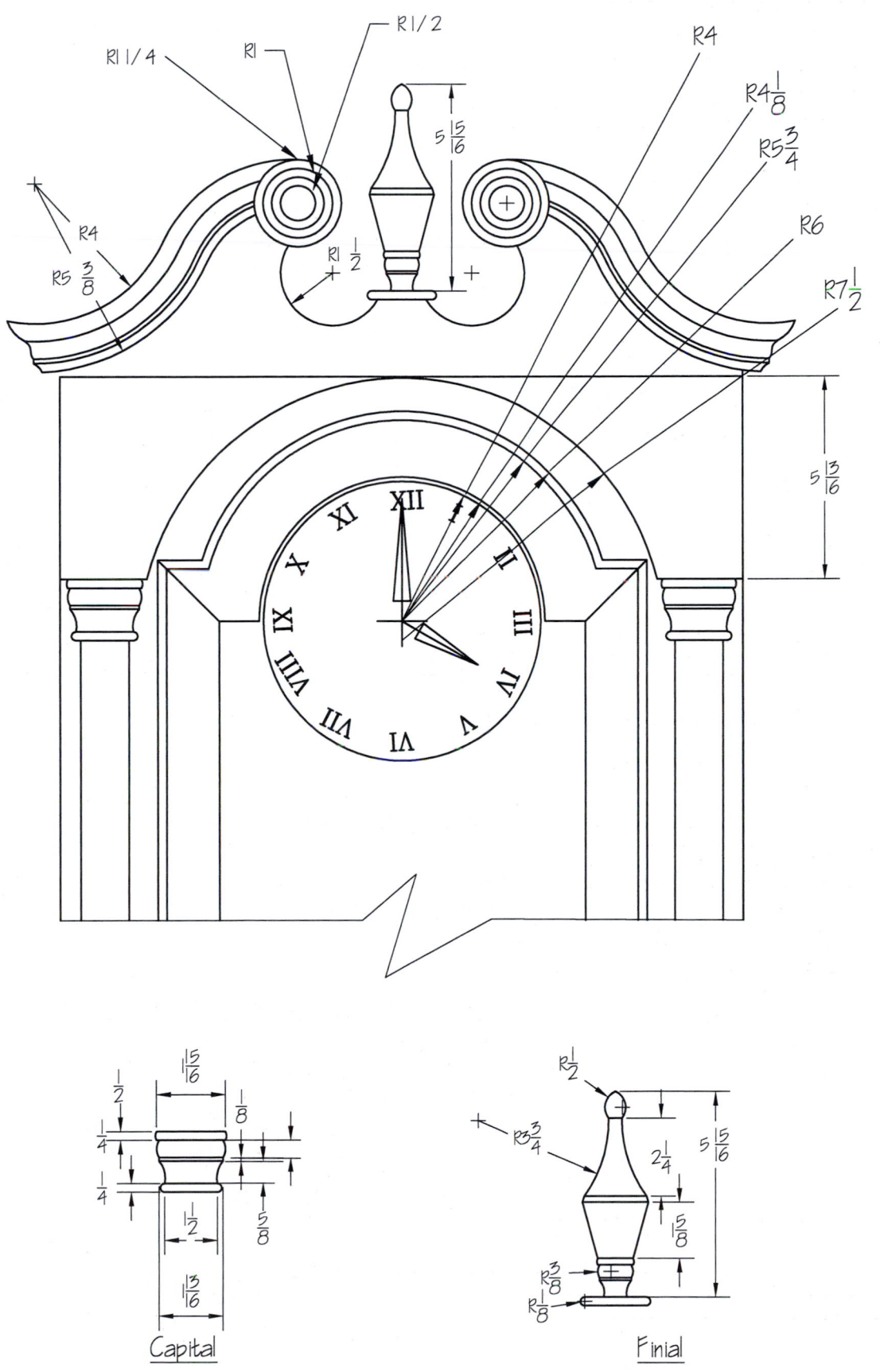

Capital

Finial

Exercise 7 Challenger

Make sure the views line up.

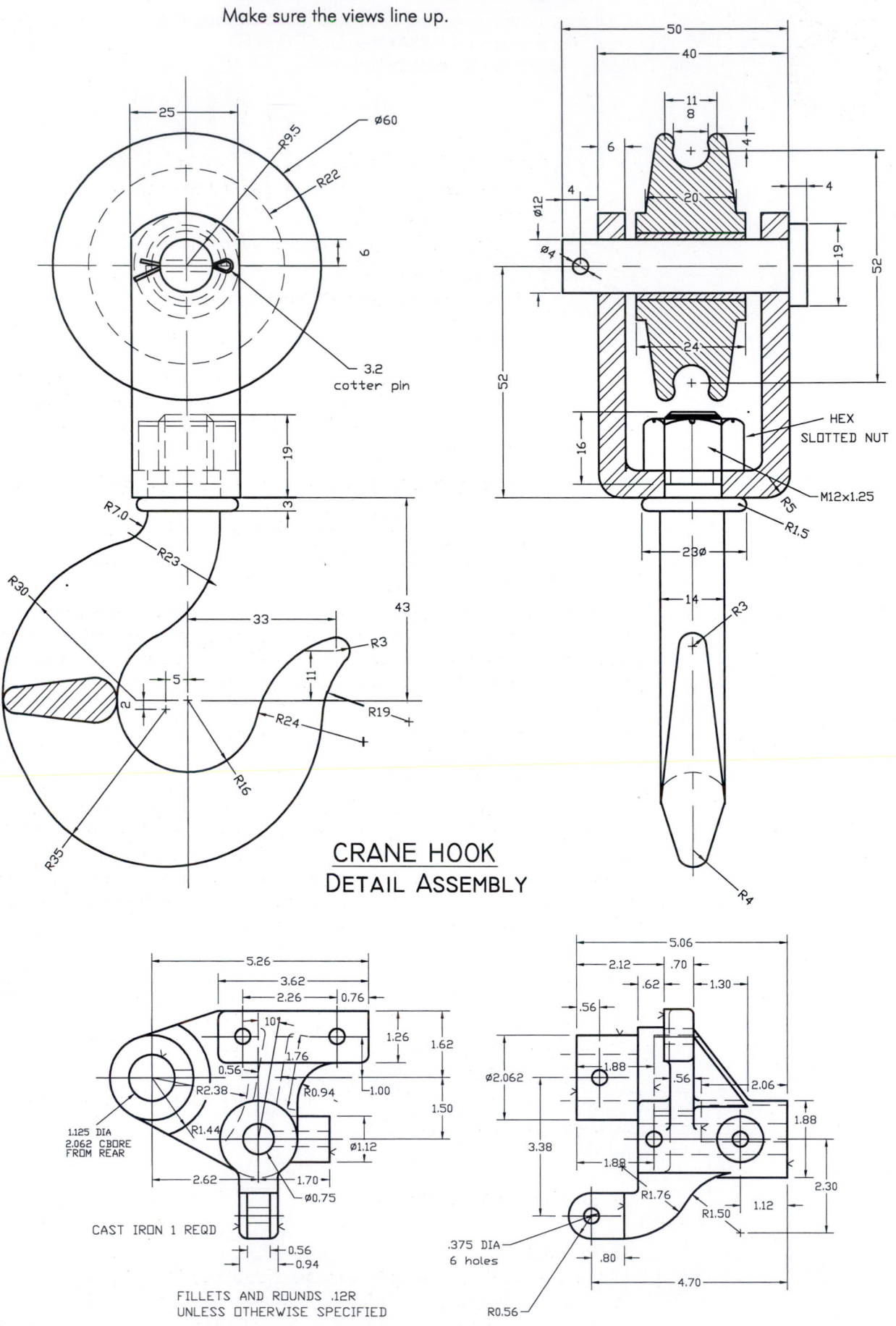

CRANE HOOK
DETAIL ASSEMBLY

HEX
SLOTTED NUT

M12x1.25

3.2
cotter pin

CAST IRON 1 REQD

FILLETS AND ROUNDS .12R
UNLESS OTHERWISE SPECIFIED

1.125 DIA
2.062 CBORE
FROM REAR

.375 DIA
6 holes

CONNECTOR

8 Text and Pictorial Views

On completion of this chapter, you should be able to:
1. Place linear text in any size at any rotation angle
2. Place paragraph text
3. Create and change collections of formats to make a text style
4. Change existing text
5. Set up isometric text and pictorial drawings.

AutoCAD provides two basic ways to create text. ***Linear text*** places simple entries of one or two lines. For longer entries ***paragraph text*** is used. Text is entered in the current text style, which incorporates the current format and font settings. Use the Text toolbar for many text entries.

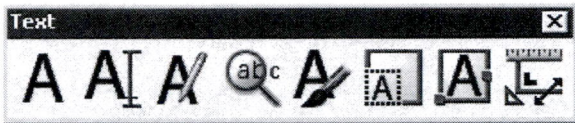

Linear Text

As stated in Chapter 5, the commands TEXT and DTEXT will place strings of characters on your drawing. When entering text, AutoCAD will prompt you to choose a height for each character, a rotation angle for the string, and a point at which to place the text string on the model or drawing. Many people prefer TEXT or Single Line Text from the pull-down menu for titles and filling in title blocks, as with it you can see the style and placement of the text as it is entered.

The *TEXT or DTEXT Command*

In Releases 2008 and 2007, the TEXT command functions in the way that the DTEXT command functioned in releases prior to 2005. Release 2006 has only the TEXT command without the multiple line capabilities. This command is very useful for filling in title blocks and adding data at a standard size and style in several spots on a drawing at once. Multiline text is very useful for paragraphs and for importing ASCII data, but the TEXT command is still the best for annotating drawings.

Toolbar From the Text toolbar choose

Pull-down menu From the Draw menu choose TEXT, then Single Line Text.

The command line equivalent is TEXT.

The TEXT command is as follows:

```
Command:TEXT
Specify start point of text or [Justify/Style]:J
Enter an option
   Align/Center/Fit/Middle/Right/TL/TC/TR/ML/MC/MR/BL/BC/BR:
```

Where:	**Justify**	= the placement of the text
	Style	= controls the style of the letters; the styles must be loaded in AutoCAD to be accessible
	Align	= an alignment by the end points of the baseline; the aspect ratio (X vs. Y) will correspond to the preset distance
	Center	= the center point of the baseline; this option will fit the text through the center point indicated
	Fit	= an adjustment of width only of the characters that are to be fit or 'stretched' between the indicated points
	Middle	= a placement of the text around the point, i.e. the top and bottom of the text are centered as well as the sides
	Right	= an alignment with the right side of the text

The examples in Figure 8.1 demonstrate the standard justifications. The default is left justification at the baseline of the text string.

Height can be chosen by picking a point to indicate the height, or by typing in a number.

The double-initialed justification options illustrated in Figure 8.2 are as follows:

Figure 8.1

TL	=	top left
TC	=	top center
TR	=	top right
ML	=	middle left
MC	=	middle center
MR	=	middle right
BL	=	bottom left
BC	=	bottom center
BR	=	bottom right

Figure 8.2

Once you have chosen a point at which to place your text, the command will prompt you for the height of the letters, the rotation angle, and the text or string of characters itself.

A text *string* is one line of text.

Figure 8.3

```
Command:TEXT (or DTEXT)
Specify Start point or [Justify/
Style]:C
Specify center point:(pick 1)
Specify height<.2000>:.5
Specify rotation angle <0>:↵ (to accept the default)
Enter text:Front Elevation↵
```

In Figure 8.3, the justify option Center was chosen, so the other options for placement were bypassed. If Justify had been chosen, the following line would have been offered:
```
Align/Center/Fit/Middle/Right/TL/TC/TR/ML/MC/MR/BL/BC/BR:
```

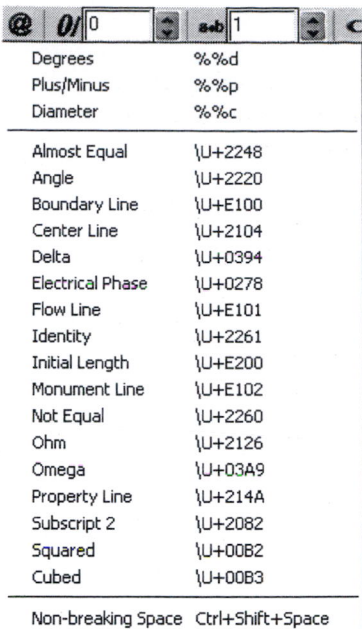

Degrees	%%d		
Plus/Minus	%%p		
Diameter	%%c		
Almost Equal	\U+2248		
Angle	\U+2220		
Boundary Line	\U+E100		
Center Line	\U+2104		
Delta	\U+0394		
Electrical Phase	\U+0278		
Flow Line	\U+E101		
Identity	\U+2261		
Initial Length	\U+E200		
Monument Line	\U+E102		
Not Equal	\U+2260		
Ohm	\U+2126		
Omega	\U+03A9		
Property Line	\U+214A		
Subscript 2	\U+2082		
Squared	\U+00B2		
Cubed	\U+00B3		
Non-breaking Space	Ctrl+Shift+Space		
Other…			

If TEXT was the last command entered, pressing ↵ at the Specify start point [Justify/Style]: prompt skips the prompts for height and rotation angle, and immediately displays the Enter text: prompt.

The text is placed directly beneath the previous line of text.

Special Character Fonts

You can underscore, overscore, or include a special character by including control information in the text string. In MTEXT, the special characters are found under the @ symbol.

% %u	=	underscore
% %o	=	overscore
% %d	=	degree symbol
% %p	=	plus/minus tolerance symbol
% %c	=	diameter symbol
% %nnn	=	ASCII characters: **%%123 %%125** = { }

When using special character fonts with the TEXT command, the special characters will be displayed as you type, i.e., **%%uFront Elevation%%u**. The entry will be updated to the desired text once the command is finished, as shown in Figure 8.4.

```
Command:TEXT
Specify start point or [Justify/Style]:R
Right side:(pick 1)
Specify height<.2500>:↵
Specify rotation angle<0>:↵
Enter text:%%u%%c25 4 holes
Enter text:↵
```

Ø25 4 holes[1]

Figure 8.4

Multiline Text

A new text string will line up with the previously entered text string, if there are no changes in the base point or justification options in both the TEXT and DTEXT commands. If Center is chosen, all of the text will be centered (see Figure 8.5b); if no option is chosen, all strings will be left-justified (see Figure 8.5a). Your last string of text will be highlighted to show where the next line will be lined up.

If you do not want your text to line up with the last string entered, simply identify a new start point. In TEXT use ↵ to reenter the command after the first text string, then press ↵ again to accept the default position, size, and rotation.

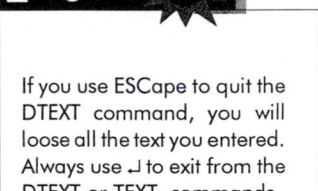

ALL FILLETS ARE RADIUS .5
BOTH SIDES

MOHAWK COLLEGE
FENNEL AND WEST FIFTH
HAMILTON ON

a   b

Figure 8.5

Danger

If you use ESCape to quit the DTEXT command, you will loose all the text you entered. Always use ↵ to exit from the DTEXT or TEXT commands.

Once your text has been entered, it is accepted as one item and can be edited using any of the edit commands, such as COPY, MOVE, ERASE, ROTATE, and ARRAY. To edit the text itself, use DDEDIT, ED, DDMODIFY, or CHANGE.

Paragraph Text

Paragraph text or MTEXT is for long, complex entries that have many lines of text. Any number of text lines or paragraphs can be entered to fit within a specified width. The paragraphs form a single object that can be moved, rotated, copied, erased, mirrored, stretched, or scaled. This is the default text command under the Draw toolbar in Release 2000 and subsequent releases.

You can apply overscoring, underlining, fonts, color, and text height to any individual character, word, or phrase of the paragraph.

Creating paragraph text is a lot more flexible, but slightly more difficult than entering single line text. Text is entered in the Edit MText dialog box. This editor can be changed using the MTEXTED system variable. Use Preferences to set up a different editor. If you are using a system with limited RAM and limited speed, the ability to have your text spell-checked does not make up for the time necessary to access the MTEXT dialog box every time you want to add a string of text. For large paragraphs, however, this is a real advantage.

Paragraph Text

The Edit MText dialog box is a very efficient way to set properties that affect the entire paragraph or selected text. As in a word processor, you should set the width before you create the text. The paragraph will be displayed in a dialog box within the specified width. The text will wrap or spill in the direction defined by the current attachment setting. The text boundary can be realigned. To create paragraph text, from the Text fly-out on the Draw pull-down menu, choose Multiline text, or the button shown on the toolbar. This will invoke the MTEXT command .

Toolbar From the Draw toolbar choose

A

Pull-down menu From the Draw menu choose Multiline Text.

The command line equivalent is MTEXT.

You will be prompted to specify the insertion base point for the text as follows:

```
Command:MTEXT
Current text style: STANDARD Text height: 0.2000
Specify first corner or [Height/Justify/Line
  Spacing/Rotation/Style/Width]:(pick a point where the text
  will start)
```

Next you specify the width of the text by using one of the following methods:

- To define a diagonally opposite corner of a rectangular text boundary, specify a point.

- To define only the width of the text boundary, enter **W** and specify a width value. Entering **0** causes the text to extend horizontally until you press ↵.

- The Properties and Find/Replace tabs offer more settings.

The Text Formatting box will appear and your text will be displayed within the box as you type it, as shown in Figure 8.6.

The width of the text that you have chosen will be reflected on the dialog box. Type in the text you would like placed at the specified location on your file, then choose OK to write it to the file. Once there, you can highlight it to change the size or properties of the text.

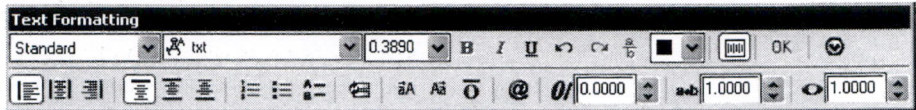

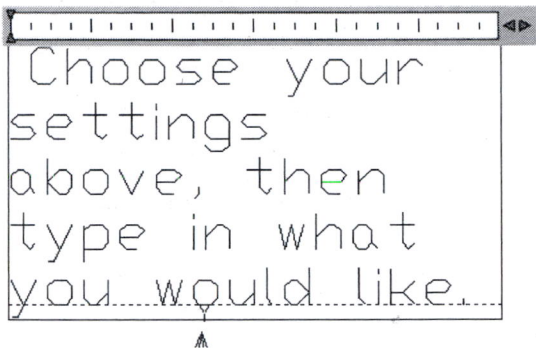

Figure 8.6

The text can be typed in as in a word processor; it will wrap according to the width chosen.

To edit the color of one word or phrase, or to have it underlined or overscored, select the text using the mouse, then choose Overscore, Underline, or Color.

Some fonts cannot be displayed in the Edit MText dialog box. If text isn't shown, select a substitute font to represent the original font, then choose OK. When you are finished editing, the original font selected appears in the graphics area.

The text will wrap according to the size of lettering that you have chosen. When you have finished typing in the information you need, pick OK to have it placed in the area on your screen that you specified earlier.

Text Styles and Fonts

Text styles are what the user names the style of the lettering chosen. *Fonts* are the style or design of the letters and numbers used to create the text string. The fonts can be supplied by AutoCAD or a third-party developer. A text style is stored with not only the lettering style but also a group of characteristic settings.

The simplest font files are shape files or .shx files. True Type Fonts can be loaded as well. These are handy, but not always appropriate for drawing notation. They also use a lot of space on your file, so be careful when loading them.

Using Text Styles

Each text style takes a font file from the AutoCAD list.

The default style is **Standard** using the txt.shx font. To use any other type of lettering you must load that style. To load a text font onto AutoCAD from another folder, find the path where the fonts are used in AutoCAD. On most systems this is C:\Program Files\AutoCAD 2007\Fonts. Copy the font file there and it will be available through the STYLE command.

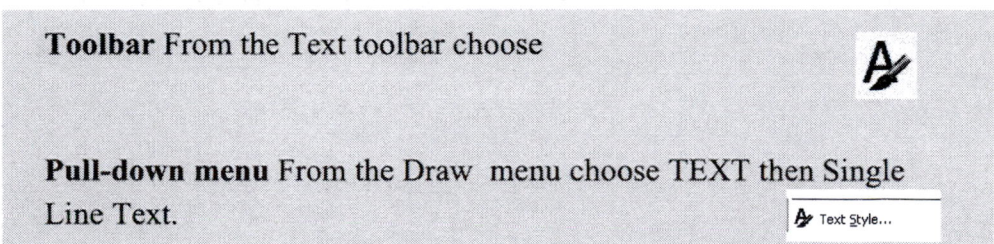

Toolbar From the Text toolbar choose

Pull-down menu From the Draw menu choose TEXT then Single Line Text.

The command line equivalent is STYLE.

Both the pull-down menu and typing will give you the dialog box shown in Figure 8.7.

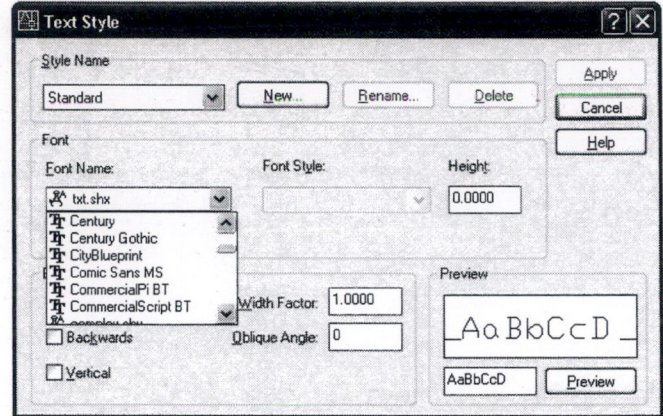

Figure 8.7

You can change the standard text font or add a new style. Next find a font that you would like to assign to this name. City Blueprint is fairly standard.

Height can be specified in this style box or within the command itself. It is much better to assign the height in the TEXT or MTEXT command than here.

To create a new text style, choose New, then pick a style from the font list. Choose Apply and it will be loaded. The word Cancel will be replaced by the word Close. Choose Close and your style will be current.

ROMANS

Width factor .5	AutoCAD	
Width factor 1	AutoCAD	Vertical
Width factor 1.5	AutoCAD	AutoCAD
Oblique angle 0	AutoCAD	
Oblique angle 30	AutoCAD	
Oblique angle −30	AutoCAD	
Upside down	AutoCAD	
Backwards	AutoCAD	

Figure 8.8

Most of the options illustrated in Figure 8.8 you will never use; keep in mind it is computer people not designers who make the software. They make these option available simply because they can.

Monotext
7 line segments

Roman Simplex
19 line segments

English Gothic
70 line segments

Figure 8.9

Using Text Styles

In Figure 8.9 you can see that the Gothic letter has 10 times the number of lines as the Monotext letter; it may be appropriate for a title, but not for a notation or a dimension. The text fonts marked TT are bitmap fonts, not made with lines. These are Microsoft True Type fonts. To make a style current, simply pick it from the list and then click Apply.

Editing Text

A text string is considered an object and can therefore be moved, copied, changed into different layers, and created in different colors. You can also alter the text string itself as well as the height, the rotation angle, and the style of the characters.

Text objects also provide grips for moving, scaling, and rotating. A paragraph text has grips at the four corners of the text boundary and, in some cases, at the attachment point. Line text provides a grip at the lower left corner and another at the alignment point. (Review Chapter 4 for more information on grips.)

When using Object Snaps, INSERT will snap to the insertion point you selected to create the text.

Editing Line Text

You can change both the text characteristics and the text string with the CHANGE command, as shown in Figure 8.10. This is useful for changing many lines of text, particularly when making charts.

DDEDIT and DDMODIFY make editing single strings of text much easier. DDEDIT can be invoked with the alias ED or simply by double-clicking the text string.

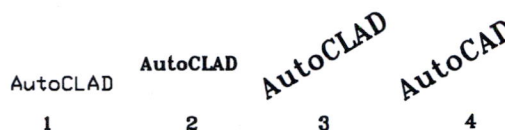

AutoCLAD 1   AutoCLAD 2   AutoCLAD 3   AutoCAD 4

```
Command: CHANGE
Select objects: (pick 1)
Select objects: ↵
Specify change point or [Properties]: ↵
Specify new TEXT insertion point <no change>: ↵
Enter new text style <STANDARD>: ROMANT (part 1, Figure8.13)
Specify new height <5.0000>: 10 (part 2)
Specify new rotation angle <0>: 30 (part 3)
Enter new text <AutoCLAD>: AutoCAD (part 4)
```

Figure 8.10

CHANGE gives you the options to change all of the variables in the TEXT command as shown in Figure 8.10.

If you pick more than one string, the command will repeat the prompts for each string.

The DDEDIT Command

In long strings of text, DDEDIT makes the editing process much easier. You can change only the text, not the formatting or properties of the text.

To edit the text string content:

> **Toolbar** From the Text toolbar choose choose
>
> **Pull-down menu** From the Modify menu choose Object, then Text.

The command line equivalent is DDEDIT or ED.

Select the line text object that you would like to edit. Within the text string offered for editing, select the text where you want it to be changed, or type over or reenter the text, then choose OK to have it updated. Pressing Backspace will delete the highlighted text.

Select another line of text or press ↵ to exit the command.

Modifying other Properties

If you want to change more than the text line content, use Properties under the Modify menu. This works on only one string of text at a time, but offers you a variety of things to change.

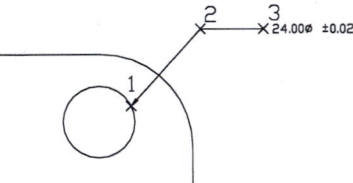

To edit the text string:

This works similarly to the CHANGE command in that you can change all of the properties associated with the line of text.

To change either the justification or the scale of the text using text paramaters choose:

   Justify text

   Scale text

Using LEADER to Create Notations

The LEADER command in the Dimension menu gives you the facility of creating text with a leader line or series of leader lines and an arrowhead as shown in Figure 8.11. The leader arrowhead emanates from the point picked. The command is as follows:

```
Command:LEADER
Specify leader start point:(pick 1)
Specify next point:(pick 2)
Specify next point or [Annotation/Format/Undo]<Annotation>:
   (pick 3)
Specify next point or [Annotation/Format/Undo]<Annotation>:↵
Enter first line of annotation text:24.00%%c %%p0.02
MText:↵
```

Figure 8.11

You can enter as many points on the leader line as are necessary.

All special text characters can be used in this dimension as well.

The arrowhead size is set in the dimension style or with DIMSCALE or DIMASZ.

To enter multiple lines of text, either keep typing at the MText: prompt or enter at the Annotation (or enter ↵ for options): prompt. If you choose the latter, you will enter the text editor and can continue entering text using it.

The options in LEADER include:

Tolerance   offers a control frame containing geometric tolerances using the Geometric Tolerances dialog box

Copy   copies text, a text paragraph, a block, or a feature control frame to the leader line

Block   inserts a block at the end of the leader line

Format   controls the way the leader is drawn and whether or not it has an arrowhead. Options include Spline, Straight, and Arrow

To enter a leader line without related text, enter a single blank space when you are prompted for the dimension text. This can be done with all dimension entries.

As in the LINE command, you can use **U** to undo the previous point entry without exiting the command. Enter multiple arrows by starting another leader, then use the space bar for the annotation.

SNAP and GRID

The SNAP and GRID Commands

SNAP allows you to indicate points or positions on the screen at preset regular integers. It also allows a rotated or isometric drawing to be entered. Set up your SNAP and GRID commands using the Drafting Settings dialog box under the Tools menu.

First set SNAP and GRID to .25 units. Then draw in Figure 8.12.

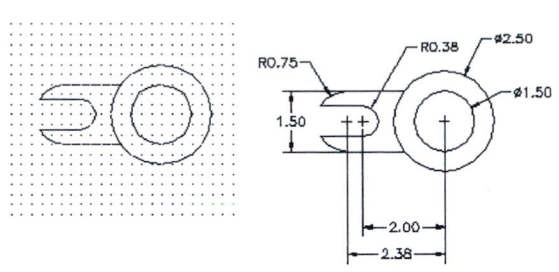

Figure 8.12

Next rotate SNAP by 45 degrees and set the base point in the center of the large circle, as shown in Figure 8.13. The GRID will follow the SNAP angle. The crosshairs will remain perpendicular, but are seen at an angle.

It is possible to enter all of this data using lines at specified angles and offset, but using GRID and SNAP can be easier.

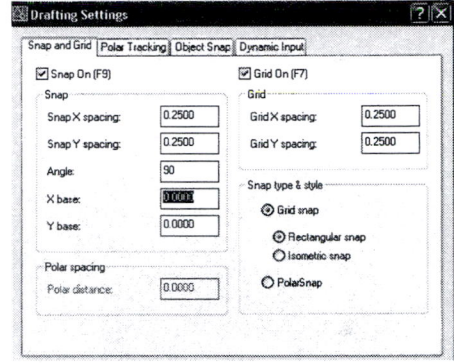

Figure 8.13

```
Command:SNAP
Specify snap spacing or
  [ON/OFF/Aspect/Rotate/Style/Type]<0.500>:R
Specify base point<0'-0.00",0'-0.00">:CEN of (pick the
  circle)
Specify rotation angle<0.00>:45
```

Draw the other parts using the new rotated SNAP as shown in Figure 8.14.

To return to standard snap and grid values, choose 0 degrees and 0,0 for the angle and base points. You can turn the SNAP, GRID, and ORTHO modes on or off by selecting or deselecting the relevant box.

Options on the bottom left will allow you to change OSNAP, and many other settings. These options are only available if you have entered via the pull-down menu.

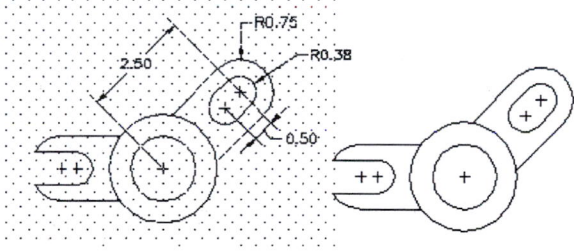

Figure 8.14

GRID and SNAP can also be useful for placing dimensions at regular or graduated intervals around the drawing. In traditional drafting there are set distances that the dimensions should be placed at. These are easier to arrange using SNAP.

In this tutorial we will create a title block.

Step 1

Use PLINE and LINE to create this title block. If you set the SNAP value to .25 to start, it will make drawing easier. Change the SNAP value as needed.

The lower left area will be for Scale, Date, etc., the central area will be for the company title, which is '3D Design Studio' and the top area will be for revisions.

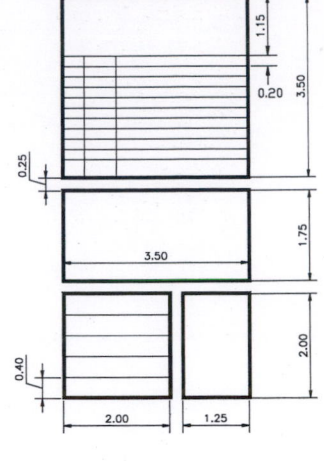

Step 2

Set up a text style.

The command line equivalent is STYLE.

Command: **STYLE**

Create a new text style. Make Futura the text font for it.

Scroll the bar up until you find Futura. Choose this font, or another if this is not available. Then choose Apply and then Close to return to the command prompt.

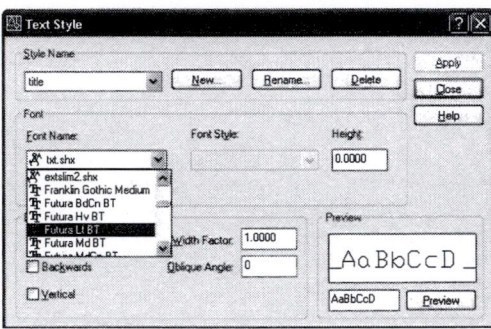

Step 3

Use the TEXT command or, if you have an older version of AutoCAD, the DTEXT command to allow multiple dynamic entries.

Make sure OSNAP is off and SNAP is set to .20. The pline distance is .40, so the text placed with SNAP will be exactly halfway between the plines.

The command line equivalent is TEXT.

```
Command: TEXT
Current text style:     "title"  Text height    0.00
Specify start point or [Justify/Style]: (pick 1)
Specify height <.5000>: .1
Specify rotation angle <0>: ↵
Enter text: Scale ↵
Command: TEXT (use ↵)
Current text ... 0.00
Specify start point or
   [Justify/Style]: (pick 2) ↵
Specify height <.1>: ↵
Specify rotation angle <0>: ↵
Enter text: Drawn By
Command: ↵
..... (pick 3) ↵
Enter text: Checked By↵
.....(pick 4) ↵
Enter text: Date↵
  ...(pick 5)
Enter text: Date of Print↵↵
```

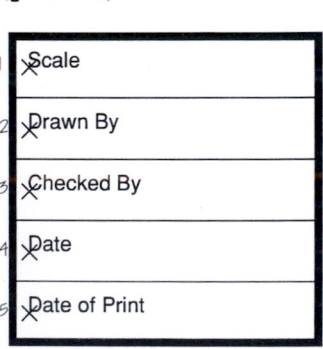

Step 4

The lettering is a bit large, so use the CHANGE command to make the letters smaller. You can change them with DDMODIFY, but for this operation CHANGE and Matchprop is quicker.

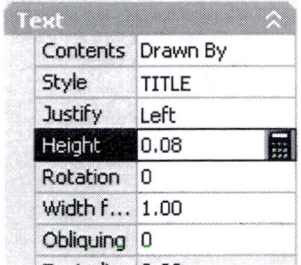

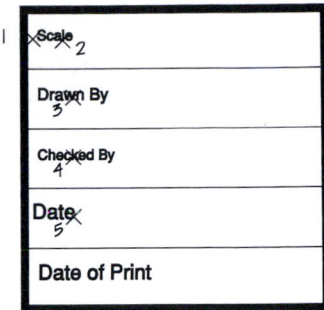

```
Command:CHANGE
Select objects:(pick 1)
Select objects:↵
Specify change point or [Properties]:↵
Specify new TEXT insertion point
 <no change>:↵
Enter new text style <STANDARD>:↵
Specify new height <5.0000>:.08
Specify new rotation angle <0>:↵
Enter new text <Scale>:↵
Command:MATCHPROP
Select source object:(pick 2)
Select destination objects:(pick 3,4,5)
```

Step 5

Now use the MOVE command to move the text up a bit.

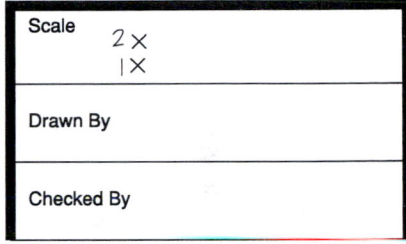

```
Command:MOVE
Select objects:P
Select objects:↵
Specify base point or
   displacement:(pick 1)
Specify second point as
   displacement:(pick 2)
```

Step 6

For the other text in this box, create a new text style and add the text as shown. The text font illustrated is CityBlueprint.

The command line equivalent is STYLE.

Then add the text as single-line text.

For title blocks paragraph text is used only for large notations.

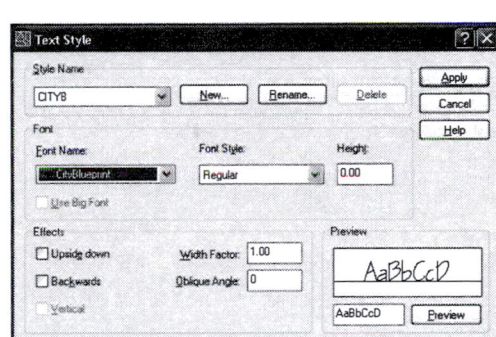

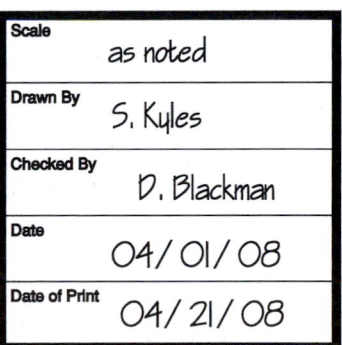

Step 7   Now make a cube with isometric lettering. Change the SNAP and GRID values to draw in the cube. From the Format menu, pick Drawing Aids. Change the rotation angle of the SNAP to 30 degrees.

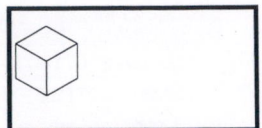

```
Command:SNAP
Specify snap spacing or
   [ON/OFF/Aspect/Rotate/Style/Type]:S
Enter Snap grid style Standard/Isometric <Standard>:I
Specify vertical spacing <0.01>:1 (will rotate at 30
   degrees)
```

The grid size will follow the snap size unless it is changed by the GRID command.

Use the line command to draw in the cube.

Step 8   Now use the STYLE command to load the ROMANT font and change the obliquing angle in order to load letters onto an isometric plane.

Note that the text font has changed. The oblique angle has also been placed at -30 degrees.

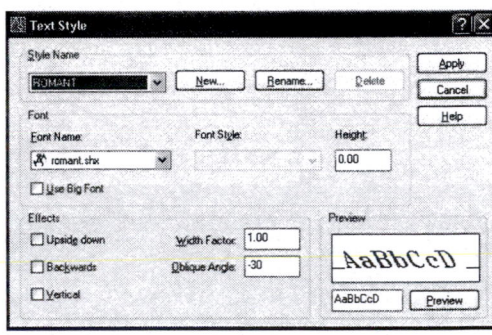

Now add the text.

```
Command:TEXT
Specify Start point or [Justify/Style]:F
Specify first endpoint of text baseline:(pick 1)
Specify second endpoint of text baseline:(pick 2)
Specify height<.2000>:.5
Specify rotation angle <0>:↵ (to accept the default)
Enter text:3 ↵
Enter text:↵
```

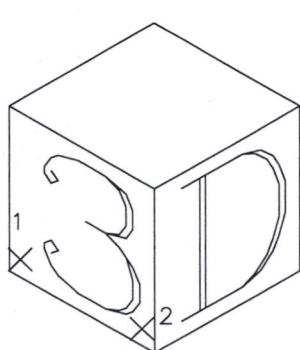

Now modify the text style to create the D. You will need to change the oblique angle.

For practice you can copy this cube and place your initials on it as part of your personal title block. The top face is either a 30 degree oblieque angle or a -30 degree oblique angle, depending on the bottom orientation.

Step 9 Change the oblique angle of the romant font to 0 and add the title and address using the Center option. Set SNAP back to normal before you start.

```
Command:SNAP
Specify snap spacing or
  [ON/OFF/Aspect/Rotate/Style/Type]:S↵
Isometric/Standard:S
Increment<1.00>:↵

Command:STYLE    Change the oblique angle.

Command:TEXT
Specify start point or [Justify/Style]:C
Specify Center point of text:(pick 1)
Specify height <.5000>:.1
Specify rotation angle <0>:↵
Enter text:Design Studio
Enter text:Slip Gate Road
Enter text:Frog Hollow
Enter text:New Hampshire
Enter text:↵
```

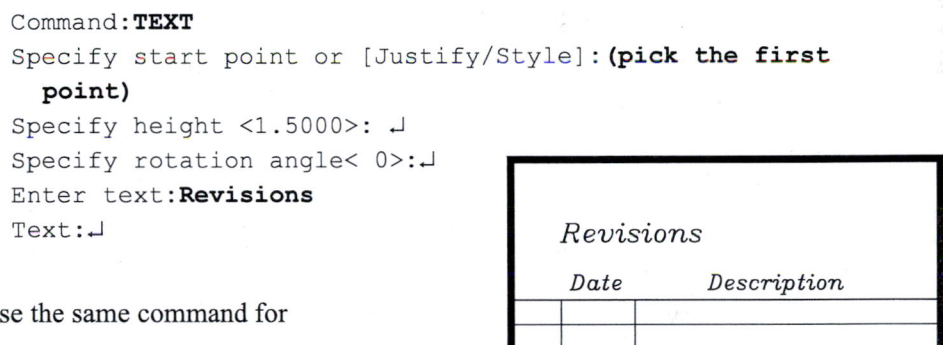

Step 10 Use PAN to move the screen down so that you can change the STYLE to Italic and add the notes regarding the revisions.

```
Command:TEXT
Specify start point or [Justify/Style]:(pick the first
  point)
Specify height <1.5000>: ↵
Specify rotation angle< 0>:↵
Enter text:Revisions
Text:↵
```

Use the same command for

```
Text:Date
Text:Description
```

Step 11 Now create another box above the Revisions box, and add a paragraph of text regarding the date of tender for the drawing. Use PLINE to quickly draw in a rectangle. Then change the style to romans. Finally add a paragraph of text using MTEXT.

The command line equivalent is MTEXT.

```
Command:MTEXT
```

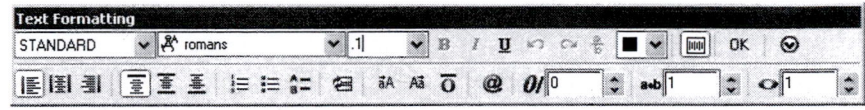

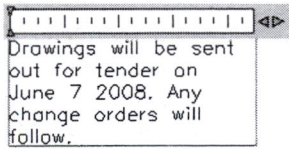

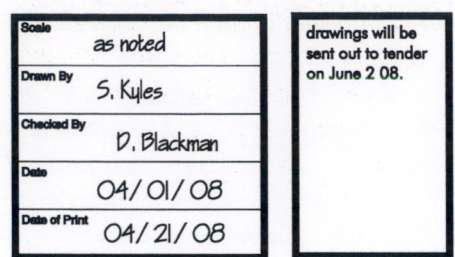

Once the area has been picked, type in the text that you want, then highlight it to change the size of the text or the text font, as shown in this example.

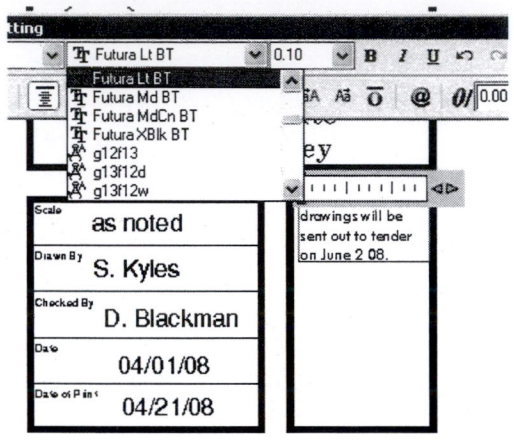

Step 12   Use ZOOM All to view your title block.

Change some of your text using DDEDIT and file it for future use under the name Titleblock.

Once you have a title block, you can insert it into drawings before printing, so don't lose it!

You can get title blocks that are made up by Autodesk on-line. Most companies, however, have their own title block.

Attributes are very useful in title blocks for setting up the parameters for the size of lettering, the text font, the format, etc. So that the information on all drawings fits a format.

Make sure that your title block fits your discipline, i.e. If you are making a mechanical drawing, don't use an architectural title block.

Step 1

Start a new file in imperial units. You are making a rabbet corner for wood. No limits are needed, but you need to make sure you are in imperial, then set an isometric SNAP and GRID. Use acad.dwt or

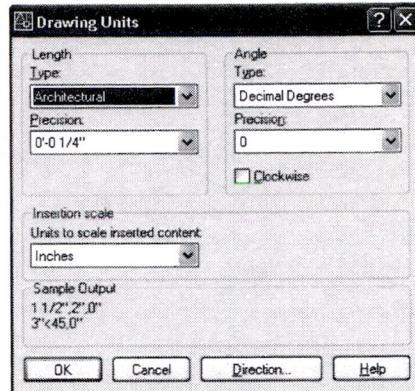

```
Command:NEW

(Choose Imperial) or  acad.dwt


Command:UNITS
  Set units to architectural.
```

Under the Tools pull-down menu, choose Drafting Settings and go to the Snap and Grid tab. Each release of AutoCAD has a different layout for the dialog boxes. The information is generally the same, but it is in a different place.

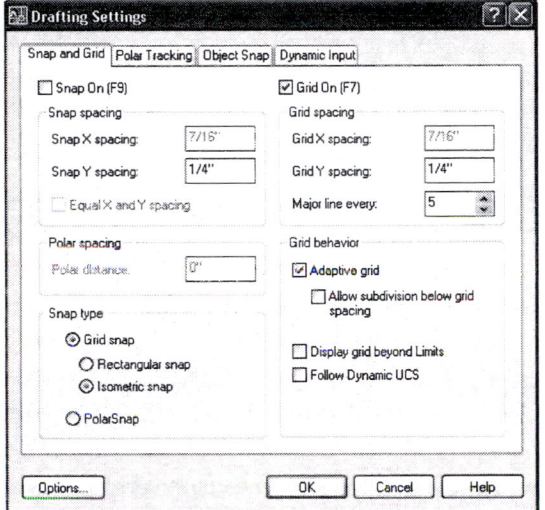

Set the SNAP and GRID values as shown. Your spacing for the *Y* value will be ¼″. The *X* value will be relative to the *Y* once isometric is chosen. Both GRID and SNAP should be on. Choose OK to exit this dialog box.

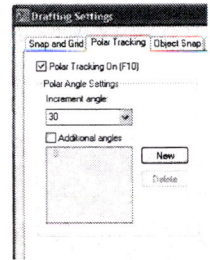

Set Polar Tracking to 30 degrees.

Make sure you have POLAR, OSNAP and OTRACK on.

Step 2

Start by drawing LINEs. Start at 0,0, then move your cursor along the direction of the grid points and enter a polar value. Follow the direction of the arrow as shown.

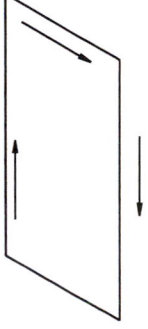

```
Command:LINE
Specify first point:0,0
Specify next point or [Undo]:6 up
Specify next point or [Undo]:3 across
Specify next point or [Undo]:6 down
Specify next point or [Undo]:c
```

Exercise 8 Challenger

The two wall sections show the traditional frame structure - R 20 and the Insulated Concrete Form - R30 to R40.

Frame

Insulated Concrete Form

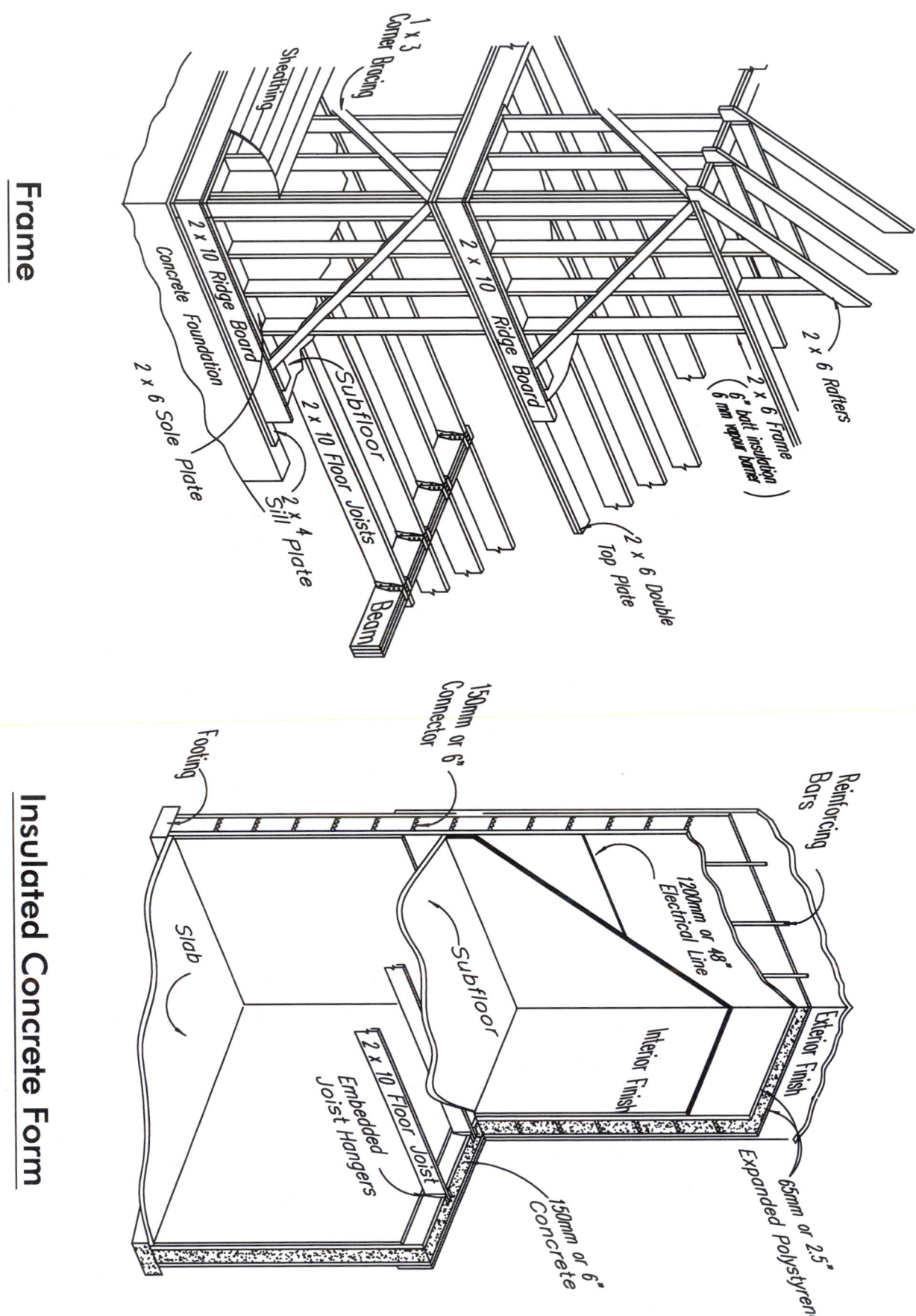

9

HATCH, SKETCH, and GRADIENTS

On completion of this chapter, you should be able to:
1. Use the BHATCH command
2. Edit existing hatches
3. Use the SKETCH command
4. Use the GRADIENT command

This chapter will deal with graphic patterns that will make your drawings look more professional. Hatches are used in mechanical and civil drawings to show cross-sections or the delineation of different materials. In architectural applications, hatches are used to show the types of material used e.g.brick/stone/ wood/concrete/sand.

The BHATCH Command

The BHATCH command fills an enclosed area with an associative hatch pattern that will update when the boundaries are modified. In addition, BHATCH allows you to preview the hatch pattern and adjust the definition or options of the hatch such as scale and angle. BHATCH is accessed through a dialog box. To access BHATCH:

Toolbar From the Draw toolbar choose

Pull-down menu From the Draw menu choose Hatch.

The command line equivalent is -HATCH for the command line, BHATCH for the dialog box.

```
Command:-HATCH
Current hatch pattern:ANSI31
Specify internal point or
  [Properties/Select objects/draw
  boundary/remove Boundaries/
  Advanced/DRaw order/Origin]:(pick 1)
```

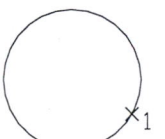

Figure 9.1

The hatch will be drawn within the area specified. The area must be a closed boundary, as in Figure 9.1.

When placing hatches, consider the boundary as containing something liquid: if the shape is not closed, the hatch will simply pour out until it meets with a boundary that will contain it. Close corners with Fillet radius 0 and try to avoid overlapping geometry.

In earlier releases there was a command called HATCH that created a hatch that was not associated with the boundary. This was useful particularly in architectural applications for showing dappled light on textured surfaces. The boundary hatch could be copied to make several regular patterns.

In Releases 2007 and 2008, this is under Options in the BHATCH menu.

The *BHATCH* dialog Box

The BHATCH dialog box offers many good options. The GRADIENT command is under the same menu but has a different tab. If you don't get the full menu below when you open the BHATCH command, use the small arrow on the bottom right.

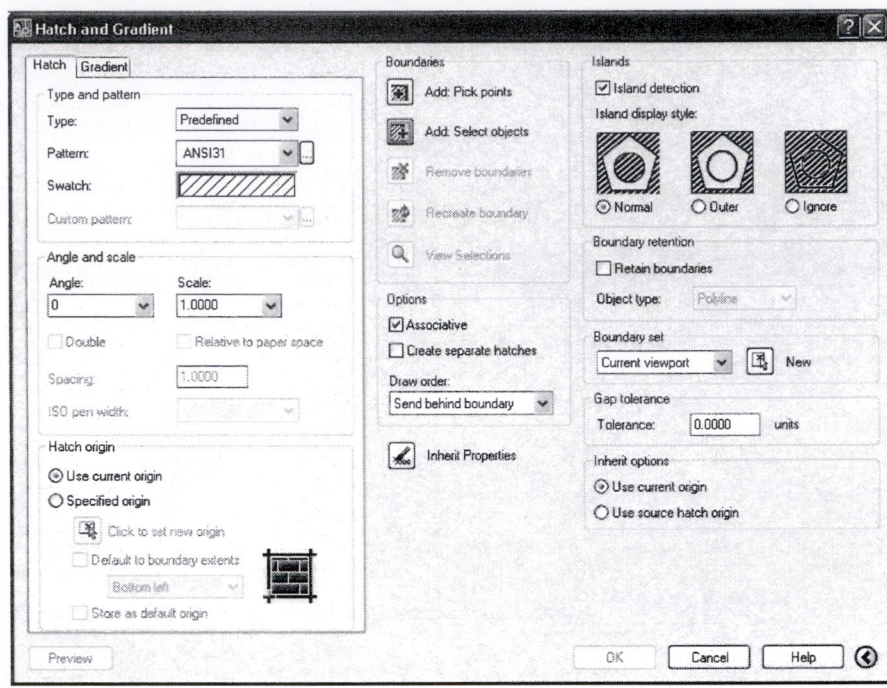

The BHATCH dialog box controls every aspect of the hatch pattern.

Type	tells you if it is a user-defined or an AutoCAD-defined hatch
Pattern	controls the ACAD standard hatch patterns. Hatch patterns are listed by name in the dialog box. Click on the down arrow to view the available patterns. This area lists only the names of the patterns
Swatch	illustrates the pattern
Preview	allows you to preview the hatch before applying it
Angle	controls the angle of the hatch patterns. Angles are generally counterclockwise
Scale	shows the scale of the pattern to be used. The default scale factor is 1. This means that the hatch is calculated to be displayed at the default screen size or an 11″ x 8 1/2″ sheet
Hatch origin	used for setting bricks or stone to start points
Boundaries	refers to the objects that are to be hatched; either closed objects, such as circles or ellipses, or sets of contiguous objects
Options	allows for the hatches to be non-associative
Inherit properties	acts like the MATCHPROPERTIES command, matching the pattern scale and angle of a chosen hatch and applying it to another hatch

Layers

In Figure 9.2 the hatches are contained by boundaries that are placed on a different layer then frozen. Only the hatch is seen on the final drawing.

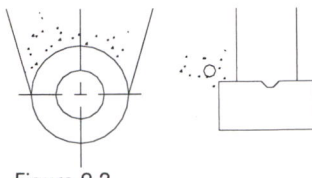

Figure 9.2

Scale

Each hatch pattern has a specific number of lines per inch. For example, the ANSI31 pattern generates three lines per inch at a scale of 1:1. If you have changed the limits and the model is larger than 11 units in X, you *must* change the scale factor.

When working in inch units, the scale factor of the hatch should be the maximum X value of the screen divided by 12, the same as the scale factor for LTSCALE. For example, limits at 0,0 and 24,18 would have a scale factor of 2.

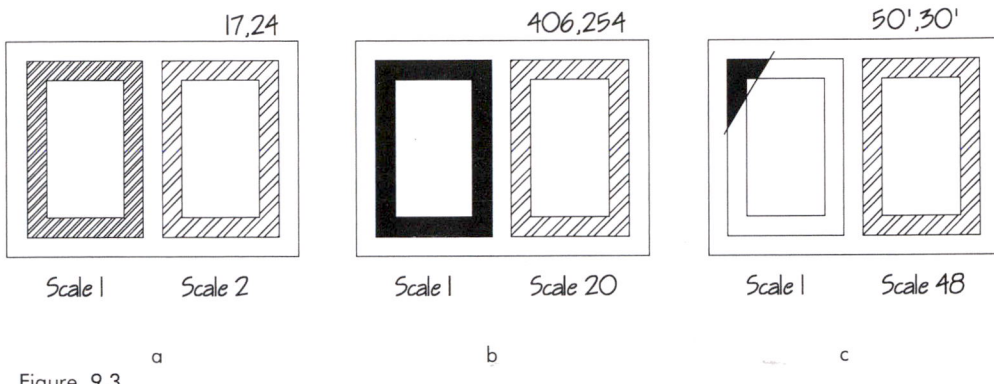

Figure 9.3

Other scale sizes are shown in Figure 9.3. Remember, if you are using architectural units or feet and inches, each unit is an inch, not a foot.

Always use the Preview option on the bottom left of the BHATCH dialog box before applying the hatch to see if it is correct.

In Chapter 11 there is a discussion on scale with regard to the size of the hatch and line-type scales and the size of the drawing. Before producing the final drawings, check Chapter 11 to see that the hatch is accurate.

Angle

If you are picking a hatch from the pull-down menu, the rotation angle will be exactly as shown. For example, Figure 9.4 a uses ANSI33, which is displayed at an angle of 45 degrees. To pick this pattern as it appears on the screen, do not change the rotation angle; leave it at zero, because that is the angle of the pattern itself.

Figure 9.4b has ANSI33 at the default rotation and at a rotation angle of 90 degrees.

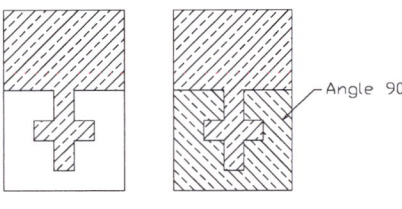

Figure 9.4

When using user-defined hatch patterns, the rotation angle is calculated at the horizontal, rotating counterclockwise.

Identifying the Boundary

For many students, this is where the difficulties begin because objects must form a perfect boundary. Lines, circles, and other objects are sometimes not entered accurately. SNAP and OSNAP are not always used effectively and consequently the lines are crooked, do not have tidy intersections with adjacent items, or are otherwise defective. This means that these objects do not provide an adequate boundary for hatching.

Preview the hatch.

If the concrete hatch doesn't look like the one opposite, use ESCape to go back to the BHATCH menu and make the necessary changes.

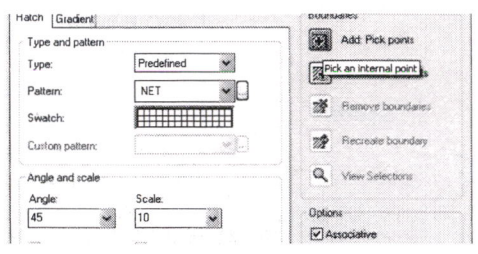

Step 5   Finally put a hatch in your wall section.

As in the previous two hatches, you will need to copy some data over to the walls layer to make a complete boundary.

In construction drawings, unless you are seeing the entire expanse of an area, there is a break line as shown at the edges of the walls, insulation and veneer. Put in the break line on your wall sections before hatching so that the ends of the walls are shown to be broken as well.

Thaw all of your layers.

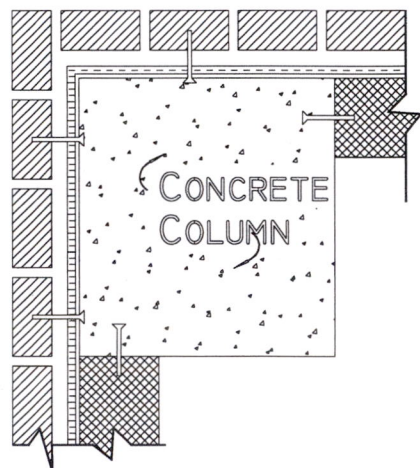

Step 6   The wall section is almost complete. All you need is to finish the drywall on the inside.

Create another layer called drywall. Put in the boundary as shown, and find an appropriate hatch for it.

This is a construction detail of a concrete column.

There is a thin layer of insulation on the outside of the exterior wall and the column.

The veneer, in this case either brick or decorative concrete block, is pinned to the column, as is the wall itself.

The inside finish is drywall.

To do this kind of detail without using layers is possible, but much more difficult.

Exercise 9 Practice

Sections are used to show what an object is made of
as well as where the indentations and cuts are within it.
In the case of the wheel, almost nothing can be gained
from looking at the front view alone. The section tells everything.

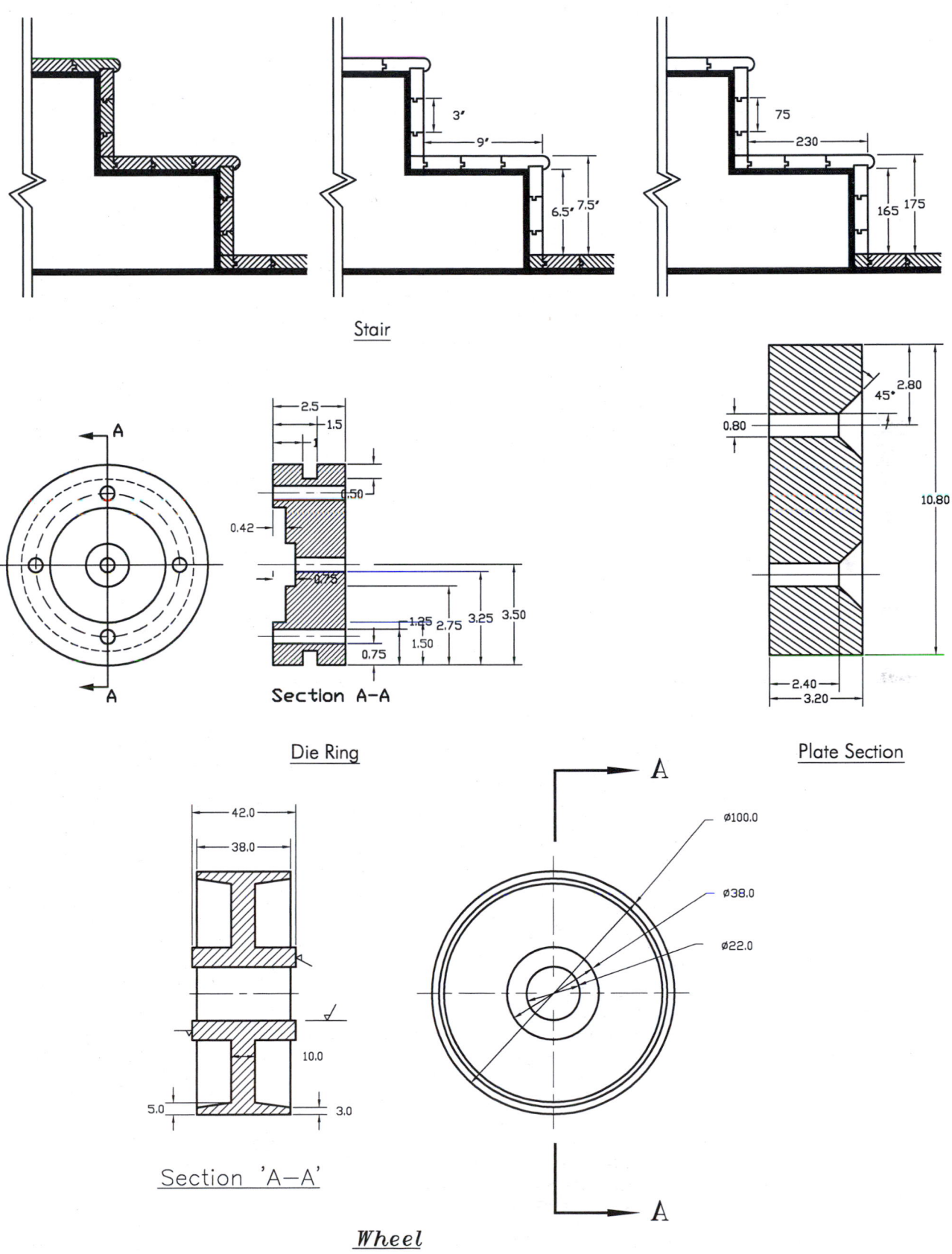

Stair

Section A-A

Die Ring

Plate Section

Section 'A–A'

Wheel

Exercise 9 Architectural

In architectural applications hatches and gradients are often used in elevations. These are elevations of the house shown in plan on pages 85 and 137. The foundation, shown in hidden lines, has a break line. Note that the elevations are lined up on the page.

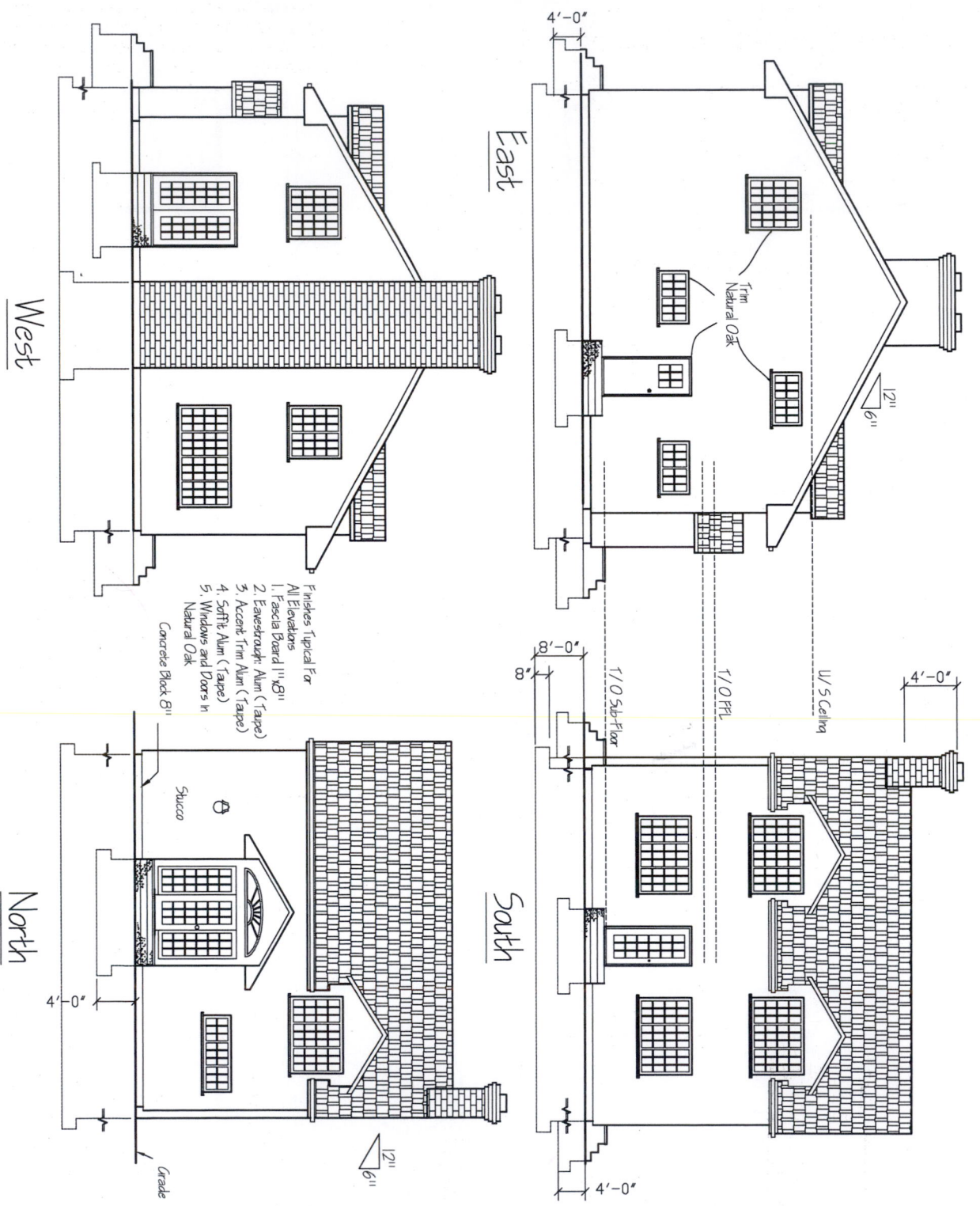

ELEVATIONS

ELEVATIONS ARE A VIEW OF EITHER THE EXTERIOR OR THE INTERIOR OF A BUILDING AS SEEN FROM "STRAIGHT-ON". THE FINISH MATERIALS ARE INDICATED BOTH WITH HATCHES AND WITH NOTATIONS. THE HEIGHTS OF FINISHED FLOORS, CEILINGS AND/OR DOORS AND WINDOWS CAN BE SHOWN AS WELL. THIS GIVES AN INDICATION OF WHAT THE PROJECT WILL LOOK LIKE WHEN COMPLETED.

Exercise 9 Wood Millwork

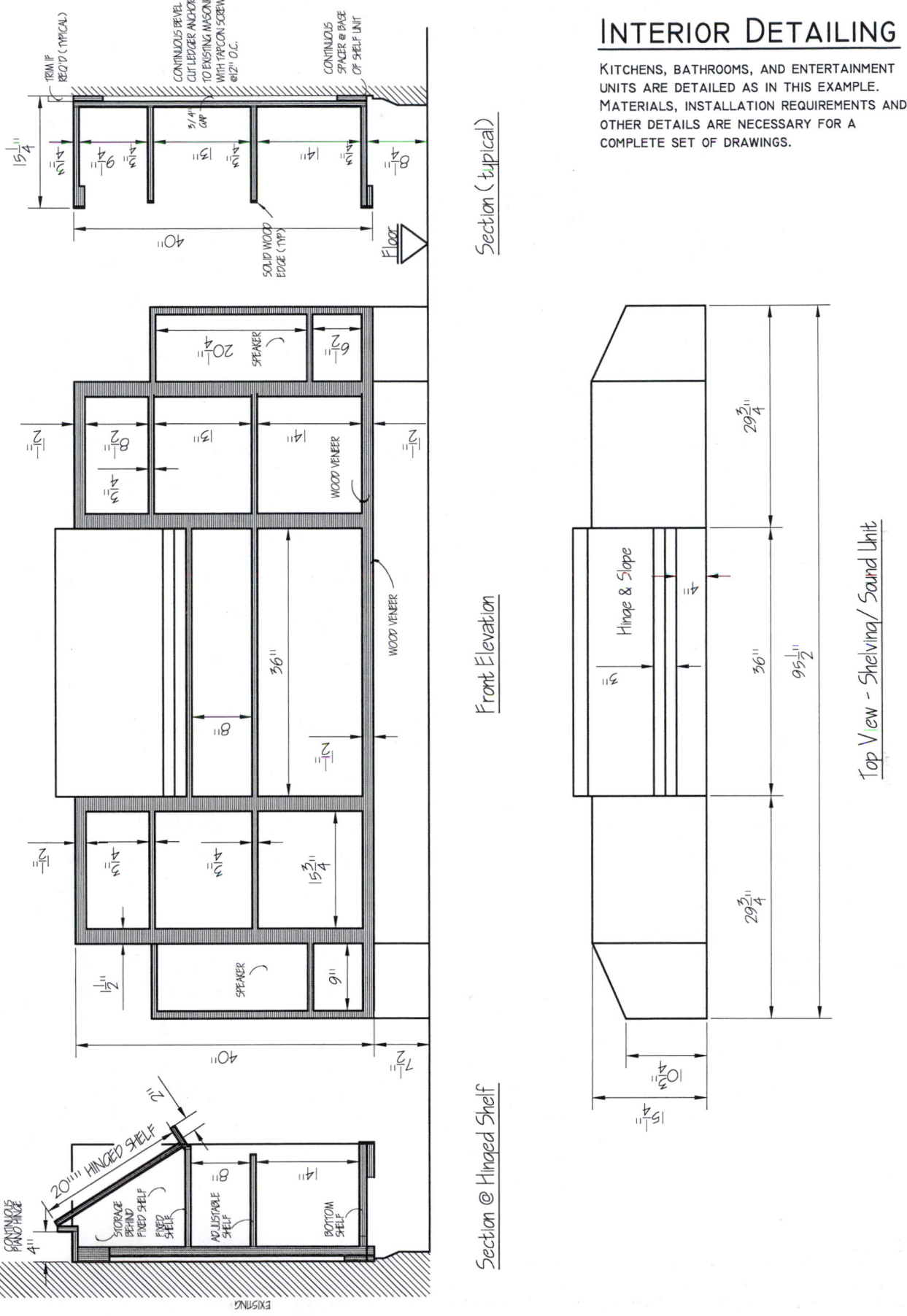

INTERIOR DETAILING

KITCHENS, BATHROOMS, AND ENTERTAINMENT UNITS ARE DETAILED AS IN THIS EXAMPLE. MATERIALS, INSTALLATION REQUIREMENTS AND OTHER DETAILS ARE NECESSARY FOR A COMPLETE SET OF DRAWINGS.

Section (typical)

Front Elevation

Section @ Hinged Shelf

Top View - Shelving/ Sound Unit

Exercise 9 Mechanical

Using the dimensions on this page, create the
assembled section for the tool post shown on page 179.

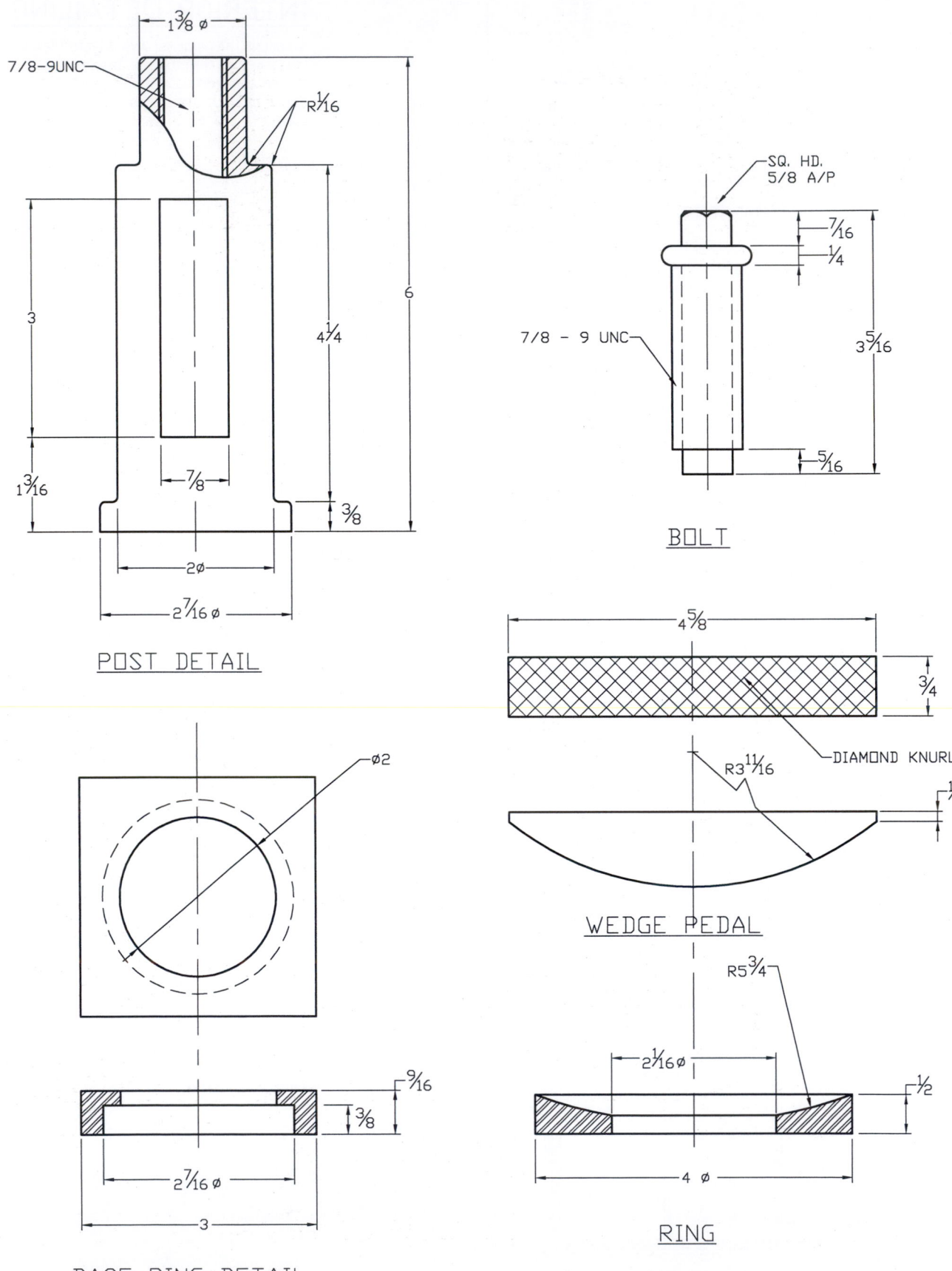

POST DETAIL

BOLT

BASE RING DETAIL

WEDGE PEDAL

RING

Exercise 9 Mechanical (continued)

Using the dimensions on the previous page create a section of the toolpost as shown.
Then try creating a section of the battery pulley.

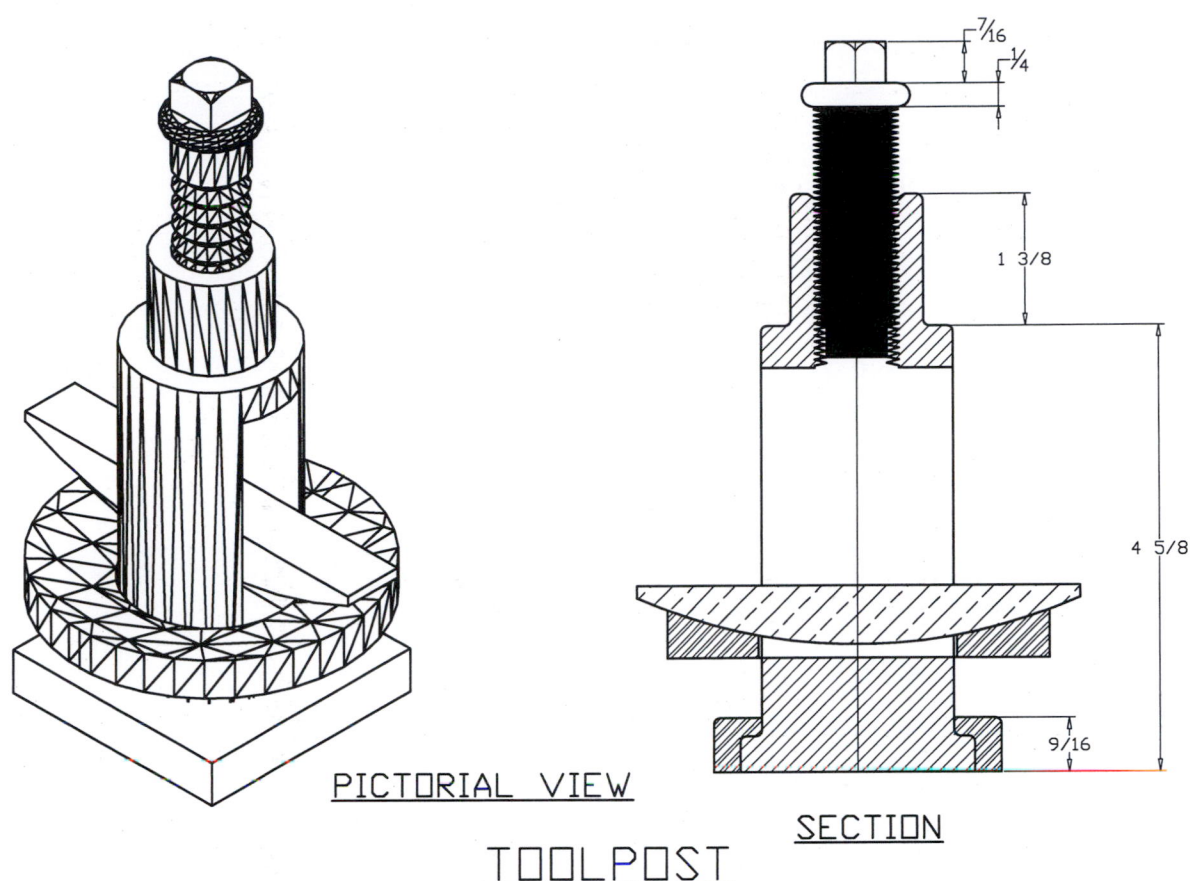

PICTORIAL VIEW

SECTION

TOOLPOST

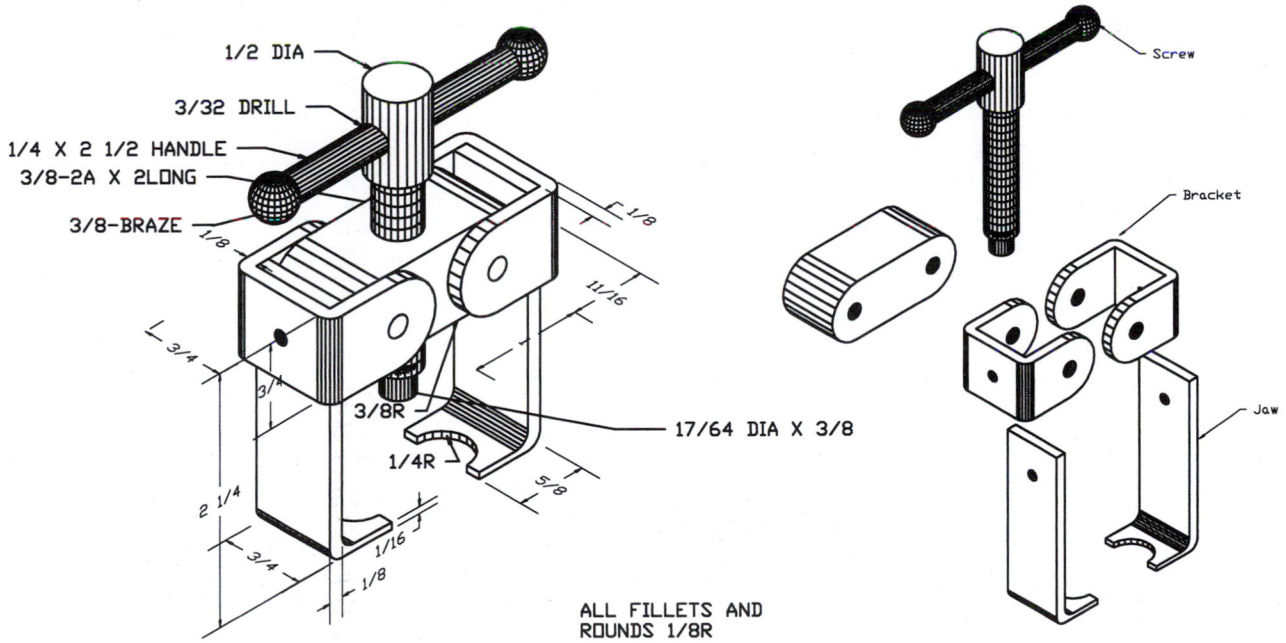

1/2 DIA

3/32 DRILL

1/4 X 2 1/2 HANDLE

3/8-2A X 2LONG

3/8-BRAZE

3/8R

1/4R

17/64 DIA X 3/8

ALL FILLETS AND
ROUNDS 1/8R

Screw

Bracket

Jaw

BATTERY PULLEY

Example 9 - Challenger

This is a stunning use of gradients. Try designing a facade yourself.

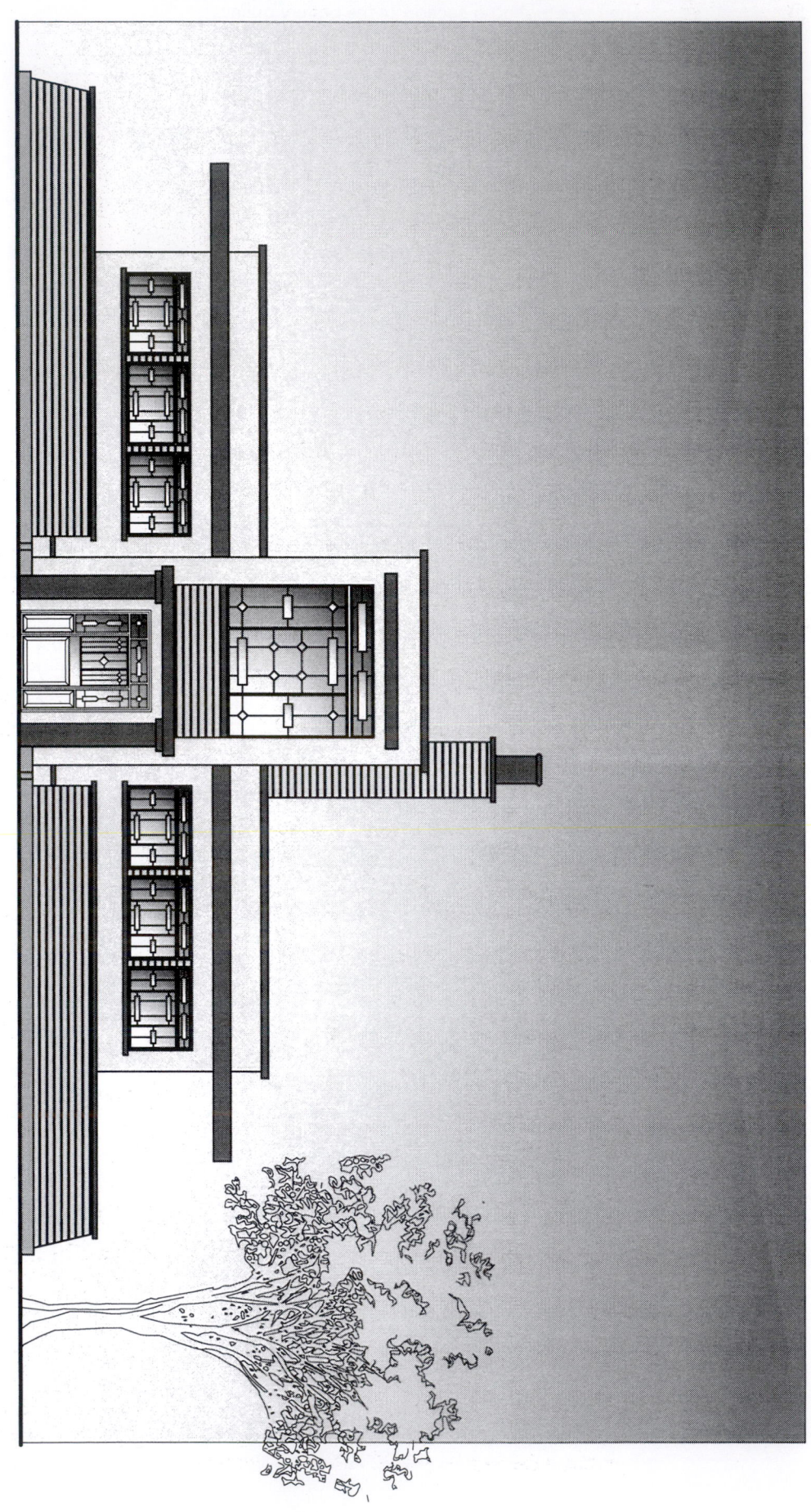

10 Blocks and Wblocks

On completion of this chapter, you should be able to:

1. Create a block
2. Create a wblock
3. INSERT both blocks and wblocks
4. Use Copy/Paste
5. Use COLOR, LAYERS, and other data with blocks
6. Use blocks form the AutoCAD Design Center.

Using Blocks

Blocks are a way of compiling many different drawings and drawing details onto one final drawing. By using BLOCKs on large projects, you can cut down design time by creating portions of the drawing separately and then assembling them on a final drawing.

There are two types of blocks used in AutoCAD. An *internal block* created with the BLOCK command is part of the base drawing and cannot be accessed except within the drawing. An *external block* or *wblock* is a drawing file.

Any drawing file can be inserted onto another drawing file at any time. Drawing files are already external blocks.

Dimensions and hatches are also considered blocks by AutoCAD because they are groups of information, but these are not made with the BLOCK commands.

The six commands that are connected with blocks are:

BLOCK	creates an internal block
WBLOCK	creates a .DWG file or external block
INSERT	inserts either a block or a wblock (external drawing file)
MINSERT	inserts blocks in rectangular arrays like the ARRAY command
BURST	reverts the blocked data back to individual objects
EXPLODE	reverts the blocked data back to individual objects

The BLOCK Command

The BLOCK command is used to create a group or set of objects that are identified by a given name. Once blocked, an object can be inserted into your drawing many times. Blocks created with the BLOCK command are internal.

Toolbar From the Draw toolbar choose

Pull-down menu From the Draw menu choose Block, then Make.

The command line equivalent is BLOCK or -BLOCK for no dialog box.

When creating a block, you must identify three things. The name of the block, the insertion base point, and the objects contained within that block. The dialog box can be accessed either through the icon or by typing in BLOCK. If you find the dialog box cumbersome, type in -BLOCK. It is much quicker.

If you want a particular wblock to always have a specific layer, linetype, and color, assign it explicitly; do not leave it on layer 0.

When in doubt, use 0,0 as the insertion point. If all of the files are created at a 1:1 scale, the data will always fit.

The **BURST** Command

The BURST command works like the EXPLODE command, but it allows attributes to remain as "values" instead of "tags". BURST is not available in all versions of Release 2006.

Removing Unwanted Blocks with the **PURGE** Command

Each time a block or drawing file (wblock) is inserted into a file, a copy of it is placed in the drawing memory or default area. If you want to clean your file, you must PURGE the blocks. The PURGE command will erase all unused blocks, layers, text styles, linetypes, etc. from a file. Any blocks, layers, etc. that have been brought into the file but never used can be purged. It erases the information from the file defaults and settings.

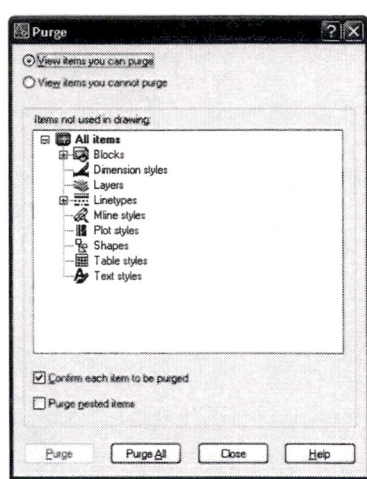

When a drawing file or wblock is inserted, then erased, the copy of the wblock remains in the memory in case it is needed again. The original file *must* be purged in order to accept a new file. PURGE removes unused named references, such as unused blocks or layers from the database.

```
Command: PURGE
Enter type of unused objects to purge
   [Blocks/Dimstyles/LAyers/LTypes/PLotstyles/SHapes/textSTyl
   es/Mlinestyles/All]: A
Enter names to purge <*>:
```

The AutoCAD Design Center

AutoCAD has an extensive database of saved blocks for your use saved in the ADCENTER. Use ADCENTER or Ctrl 2 to access the design center. Under the Folders tab you will find a variety of categories for saved blocks.

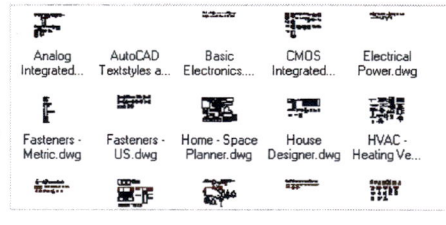

Pick the category that you would like to choose from, then pick the block that you need. Each block has a logical insertion base point. They are all drawn to a 1:1 scale.

Some of these blocks are dynamic blocks.

Figure 10.12

Dynamic Blocks

Autodesk has provided seven new software upgrades for AutoCAD since 2000. In 2006, the big feature was dynamic blocks. For those who need such things, dynamic blocks provide flexible blocks that can be placed within drawings. There are tutorials on how to make and use these blocks under the Block Edit icon. Essentially they allow you to create a door, window, bolt, etc. that can be sized dynamically when placed on a drawing.

In this tutorial we will draw a standard chair, block it, insert it, insert a title block onto it, and then see the difference between this and Copy/Paste.

Use STARTUP or acadiso.dwt to start a metric drawing. Create a layer called Dims and another called Objects.

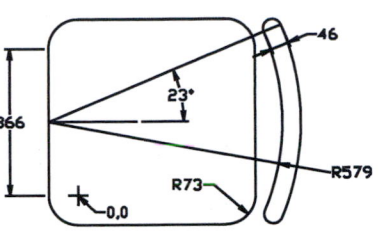

Step 1

Draw in the chair as shown on the Objects layer.

The insertion base point is REALLY important!

Make sure the center of the bottom left circle is at 0,0 - the origin.

Create a dimension style called Furniture. Set it to an overall scale of 10 under the Fit tab and two decimal places of accuracy under the Primary Units tab.

Dimension the part on the Dims layer.

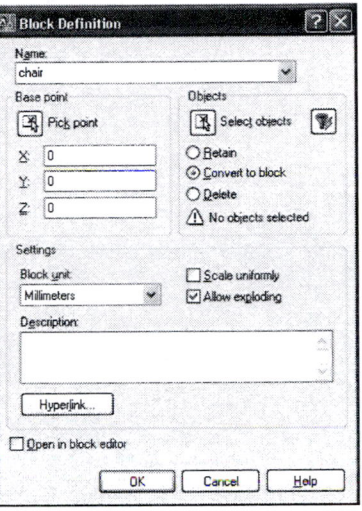

Step 2

Make layer 0 current, then use BLOCK to create a block of this data.

The command line equivalent is -BLOCK.

```
Command:-BLOCK
Enter block name or [?]:chair
Specify insertion base
   point:0,0
Select objects:(pick the chair and dimensions)
Select objects:↵
```

If you use the command line, the chair will disappear from the screen. Re-insert it.

Step 3

Now draw a table 1800mm by 1250mm and insert the chairs around it.

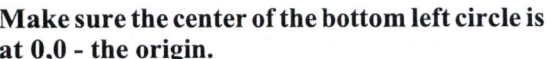

```
Command:-INSERT
Enter block name:chair
Specify insertion point or [Scale/X/Y/Z/Rotate/Pscale/PX/PY/
   PZ/PRotate]:(pick )
Enter  X  scale  factor,  specify  opposite  corner,  or
   [Corner/XYZ]:↵
Y scale factor <use X scale factor>:↵
Specify rotation angle <0>:90↵
```

You can use MIRROR and ROTATE to place the chairs. Notice that the chair on the left has the dimensions in backwards – it was inserted with a 180- degree rotation. The bottom two chairs were mirrored.

Freeze the Dims layer.

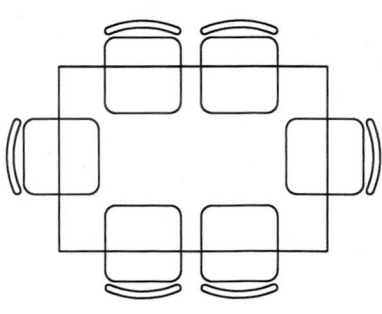

Step 4   Save the file as Dining room.

Step 5   Open a new file in metric units. Create the title block below and name the file Chap10title. (Draw the lines and the text of the title block, but do not dimension it.)

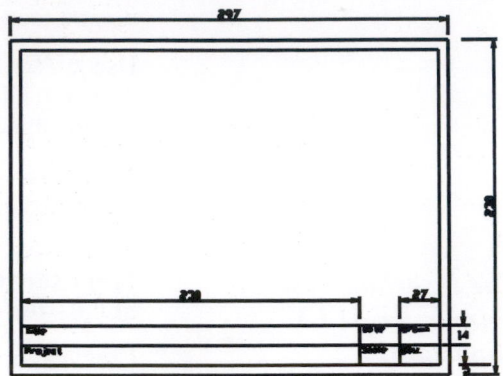

The outside line represents the outline of a 297 × 210 mm paper. The title block fits inside it.

Save the file as Chap10title.

Step 6

Now go back to the chairs and INSERT the title block in. Make sure you have layer 0 current. If not, the title block will be placed in the current layer and will take on that layer's properties. If layer Dims is current (now frozen), the title block will simply not appear.

You will notice that the title block is much too small.

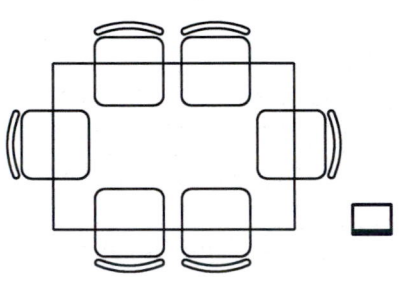

The scale that you will print the file at on a 297 by 210 sheet will be determined by the scale that you must use to expand the title to fit the drawing.

When scaling, make sure that the scale is reasonable; 1:10 or 1:20, not 1:17. For this drawing use 20.

Step 7   The title block here is scaled by 20. This means that when the title block is printed at its original size, the table and chairs will be 1/20 of their size or a scale of 1:20.

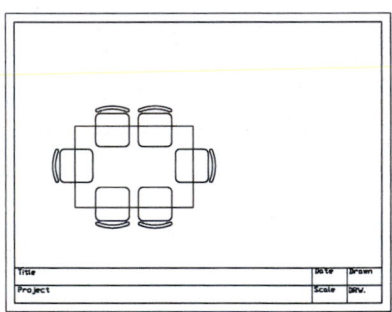

Scales are discussed in Chapter 11.

Freeze the Dims layer. Copy one chair over to the side away from the table setting.

Step 8   Now explode the chair that is not at the table, and add arms as shown. Then block the chair under the original name.

```
Command:EXPLODE
Select objects:(pick the chair)
```

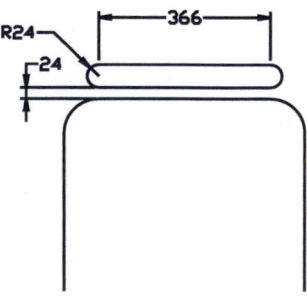

Draw in the arms as shown.

```
Command:-BLOCK
Enter block name or [?]:Chair
Block Chair already exists. Redefine
  it?<N>:Y
Insertion base point:CEN of (pick 1)
Select objects:(pick everything)
```

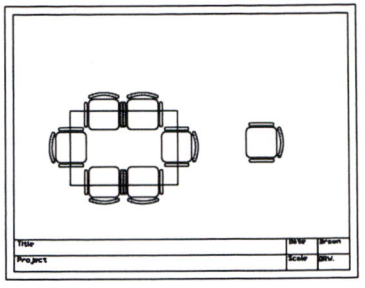

Your chairs will now have updated to include the arms.

This can be a very dangerous function. Always backup your files every few hours in case you really make a mess. Three copies is always best.

Step 9

We are now going to take a look at the difference between WBLOCK and Copy/Paste.

Zoom into the chair placed away from the table or insert one away from your drawing. Thaw the Dims layer. EXPLODE the block.

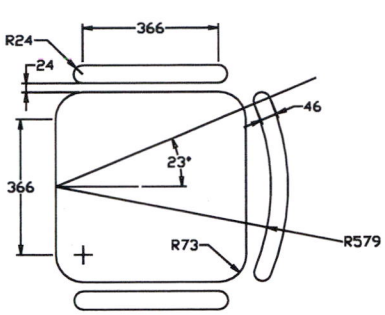

Make Dims current and add all the dimensions as shown here. Make sure you use the Furniture dimension style.

Make a text style called Label and set the font to GothicE - (English Gothic). Add the text 0,0. Then WBLOCK it. Use U↵ to bring it back.

```
Command:WBLOCK
Enter name of existing block or [=(Block=output
    file)/*(whole drawing)] <define new drawing>:TESTCHAIR
Block name (or ?):↵
Specify insertion base point:(pick the 0,0 area)
Select objects:(pick all the objects show in Step 9)
Select objects:↵
Command:U.↵
```

U will bring back your data but maintain the wblocked file.

Step 10

Finally we are going to open a new file, INSERT one chair from the hard drive and Copy/Paste another from the existing file. You will have two chairs on screen.

Open a new file in metric units. Make a Dimension style called Furniture. Set the Fit size to 10. Under the Primary Units tab, have no decimal places under precision. Under the Alternate Units tab, turn on the Alternate Units. Make it current.

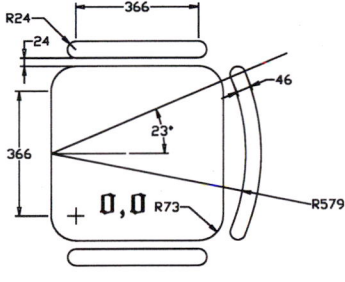 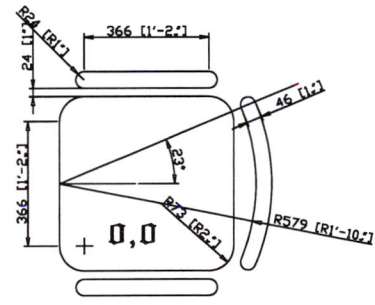

Make a text style called Label and set it to Gothic E.

INSERT the file Testchair.

Now go back to your Dining room file and under the Edit menu pick Copy. Select the objects shown in Step 9.

Go to the new file and pick Paste from the Edit menu. See the difference? The dimensions are updated to the current file with Copy/Paste, the text style is updated in both.

Remember: *A file is not a block until it is inserted into another file.* You do not need to block a drawing to insert it into another file.

Chart 1: Determining Scale for Drawings

Scale Factor		Decimal Value	Fraction
Architectural or Imperial			
3″ = 1′-0″	3:12	.25	1/4
1″ = 1′-0″	1:12	.0833333	1/12
1/2″ = 1′-0″	.5:12	.0416666	1/24
1/4″ = 1′-0″	.25:12	.0208333	1/48
3/16″ = 1′-0″	.1875:12	.0156246	1/64
1/8″ = 1′-0″	.125:12	.0104166	1/96
1/16″ = 1′-0″	.0625:12	.0052083	1/192
Mechanical			
3/4″ = 1″	.75:1	.75	
1/2″ = 1″	.50:1	.5	
1/4″ = 1″	.25:1	.25	
Metric			
1:10		.1	
1:50		.02	
1:100		.01	
1:1000		.001	

The first step is to determine the scale factor for the plot, and change the overall dimension scale factor under the Fit tab of the Dimension Style dialog box. Remember that it is not only the dimension text but the arrowheads, overshoot, dimension text gap and other variables that need to be adjusted.

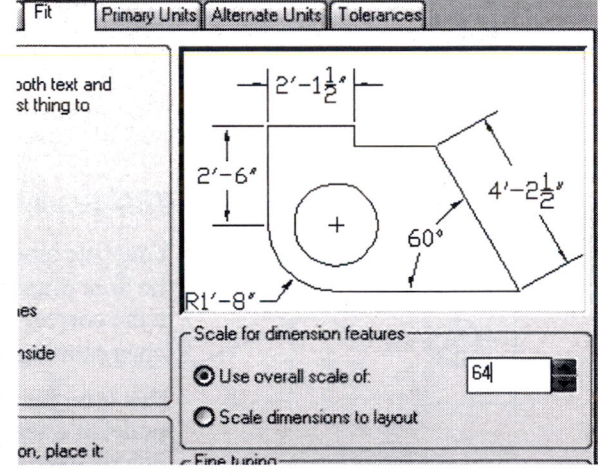

Always change the fit size not the dimension text size.

On a simple drawing with only one scale factor, Chart 1 can be used to determine the plotting scale factor and the block insertion factor for title blocks and other blocks.

For example, if you are creating a plot or drawing that will be 3/16″ = 1′-0″, draw the floor plan at a scale of 1:1. Add the dimensions at an overall scale, under the Fit tab, of 64. Add the hatch at a scale of 64 (unless it is an architectural hatch), then insert the title block at a scale of 64. Finally, plot the drawing at a scale of 1/64 or 3/16″ = 1′-0″. Chart 1 will help you to scale the model correctly. The dimensions, text, hatch, and linetype must be scaled to fit the drawing.

Scaling the Text and Annotations

When creating text and dimensions, you must change the scale factors in order to see them on the screen. This is even more of a difficulty when working with very large drawings.

If, for example, you want to make sure the text on a floor plan is at 1/8″ on the final drawing, you would need to find the relationship between 1/8″ on the paper and the size of the text on the floor plan. If 1/4″ on the paper will be equivalent to 12 inches on the model, then 1/8″ or half of that value will be 6″on the model. If you create your notations and dimensions at 6″, as shown in Figure 11.1, the final drawing plotted at 1/4″ = 1′-0″ will show text that is 1/8″ in height.

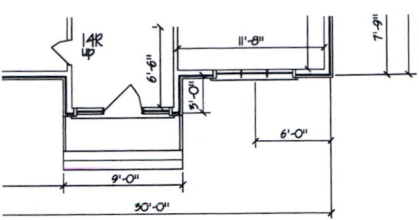

Figure 11.1

Text, dimensions, and related notations must be scaled to attain the appropriate size on the paper. You can always tell a novice AutoCAD user by the size of their text.

If you can read the text when you ZOOM All, the text is too big.

When creating metric drawings, use the same formulas. For 3 mm text on the final drawing, you will need to create the text at 300mm high in order for it to print at the proper size when plotted at a scale of 1:100. See Figure 11.2.

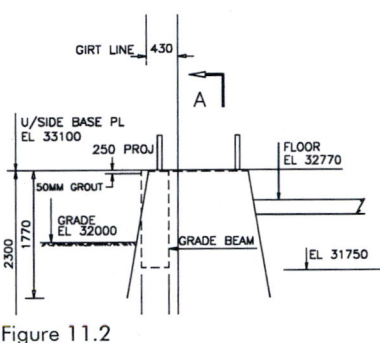

Figure 11.2

Chart 2 (see page 200) shows you at what sizes to scale your text to obtain the proper size on the final drawing. Find the size that you want your text to be, then the scale you intend to use, and make the text the suggested size.

If you take your final plot scale and multiply it by the default sizes and scale factors, you will arrive at workable scales and heights.

LTSCALE and HATCH Scale

While working with views to be placed on a drawing, you must also consider hatch and linetype. In creating these drawing aids for display on a screen, we determined that the scale for these functions should be the furthest value of *X* on your screen divided by 12. This provided a working area that was visible on the screen and easy enough to work with.

To place views onto a sheet of paper with the linetype and hatch at the proper paper format, use the same setting that you would for dimension scale or final plot scale. Architectural hatches such as arconc and arroof are set to be viewed on floor plans, so a scale of 4 can be used for a drawing to be plotted at 1/4″ = 1′-0″.

Using Blocks to Compile Drawings

For plotting different drawings at different scale factors, you can insert drawings as blocks on a title block or another accepted drawing sheet. However, although this will work, it is not recommended because any changes will have to be made in the original file.

In Figure 11.3 there are two separate files inserted onto a drawing sheet that is 36″ × 24″.

Chart 2: Dimension and Text Size

Plotted Text Size (Architectural or Imperial)	Scale To plot drawing at.	Text on the Model	SCALE for Fit tab
1/8″	1/16″=1′-0″	24″	192
	1/8″=1′-0″	12″	96
	3/16″=1′-0″	8″	64
	1/4″=1′-0″	6″	48
	1/2″=1′-0″	3″	24
	1″=1′-0″	1.5″	12
1/4″	1/16″=1′-0″	48″	192
	1/8″=1′-0″	24″	96
	3/16″=1′-0″	16″	64
	1/4″=1′-0″	12″	48
	1/2″=1′-0″	6″	24
	1″=1′-0″	3″	12
3/16″	1/16″=1′-0″	36″	192
	1/8″=1′-0″	18″	96
	3/16″=1′-0″	12″	64
	1/4″=1′-0″	8″	48
	1/2″=1′-0″	4.5″	24
	1″=1′-0″	2.25″	12
Mechanical			
.25″	1:20	.5	2.0
	1:10	25.0	10.0
	2:10	.125	0.5
.125″	1:20	.25	2.0
	1:10	12.50	10.0
	2:10	.0625	0.5
.1875″	1:20	.375	2.0
	1:10	18.75	10.0
	2:10	.09375	0.5
Metric			
3 mm	1:10	30	10
	1:25	75	25
	1:100	300	100
	1:250	750	250
	1:500	1500	500
	1:1000	3000	1000

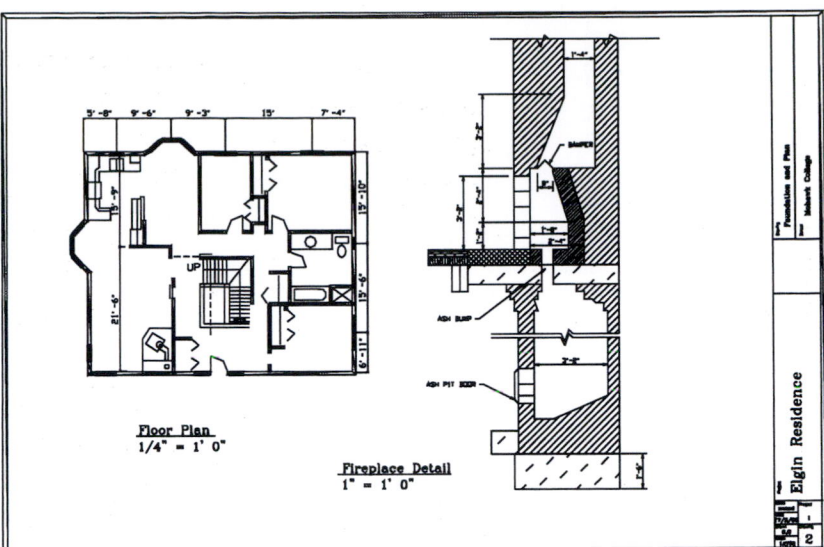

Figure 11.3

The drawing shown in Figure 11.3 consists of three separate stored files:

1. a title block at 36″ × 24″

2. a fireplace detail at 16′ vertically

3. a floor plan 35′ × 30′.

Since the floor plan is the most important view and the one most likely to have changes made, it is the base file. The other two files are inserted onto it. The title block was inserted onto the floor plan at a scale of 48. The drawing will be plotted at 1/48, so the title will get back to its original size. The fireplace, being a detail, was inserted at a scale factor of 4, four times its original size. It will be plotted at a scale of 1″ = 1′-0″.

```
Command:-INSERT
Enter block name or [?]:Fireplace
Specify insertion point or [Scale/X/Y/Z/Rotate/PScale/PX/
PY/PZ/PRotate]:(pick a point)
Enter   X   scale   factor,   specify   opposite   corner,   or
   [Corner/XYZ]:4
Y scale factor <use X scale factor>:↵
Specify rotation angle <0>:↵
```

If any dimensions need to be added at this time, the fireplace is four times its original size, so the Measurement Scale factor under the Primary Units tab of the Dimension Style dialog box will need to be changed to .25.

If a drawing is to be plotted at many different scale factors, create different dimension styles for each different size.

You can insert drawings at the fractions noted in Chart 2 onto the title block and plot at a scale of 1:1, but this allows for no changes in text or dimensions.

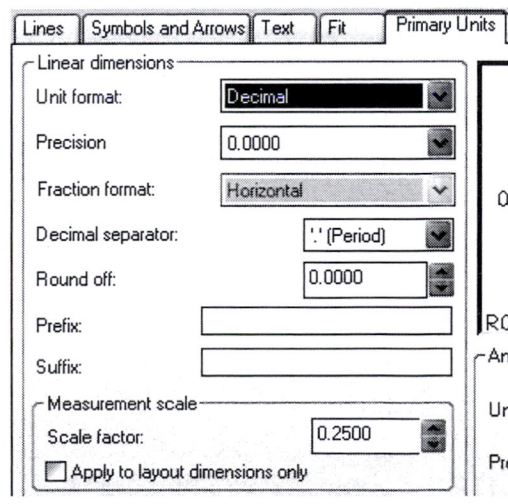

Once you have created your drawings and are ready to plot, you will want to add the final drawing-related information. Each view should have a title. If there are section drawings, these sections must be noted on the plan, front or elevation view. Finally, if you are showing details, the place where the detail attaches to the main portion of the object must be shown with a break line.

View Titles

View titles are always shown below the view. These can be slightly to the left or the right, but never above.

In applications with orthographic projection, as shown in Figure 11.4 (from page 138), the views are lined up and the titles are directly beneath each view.

This is fairly straightforward in a simple part like this, but there are times when the parts are difficult to distinguish if they are not properly labeled.

The same orientations are used in civil and architectural applications, but the views are labeled differently.

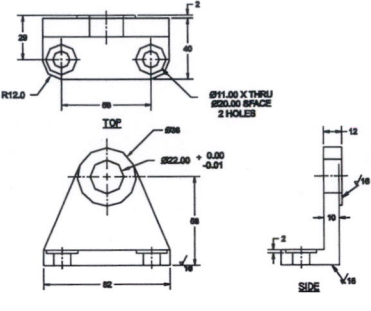

Figure 11.4

The Top view becomes a Plan or Floor Plan, the Front view becomes an Elevation, and the Side is also an Elevation. These are labeled North Elevation, West Elevation, etc. for those buildings where the designer has actually taken the trouble to orient the building on the lot. Otherwise they are called Front Elevation, etc.

Section views are for all types of drawings, and these show a vertical or horizontal slice through the object to show how it fits together and what is used in the construction or manufacture of the object in question.

In Figure 11.5, the view title indicates where the object is cut, in this case through the hinged shelf (page 177).

When the title for the section view is not sufficient to describe where the section is taken from, a section line is used.

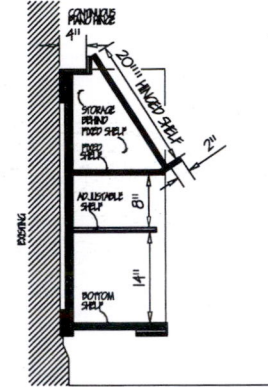

Figure 11.5

Section Lines

Section lines are used to show where a cut goes through an object. Figure 11.6 shows a simple mechanical part cut through to show where the section is. A more complicated mechanical part with several section lines is shown on page 238.

In civil and architectural applications, section lines are most often shown on floor plans. The section is often jagged so that the section of the building (page 117) is taken through all the 'difficult' parts, or the parts where the builder may want more clarification. Note the section lines on the floor plans on page 192. The section in architectural drawings is often used to find the height of objects on the elevation since these are not generally added to the elevation views. To see how this works, look at pages 194 and 195.

Section lines are made with PLINEs at varying widths (page 91).

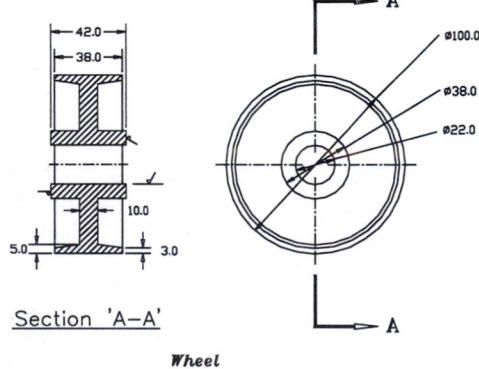

Section 'A—A'

Wheel

Figure 11.6

Break Lines

Break lines are used at the edge of objects to show that there is some material extending beyond what is being shown. In Figure 11.7a the stone at the top of the fireplace extends beyond the detail up to the roofline.

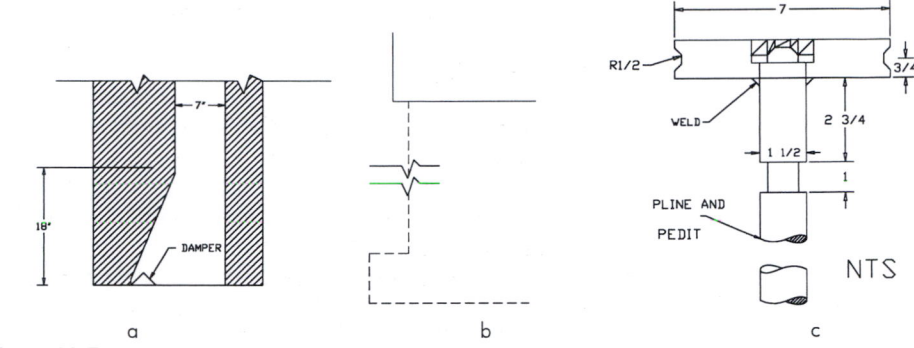

Figure 11.7

Break lines are also used to show that the full length of an object is not being shown, as in Figure 11.7b. Here the foundation of the building, under ground, is a full 8′, but only a portion is shown. Both of these applications are used in every discipline.

In mechanical applications, the break shown in Figure 11.7c is typical because it illustrates that the object being broken is cylindrical.

Importing Notations

Often you have notations on drawings that are standard or typical of many drawings such as manufacturing requirements, notations concerning adherence to building code or national code requirements, or materials requirements. If this is the case, you can type these up in a word processor and import them onto your file using MTEXT.

To test this out, type up a few sentences in Wordpad or Notepad. Open either of these programs under the Programs area of the start menu on your computer.

Save the data as a .txt file or simple ASCII text file format, as shown in Figure 11.8. Make sure you are aware of what directory the text is being saved to. If you have a flash drive or floppy disk, that is the best place to put it. The extension will be .txt. The .rtf (rich text format) files work as well.

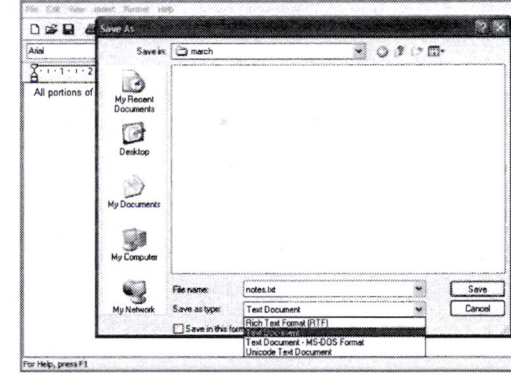

Figure 11.8

Invoke MTEXT or multiline text, pick the location for your notation, and then right-click to get the Import Text function, as shown in Figure 11.9.

Find the directory that contains your .txt file, and bring the file onto the current drawing. The text will be imported onto your file where required. You can change the size and font once it is in by highlighting it and adjusting the properties on the MTEXT menu.

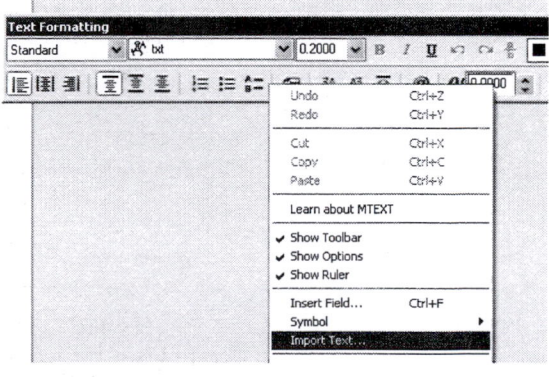

Figure 11.9

Using Lineweights

To have a professional-looking plot, you must have different line weights for each type of information. The object lines will be the darkest. The dimension lines will be lighter, and the hatch lines should be lighter still. The lineweights of your objects are best set up within the Layer Properties Manager.

Lineweights and the Layer Properties Manager

You will want to spend some time finding out what lineweights work best with your plotter. Do a few test runs to make sure the settings are correct, then make note of them. Each plotter will plot at a slightly different scale.

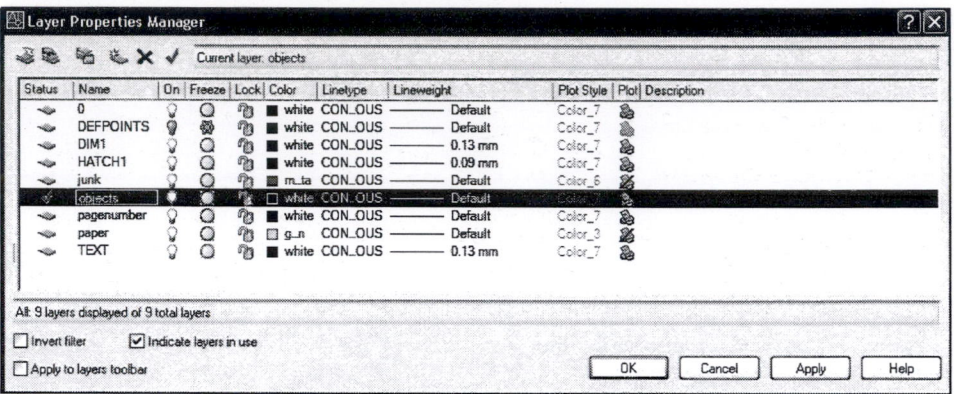

Before you plot, in the Layer Properties Manager, set all your layers that you want to print to black. The Defpoints layer, which appears automatically with dimensioning, will not print. The other layers used for page set up you can toggle to unprintable as shown on the right above.

Leave the object lines and text as default or increase them to .3, as in the examples in this book. Change the lineweight of your dimensions and hatches to be much smaller. Try 0.13 for dimensions and 0.09 for hatch. By making a .pdf of the page you can test these lineweights out before sending them to paper (see page 206).

Plotting

Plotting is only easy after you've done it a few times. If your text sizes, notations, and lineweights are all good, you should get a good plot.

There are three main ways to plot: directly to a plotter or printer, plotting to a .plt file, then taking this .plt to a plotter, or printing to a .pdf file, which can then be sent over the Internet. The way you plot will depend on how your system is set up.

Plotting Directly to a Plotter

If your printer or plotter is hooked up directly to your computer and working properly, you should have no trouble printing. Simply pick the PLOT command from the File menu or type in the word PLOT.

```
Command: PLOT
```

If you are working with an older version of AutoCAD, the command listed under the File menu may be Print. It changes every few versions. Both Print and Plot do the same thing.

Once you have accessed the PLOT command, you will get the Plot menu. The title on this dialog box will say Plot-Model unless you are in paper space (see Chapter 12).

Danger

Do not leave the plotting of your drawing to the day it is due. Any system will have difficulty printing the first time. You need to know the final plotter destination and have the necessary driver before you can even start to plot. Practice some plots first.

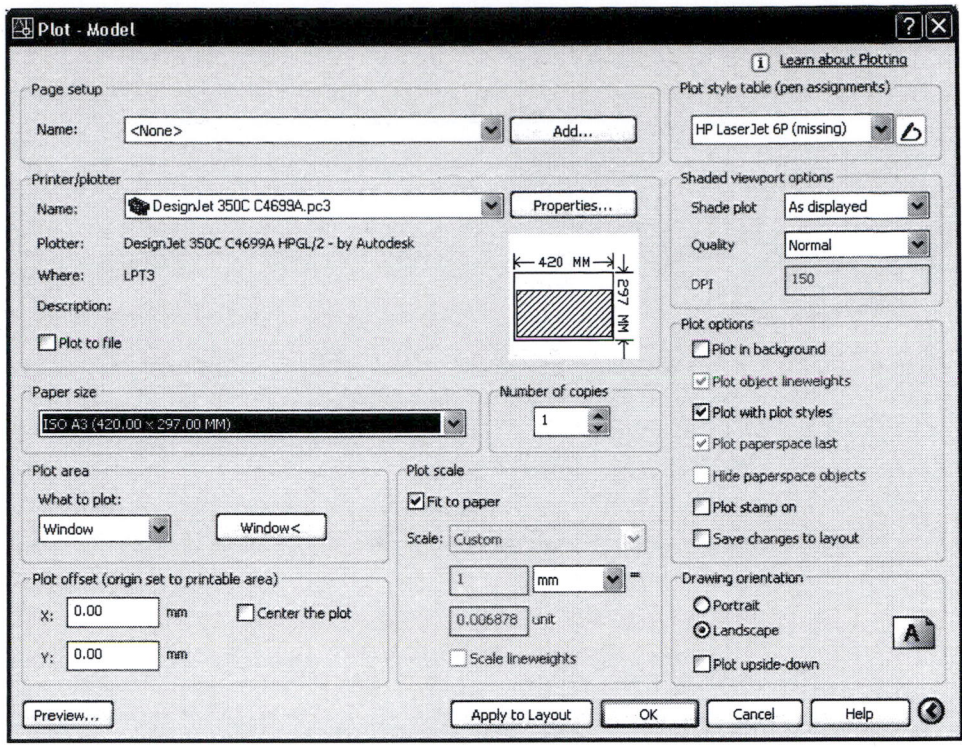

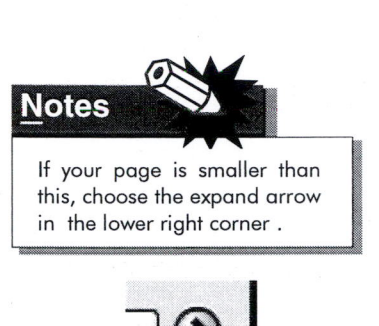

If you want to accept all of the defaults, press OK and your plot, if the plotter is set up, will be produced. The system defaults to printing the display from the screen. If your system is set up to a laser or inkjet printer, then you should be able to simply pick up your print. Should you wish to change any of the defaults or print to another plotter, another size, or another area of the screen, pick the related button. Always Preview before printing.

Printer/Plotter Selection

The Printer/Plotter tab lets you direct your plot to a series of plot devices. The one you are set up to should plot directly. It also allows you to have the plot written to a file as opposed to being sent straight to the plotter. The file types are .plt, which will allow you to take this to another printer and get a hardcopy, or .pdf, which is a portable document file for Internet use.

Use the down arrow beside the Printer/plotter name to choose a plotter. If the one you want to use is not configured, you need to add it to the list.

Configuring a New Plotter

If the plotter you need is not on the list provided, use the CONFIG command to load a new one. At the command prompt, type in CONFIG, or go to the Tools menu and choose Options at the bottom of the list.

Add-A-Plotter Wizard
Shortcut
1 KB

Figure 11.10

```
Command: PLOT
```

Then choose Add or Configure a Plotter, as in Figure 11.10.

You will get a choice of plotting options. Choose the Add-A-Plotter Wizard.

Choose the plotter you want from the many listed (Figure 11.11) and work your way through all the dialog boxes. If the plotter you need is not there, you can generally get the plot driver from the Internet.

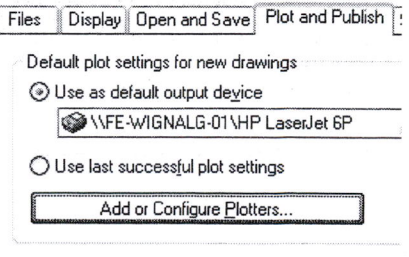

Figure 11.11

Once the plotter driver is loaded, go back to the Plot menu and change the Printer/plotter to the one you have just loaded.

Plot to File

Plot to File will create a .plt file for transfer either into a word processing file using a HP driver or to a plotter not connected to your machine. Always take a copy of your drawing file with you if you are plotting off-site in case there are errors.

Paper Size

The paper size can only be changed once the plot device has been chosen. If you have a laser printer as your plot device, you can only get a small paper size. Change to another plotter to get a larger size of paper.

Plot Area

The system is set to Display. Window gives you more control. Choose the window, then Preview.

Plot Preview

This will give you an idea of how big your plot is, and how it fits on paper. Use either the Full Preview or the Partial Preview before you spend the time or money on a paper plot. Always preview your plot before clicking OK.

Plot Scale

The drawing will be set in terms of plotted inches = drawing inches or plotted millimetres = drawing units. Use Fit only when the actual scale of the drawing is not needed.

Use Chart 1 to determine the plot scale. For a drawing at 1/4″ = 1′-0″, make the scale 1/48. To make a drawing that is 1:50, use a scale of .02. Always preview before you plot.

The plot origin is always assumed to be the bottom left corner of the paper. If you want to relocate the origin of the plot to another portion of the paper, reset this parameter by changing the X and Y values in the edit box. This is usually done to place a title block onto an existing drawing or to place a missing view onto an otherwise completed drawing. Some skill is needed here, so try placing the view or plot onto a test sheet before placing it on the final papere, especially if there are drawings on the paper that could be ruined.

PDFs

PDF files (Portable Document Files by Adobe) are a quick and easy way to both check the lineweights and quality of your plots prior to printing, and send plots over the internet.

In Releases 2007 and 2008 AutoCAD has added a built- in PDF driver. Configure this driver using AutoCAD e-PLOT under the Add-a-Plotter wizard Manufacturer's list, then from the Models list choose PDF.

In Release 2006 and earlier, try the PDF995 software available on the internet.

In this example we will take a floor plan of a house, copy a window, and plot so that the floor plan is at a scale of 1/8″ = 1′-0″ and the window is at a scale of 1/4″=1′-0″.

Step 1

Retrieve the file for Tutorial 3b.

If you haven't got it, quickly draw it up. It is only about 20 commands, see page 61.

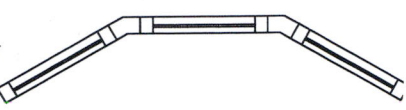

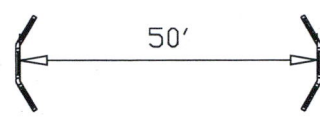

Once in, draw a line from the MIDpoint of the central window to the right 50′. ARRAY this window unit around the central spot. This will make a nice second story or well-lit tower.

Now just add some lines to place walls between these windows, and a small opening for a staircase.

Make sure all of your lines are on Layer 0.

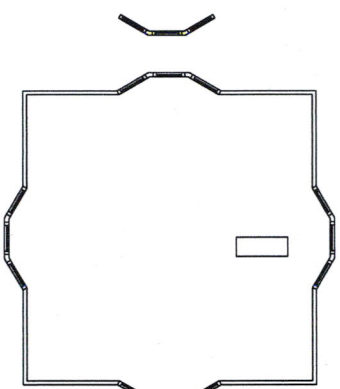

Step 2

COPY the window unit on the right over to the right.

SCALE it by 2.

The main object that you are concentrating on is the floor plan. The window is a detail added to it.

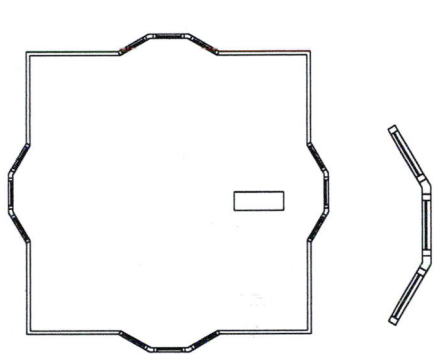

Step 3

First dimension the floor plan. Make a new style called Eighth for 1/8″= 1′-0″. Set the Overall scale to 96. Just add a few overall dimensions (the dimensions in this image are larger than on your drawing).

Make sure you put the dimensions on a separate layer.

Now we are going to add some dimensions on the detail view. The view has been scaled by 2, so you want the Measurement scale to be .5.

Make a new dimension layer called Quarter, and change the measurement scale to .5.

Measurement scale	
Scale factor:	.5
☐ Apply to layout dimensions only	

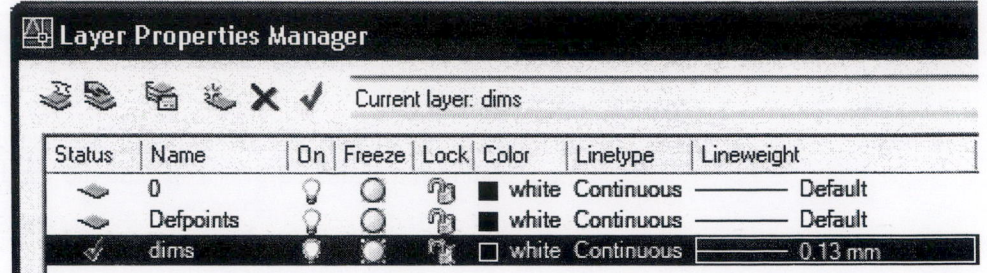

When you are setting up your layers, change the lineweight of the Dims layer to 0.13.

Notice that your dimensions for the detail show up at the same size as your other dimensions, but the readout reflects the changed measurement size.

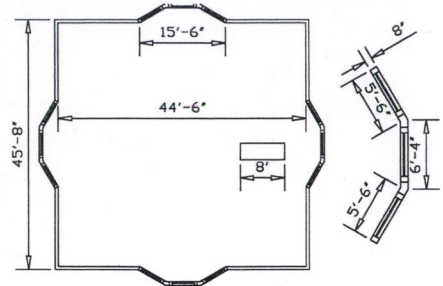

Step 4   If you have an 8.5 × 11 title block, use it. If not, just make a simple title block 10.75 ×8.25 in size. INSERT it onto the drawing at a scale of 96.

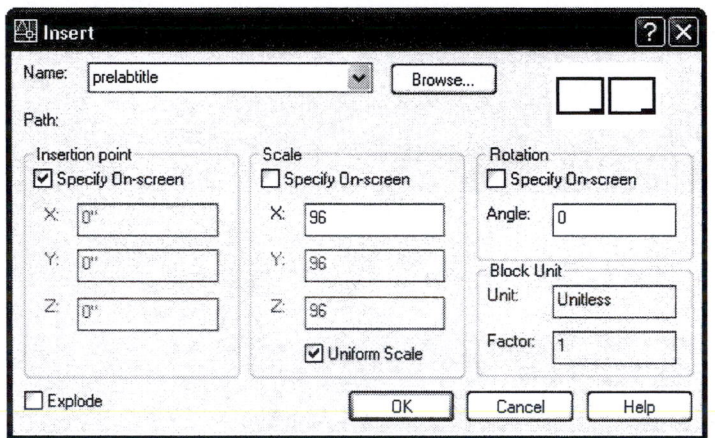

The views fit well within the title block. If they don't, you have the imperial/metric problem. If you have started in metric, all the numbers will be wrong, regardless of the fact that you may have changed your units.

Scale the title block by metric numbers - 25.4, 2.54, etc. and it should eventually fit.

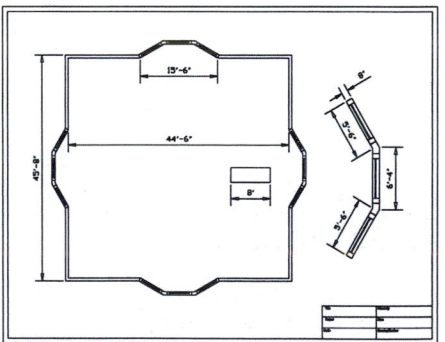

Step 5   Look on Chart 2 for the size of text that you want for your titles. Since you will be plotting at a scale factor of 1/96, you will want text that is 12″ high to show up as 1/8″ on the final drawing.

The title should be underlined. Under the title should be the scale of the view:

<u>Floor Plan</u>

Scale 1/8″ = 1′-0″

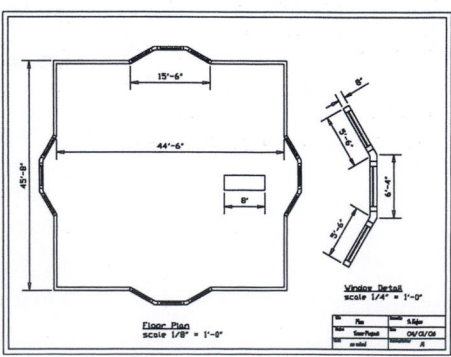

Use the default printer that is set to your machine. This will be plotted at a scale of 1/8″ = 1′-0″. Set up the plot file as shown below. Using the option Window in the Plot area section will allow you to pick which area you would like plotted.

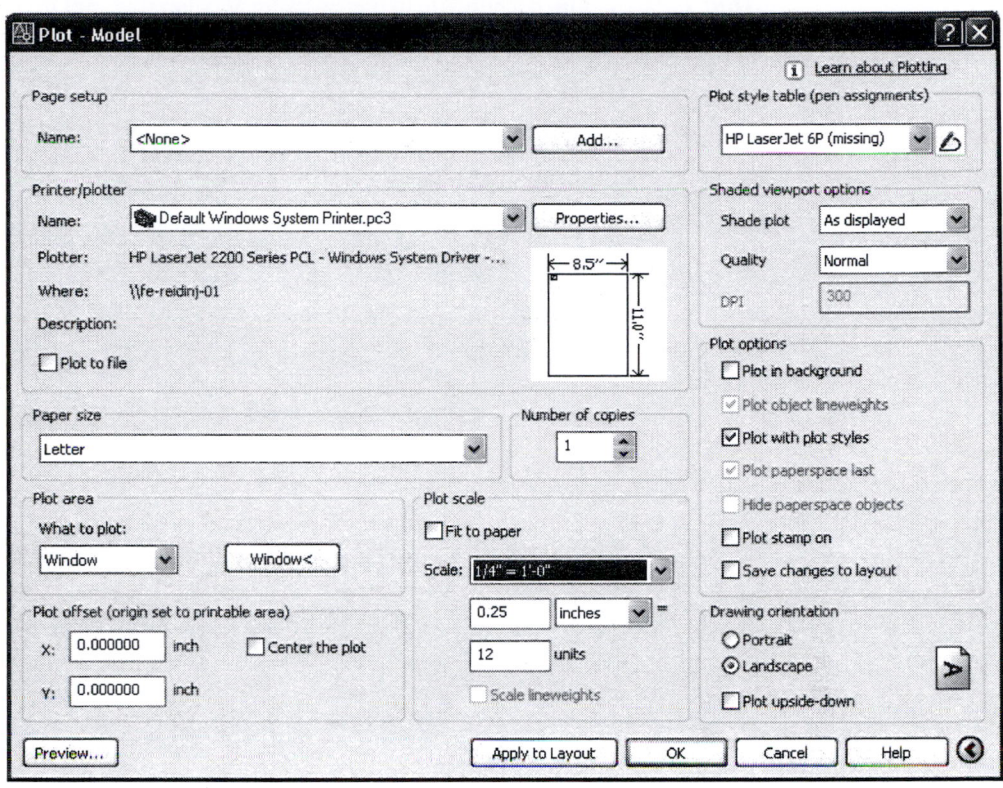

Once everything is set up, be sure to use Preview to get a preliminary view of the plot.

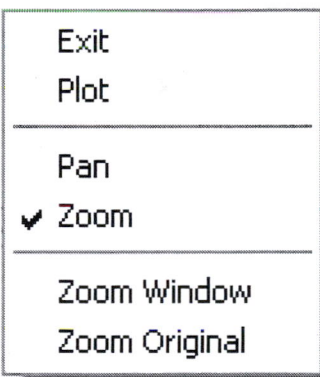

At this point you may notice any errors in placement, line quality, or size that have been overlooked in the menu. Use the roller ball on your mouse to zoom into the drawing and check out the lineweights.

If there are errors pick Exit to get back to the Plot menu.

If there are no errors pick Plot to continue with the plot.

The menu choice Plot to file can be used to make a .plt file that can be taken to a plotter for printing. Not every plotter accepts .plt files, so unless you are certain that the plotter will accept these files, making a .pdf is probably a better choice.

In this example we will take a mechanical part and plot it at a scale of 1:40. The final drawing will be plotted on a letter-size sheet (8 1/2″ ×11″ or 297 × 210).

Step 1

Open a new file in metric using acadiso.dwt. Set up your layers as shown below. Don't forget about the lineweight. Different plotters print at different weights. These lineweights are good for some printers. Others print well at a larger lineweight.

The standard is minimum 0.30 mm, with thicker lines at 0.60 mm. The plots in this book are printed with objects and 0 layer at 0.30 mm. Try this, but if it's too dark, set up as shown below.

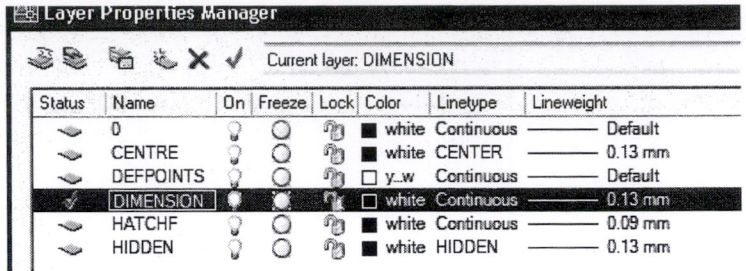

Now draw up the objects below.

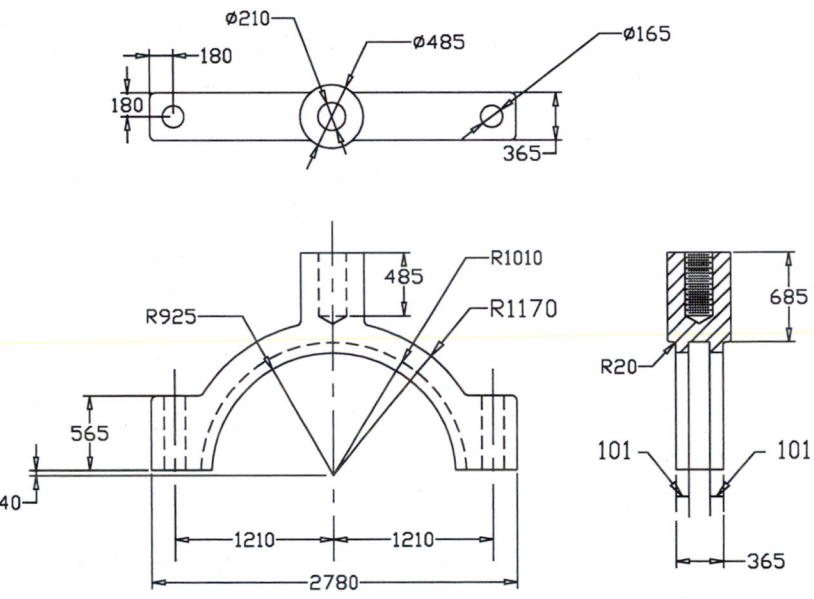

Step 2

Make a new dimension style for these dimensions called Forty because it is being plotted at a scale of 1:40. When you open the Dimension Style dialog box, if the basic dimension style is Standard, you have started the drawing in imperial units.

The base style should be ISO-25.

If you are in the wrong units, open a file in metric using acadiso.dwt. You might need to SCALE the drawing until the dimensions are correct.

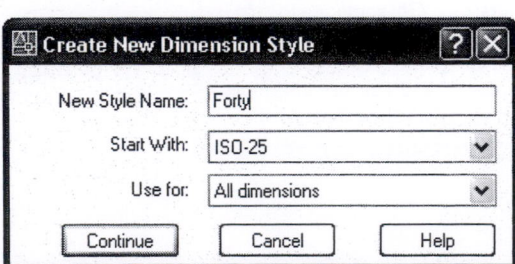

Now change your Primary Units so there are no decimal places.

Change the overall scale under the Fit tab to 40.

Many people set their text height, arrowhead size, gap, overshoot, etc. If you have a template file – a base file that always has your settings – this is a good idea. It is difficult, however, to remember to change all the settings each time, so an overall scale is a lot simpler. The graphic image will reflect what settings you have.

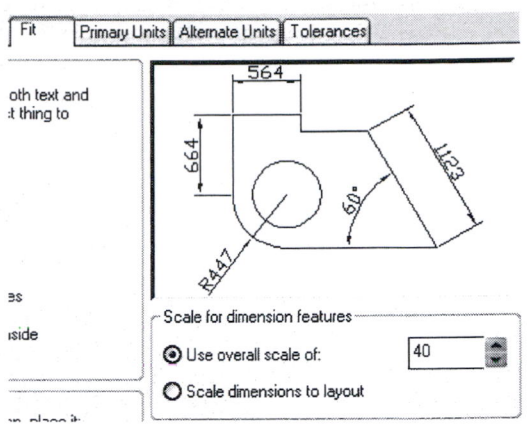

Dimension the object as shown above.

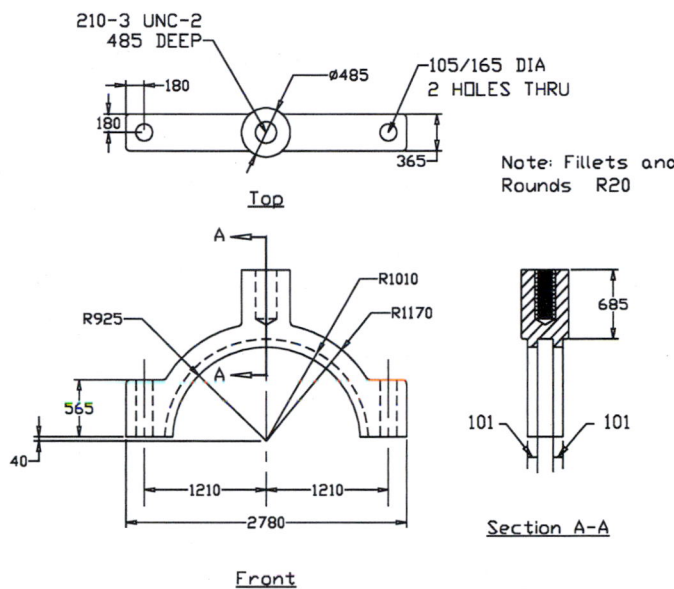

Step 3 In mechanical applications, often the notations for threaded fasteners:screws, bolts, etc., are not shown with a diameter. Instead they are shown with class notations.

For example, the top left thread is shown in this diagram. The diameter and depth of the fastener are thus not needed in the drawing. This notation says it all.

210 - 3 UNC - 2A

Update your drawing using the notations above. The height of the text should be 120. Then add:

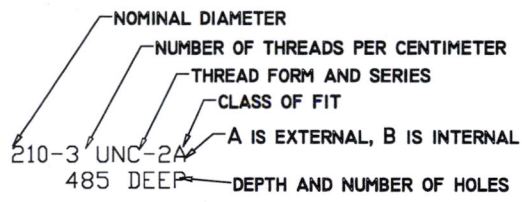

View titles

Section A-A using PLINE and TEXT

Notations for both sets of 'holes', as shown

The notation for all fillets and rounds.

Erase the dimensions that are better shown in notation

Step 4   Zoom into a small portion of your screen. Make a PLINE rectangle 265 × 185 units which will fit onto a 297 × 210 paper.

TITLE	MECHANICAL PART	DATE	17/03/08	SCALE	1:40
PROJECT	SECTION	DRAWN	BOB EDMONDS	NUMBER	M1

Zoom into the right corner and create the title block shown at the side. The distance between the lines should be 6 mm.

Now scale this block by 40. You drew it at the size it will be when plotted. Then you scaled it to fit the image. The objects will be plotted at a scale of 1:40.

Move the title block over so that it fits the image.

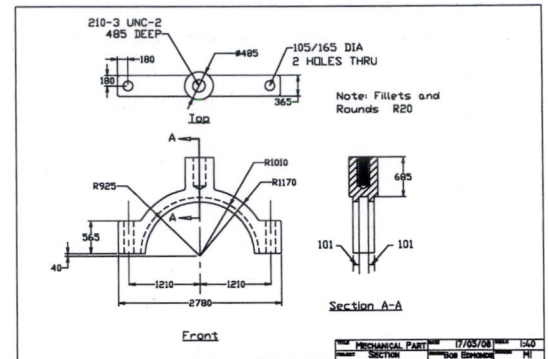

Step 5   If your linetypes don't show up hidden and centered, use LTSCALE to change the display. Just change it until it looks good.

Double check your layer lineweights as in Step 1.

Step 6   Set to the default printer, use a scale of 1:40 and a paper size of A4, and you should be fine. You may need to Window and Preview a few times to get the image properly on the page.

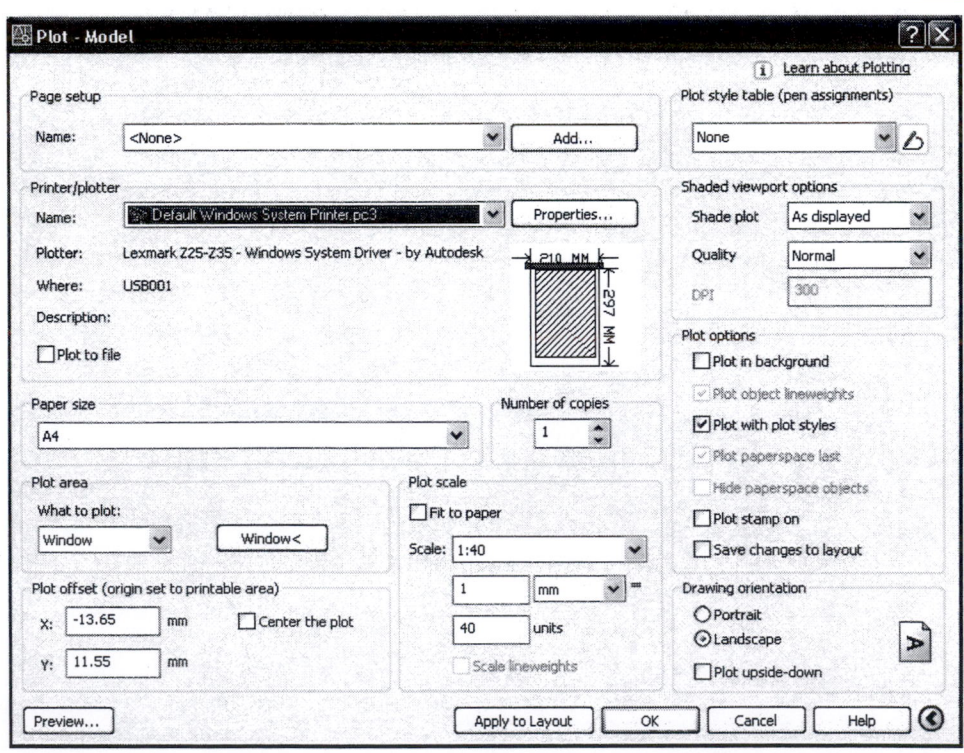

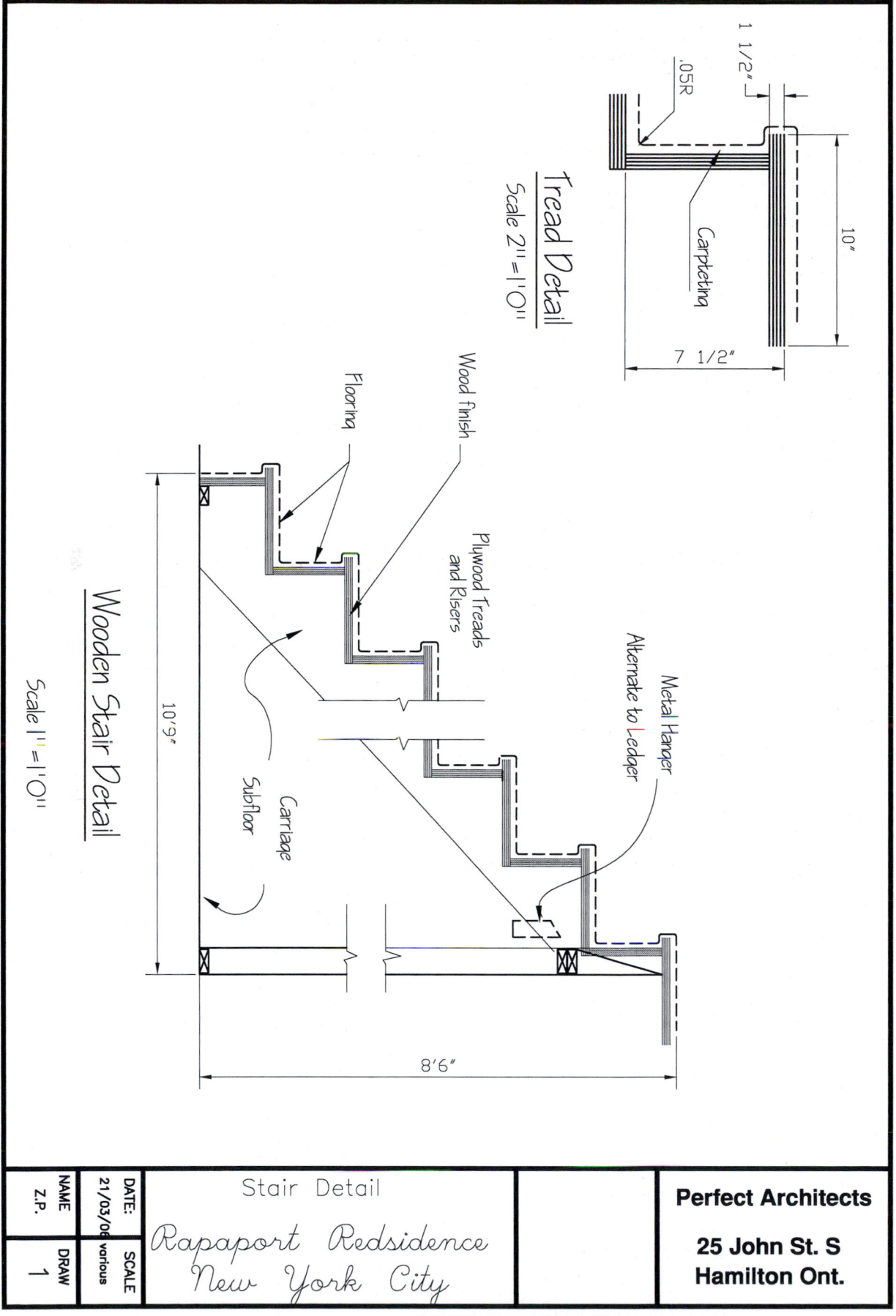

Tread Detail
Scale 2" = 1'0"

.05R

Carpeting

1 1/2"

10"

7 1/2"

Wooden Stair Detail
Scale 1" = 1'0"

Flooring

Wood finish

Plywood Treads
and Risers

Metal Hanger
Alternate to Ledger

Carriage
Subfloor

10'9"

8'6"

Perfect Architects			
25 John St. S **Hamilton Ont.**			

Stair Detail

Rapaport Redsidence
New York City

NAME	DATE: 21/03/06	SCALE various
Z.P.	DRAW 1	

Note the break line in the stair.

Draw the stair and thread details, add a title block, and PLOT.

Exercise 11 Practice

Exercise 11 Architectural

This illustrates four types of drawings made for commercial applications.
Use ARRAY for the curved text.
Print it out with differing lineweights on a letter sized or A4 sheet.

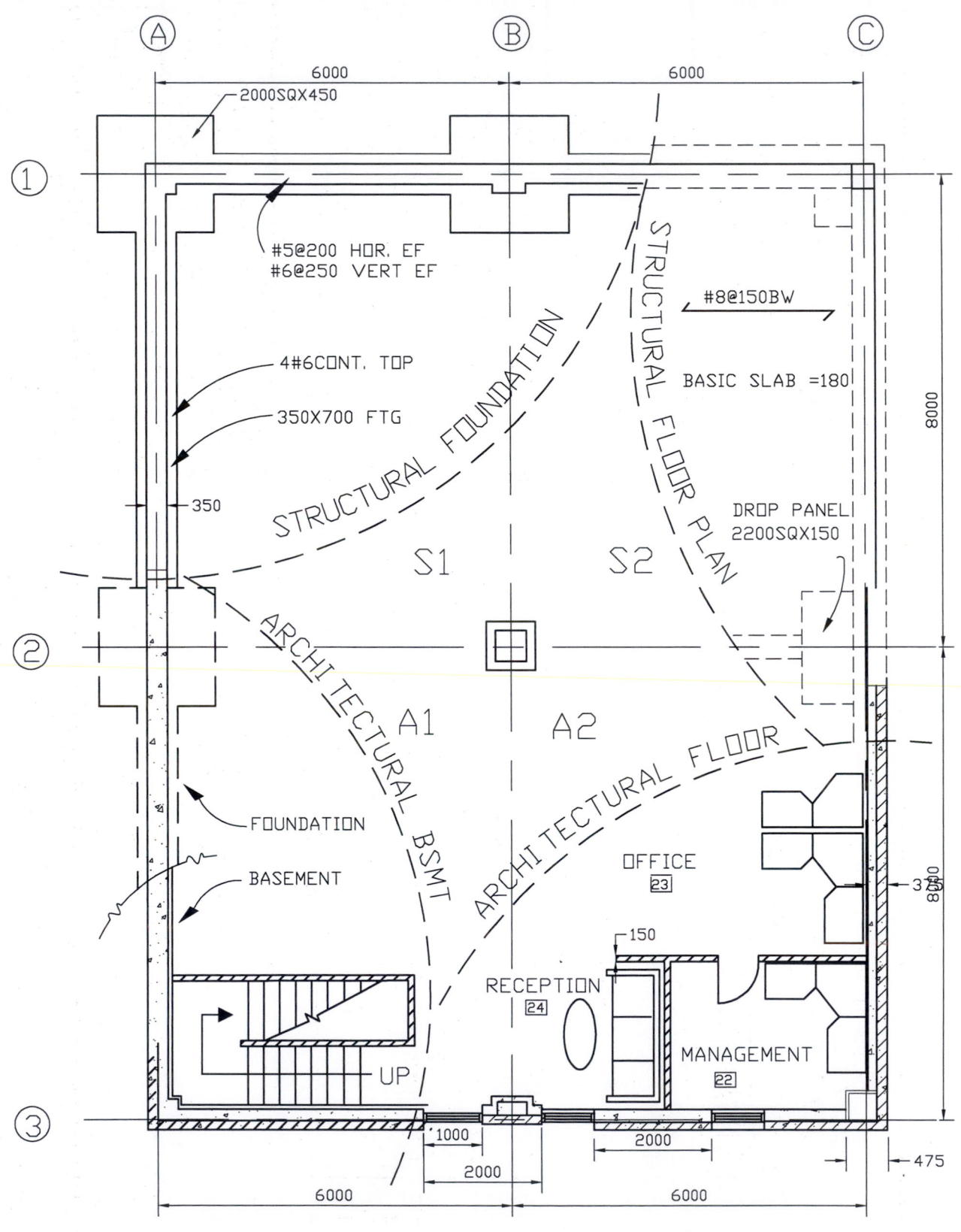

Exercise 11 Mechanical

This part should be drawn at a scale of 1=1.

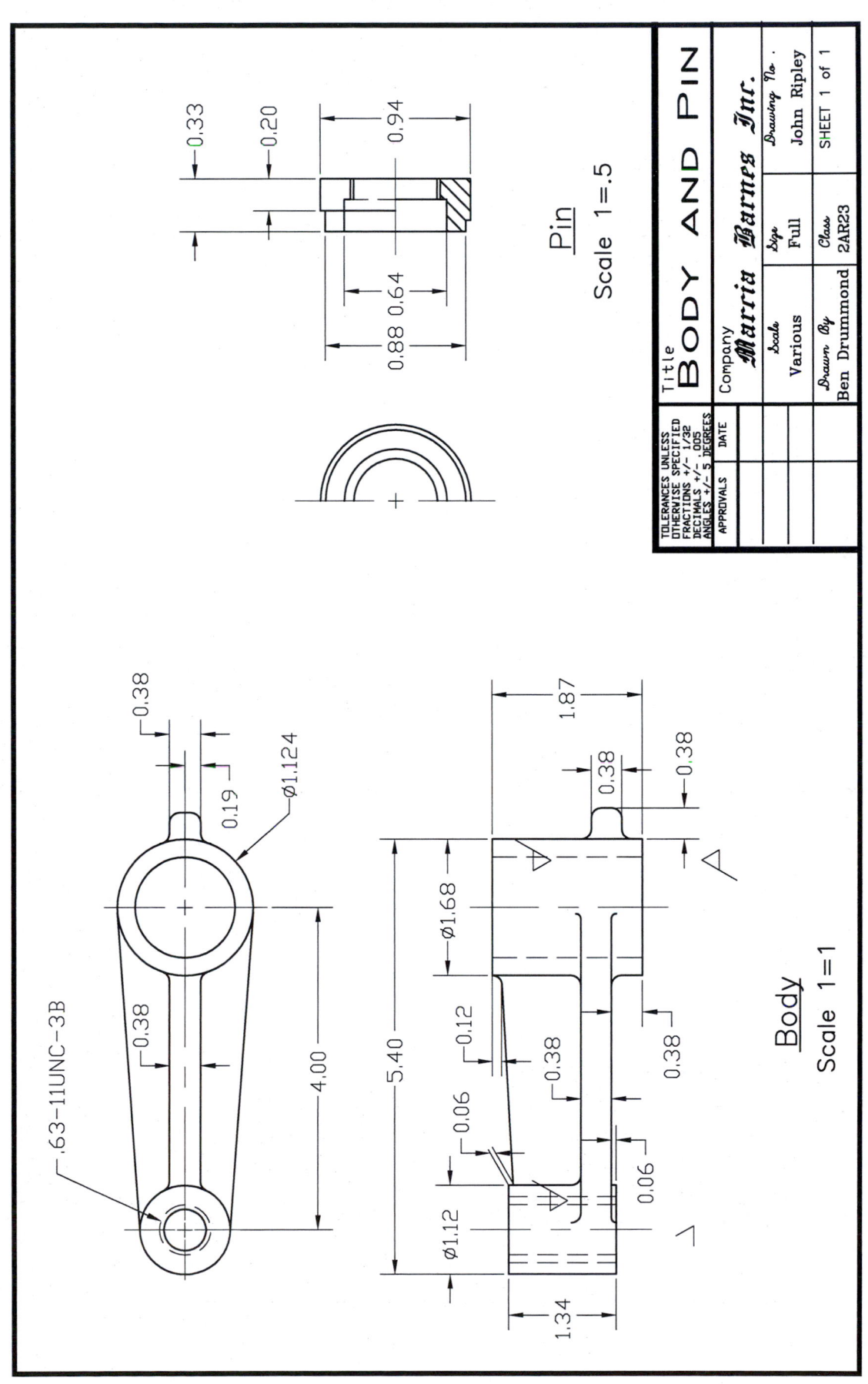

Exercise 11 - Challenger

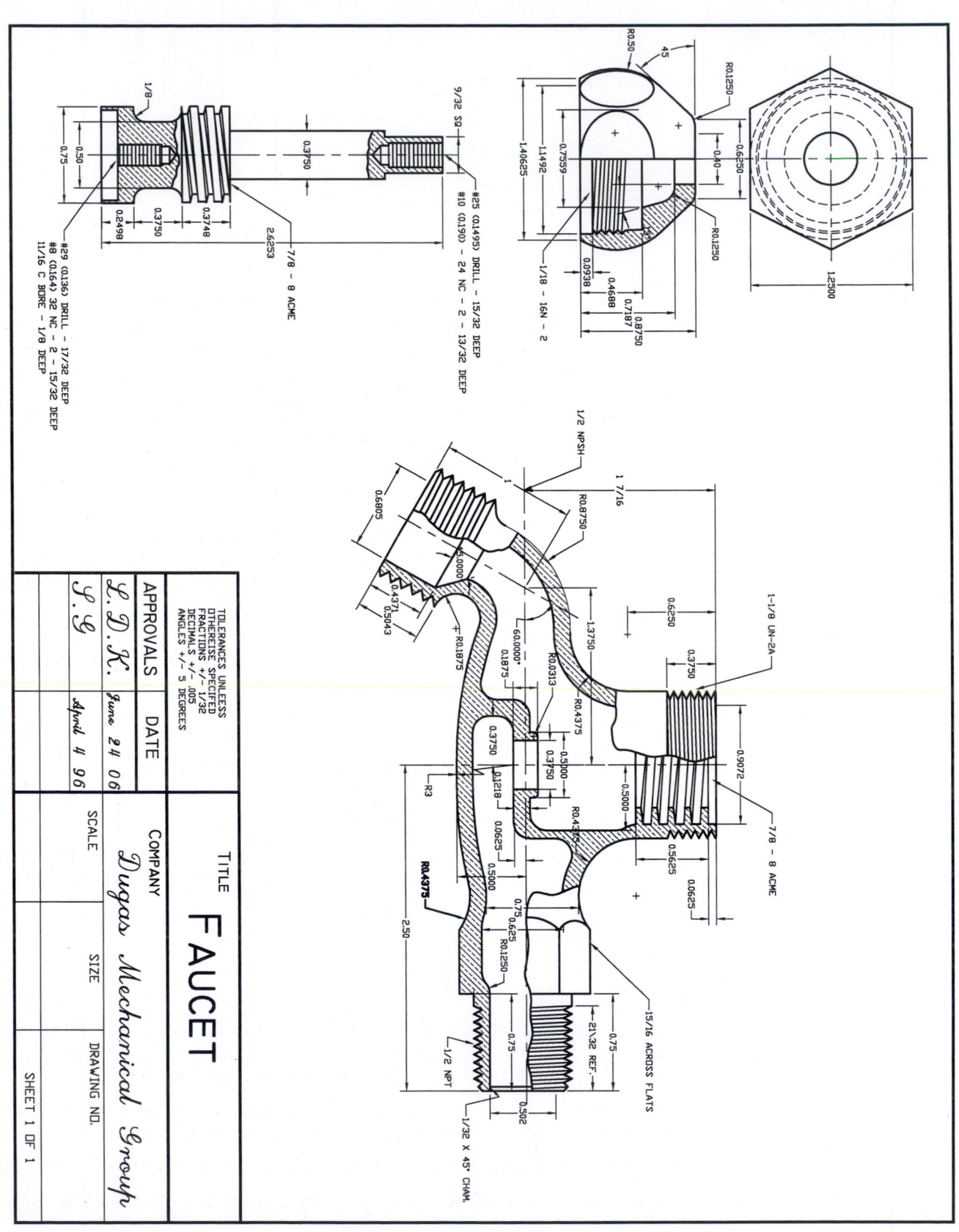

12

Paper Space for 2 Dimensional Drawings

On completion of this chapter, you should be able to:

1. Use paper space to access a paper environment
2. Use MVIEW to set up a drawing with four views on a standard title sheet
3. Use VPLAYER to have layers visible in selected viewports only
4. Use AutoCAD's stored templates to lay out a drawing.

So far you have been working in model space, the part of AutoCasd that creates designs or models. In Chapter 11 we discussed creating drawings in model space by adjusting the title block size. In this chapter we will use paper space to create drawings. Paper space is much easier to use when you are dealing with multiple views.

Paper Space

The most efficient way to compile a drawing that has details to be shown at different scale factors is through paper space.

Essentially, *paper space* is a 2D document layout facility. Once the views are completed with all the dimensioning required, paper space takes these views and places them on a 'paper' much like a cut-and-paste routine, so that they can be compiled as a drawing.

Paper space makes use of the multiple viewport facility. Multiple viewports are used to place the various views of the objects onto different portions of paper layout both in 3D and 2D.

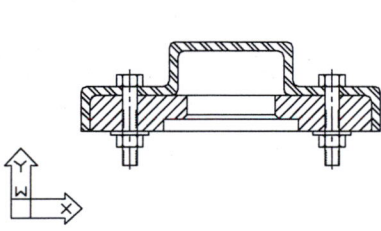

Model space   Paper space

Figure 12.1

In Figure 12.1, the object on the left is in *model space* and the layout on the right is in paper space. Both are filled with the same drawing. One has one view, the other has three views at different scale factors and different dimension styles.

The image on the left is in a tiled space, the TILEMODE system variable is on. The image on the right is in a non-tiled environment: the views can be anywhere.

Note that the view title and scale are always shown beneath the view in a drawing.

In paper space the crosshairs cover the screen. On the left in Figure 12.1 is a regular UCS icon; on the right is the paper space icon.

Accessing Paper Space

Getting from model space to paper space is quite simple. Pick Layout1 from the Layout tab, shown in Figure 12.2, at the bottom of your graphics screen. If this tab is not at the bottom of your screen, type in the word TILEMODE and then enter 0 (see bottom of page).

Figure 12.2

Once in paper space, right-click the Layout1 tab for the Page Setup Manager. This will allow you to set up the page that you are working with. Either make a new page or accept the current layout.

Similar to the Dimension style and Text style menus, it is always a good idea to set up a new page so that you never lose the defaults of the original page setup.

The next screen is the Page Setup dialog box. It looks much like the Plot dialog box. You must set your paper size before you begin. If you do not, you will waste a lot of time later adjusting the views.

First pick your plotting device from the Plot tab, then pick your paper size from the Layout tab. If you are trying to set up a drawing, and all you have loaded are laser printers, you will need to get another plotter driver loaded before you can access any paper bigger than 297×420 mm (11×17 imperial). In AutoCAD, at the command prompt, type in CONFIG, then choose Load a Plotter to add a plotter .

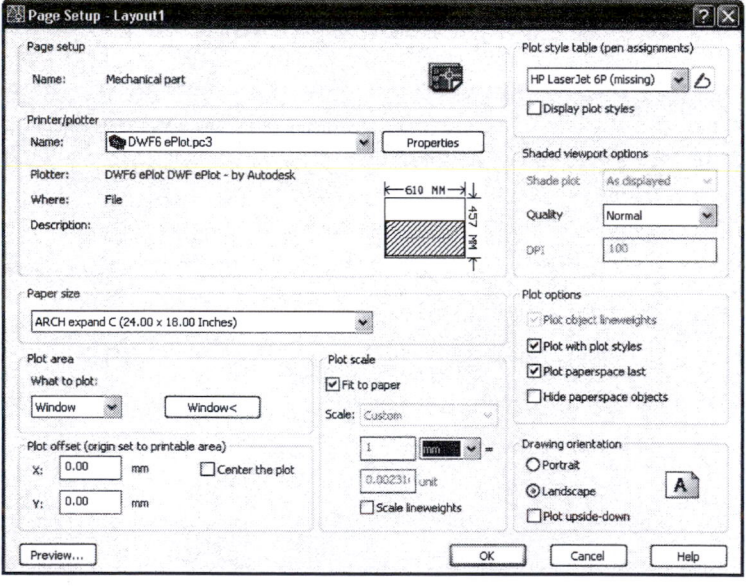

Your page will now have the correct size. Return to the paper space layout to compile the various views in your drawing.

Setting up a paper space environment toggles the TILEMODE system variable to off.

You can access the paper space mode through the command TILEMODE as well as the Layout tab. You will still need to set up your paper size in the Page Layout Manager.

```
Command:TILEMODE
Enter new value for TILEMODE <1>:0
Regenerating drawing.
```

Floating Viewports or MVIEW

The MVIEW or Floating Viewports command is used to identify views that will be used in paper space. The Page Setup Manager dialog box forces you to pick a paper size, then fits whatever is currently on screen within a border on that paper. You can adjust the size of the view using the corner grips, then add more views with MVIEW.

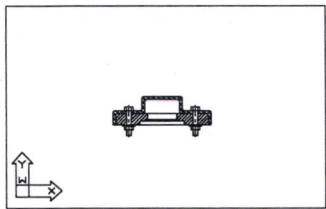 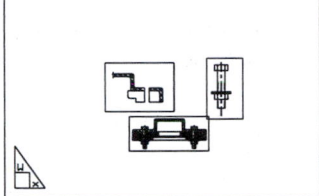 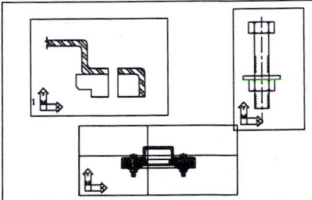

Figure 12.3

In Figure 12.3, the left screen has the model space object. The central screen has the paper space views, two new ones have been added with MVIEW. The third screen shows the three views in model space with some layers frozen in two of the views. To add new views use MVIEW.

> **Toolbar** There is no button for MVIEW.
>
> **Pull-down menu** From the View menu, choose Viewports.

The command line equivalent is MVIEW.

The pull-down menu offers the same choices as the command line: 1 Viewport, 2 Viewports, Restore, etc. The border for the viewport will be in the current layer. Make sure you have a new layer for the borders.

Once paper space has been accessed and the views have been made, you can toggle between model space and paper space quite readily. The Layout tab means the system variable TILEMODE is off, you can place your views anywhere, they don't need to be placed like ceramic tiles, side by side. When you return to TILEMODE on, AutoCAD restores the drawing as it was before paper space was entered.

When using the Floating Viewports or MVIEW command, you can create various con-figurations of viewports and have different layers active in specific viewports.

Using Model Space and Paper Space

When the Layout1 tab is on, you are creating a drawing in pa-per space. The TILEMODE system variable is off. The model is still accessible.

When the model tab is on you are working in model space. TI-LEMODE is on. This means either a paper or non-paper environ-ment.

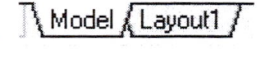

When working in paper space (TILEMODE is off), each view can be accessed as if in model space by using the MODEL/PAPER toggle.

In the Layout mode you can toggle from paper space to model space in order to access your model for dimensioning. You can type in MSPACE or MS and PSPACE or PS, or use the toggle switch.

LWT MODEL   Command:**MS** (toggles back to model space)

LWT PAPER   Command:**PS** (toggles back to paper space)

Paper space (PSPACE) only lets you access the paper view of the model. Model space (MSPACE) lets you access the model. These can both be used in the Layout1 tab.

When you use the ZOOM command in paper space, it affects the entire page or paper. A ZOOM .5X will result in the paper with the views intact at half the size that it was before.

When you use the ZOOM command in model space, the model or drawing within the view will be affected. Be careful not to ZOOM the views in model space after you have scaled them with ZOOM XP.

In model space you can PAN the objects within the viewports; in paper space the PAN command will affect the view of your paper environment.

In model space the MOVE command will move the objects within the views. In paper space the MOVE command will move the views relative to the other views.

Manipulating the Views

Any of the Modify commands will work on the viewports in paper space in the same way that they work on the objects in model space. You can use MOVE, COPY, STRETCH, etc. to place the viewports where you want with the necessary information.

Scaling Views Within a Drawing

In model space, the command ZOOM All causes you to lose any relation to the actual scale of the part being designed; the image expands to fit the space provided. Use the ZOOM "times paper space" XP option to set the drawing size relative to the paper.

```
Command:MS
Command:ZOOM
Specify corner of window, enter a scale factor (nX or nXP)
   or [All/Center/Dynamic/........Object] <real time>:.5XP
```

If you are creating a standard page setup for use with many objects of similar layout, this page layout zoom factors can be saved with the layout.

In paper space you want the views to be scaled to a relative size. Use the ZOOM XP option to scale the object relative to the paper scale units.

While the ZOOM X option scales the object relative to its current size within the viewport, the scale factor of XP gives you a zoom factor times the paper scale or relative to the paper and also relative to the actual part. Zoom XP scales can be seen in Chart 12.1.

LWT PAPER

You can also scale with the Properties menu as shown on the right.

Pick the Viewport border, then Modify, Properties. You must be in paper space to do this.

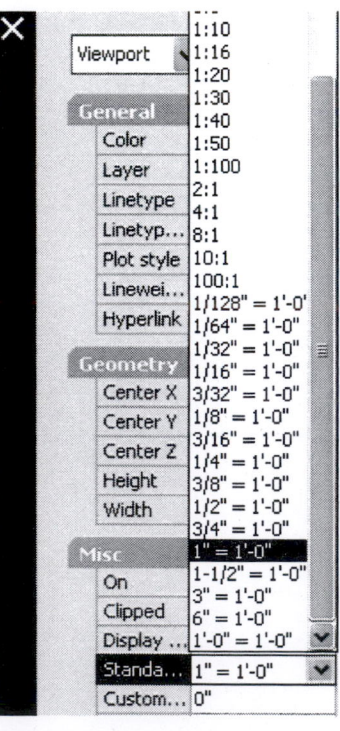

Chart 12.1

For a Scale of	*Use*	*Size Relative to Actual*
1:1	ZOOM 1XP	1
1/2 = 1	ZOOM .5XP	1/2
1:50	ZOOM .02XP	1/50
1/4 = 1 -0	ZOOM 1/48XP	1/48 or (¼ x 12)

Layers Within Viewports

The easiest way to control your layers is within the active viewport. Viewport-dependent layers are only possible in a non-tiled environment, with the Layout tab showing that you have TILEMODE off.

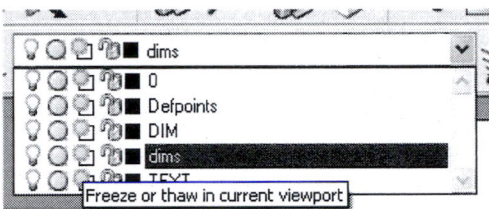

Both in the layer pull-down menu shown on the left and the Layer Properties Manager dialog box shown below, you can set the layers to be viewport dependent. Some layers are visible in some viewports but not in all.

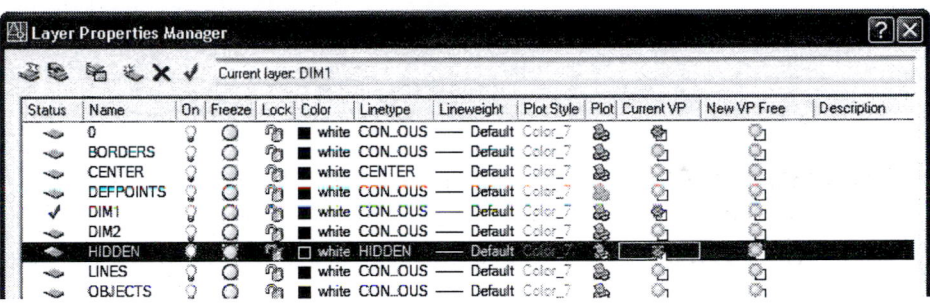

This allows you to have dimensions in different scale factors as well as notations and geometry that are not visible in all views.

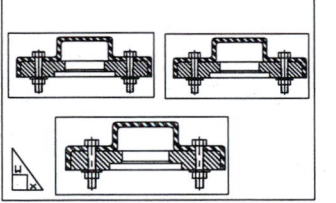
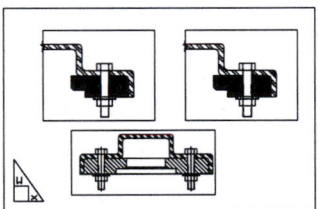
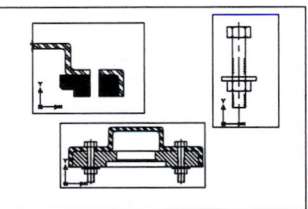

Paper Space
Three viewports

Paper Space
Zoomed views

Model Space
Layers frozen in viewports

Figure 12.4

In Figure 12.4, the first view shows three viewports in paper space. In the middle view, the two details have been zoomed with 2XP to be twice the size of the full part. The third view shows the non-relevant layers frozen in the two detail views.

Once the layers are frozen, a new layer for dimensioning is added and the dimensions are added to the various views. This dimension layer will then be frozen in the other two views. Dimensions are added in model space because it is the actual size of the model that is to be recorded with the dimensions.

The VPLAYER Command

VPLAYER is useful for manipulating many viewports. The Layer functions are easier for just one or two viewports. You can set the visibility of layers with the Layer dialog box and the Layer pull-down menu or you can use the VPLAYER command for a large number of views.

As with the Floating Viewports or MVIEW command, this command only shows up in paper space.

Toolbar and pull-down menu There is no button or menu choice for VPLAYER.

The command line equivalent is VPLAYER.

```
Command:VPLAYER
Enter an option [?/Freeze/Thaw/Reset/Newfrz/Vpvisdflt]:
```

Where: **?** = lists the LAYERs that are frozen in any selected viewport

Freeze = freeze a LAYER or LAYERs in selected viewports

Thaw = the reverse of freeze; turns on selected LAYERs in selected viewports

Newfrz = creates a new LAYER that is frozen in all viewports, and is primarily used for creating a LAYER to be viewed only in one viewport ever.

Reset = the default display for layers created in the LAYER command is thawed; with the Reset option, the layers are returned to their original default setting

Vpvisdflt = set a default visibility for any layer in any viewport.

```
Command:VPLAYER
Enter an option [Freeze/Thaw/Reset/Newfrz/Vpvisdflt]:F
Enter Layer names to freeze or select objects:DIM1 (name of
   the LAYER)
Enter an option [All/Select/Current]<Current>:S
Select  objects:(here you can pick any number of viewports
   by their frames)
```

Dimensioning in Paper Space

Your final drawing, regardless of the scale factors of the views, should have all text and arrowheads the same size.

In Figure 12.5 the dimension arrow and text size are the same in each of these views. The part is shown in four different viewports with four different scale factors. The dimension text reads the same dimension regardless of the dimension scale factor.

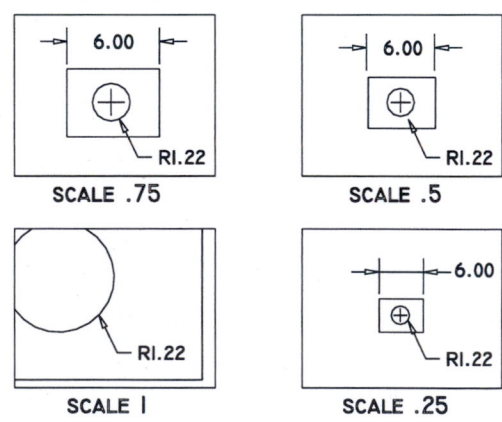

Figure 12 .5

To use different scales in different viewports on the same drawing, set the DIMSCALE value to 0.

In the Modify Dimension Style dialog box, set the overall scale in the Fit tab to 0.

This paper space scale will automatically cause all dimensions to be shown in the default size regardless of the scale of the view. A dimension scale or fit size of 0 is useful if you are happy with the defaults of the system, ie. the size of text/arrows/gaps and offshoots.

There are some cases where another size of dimensioning would give a better result. In this case, *always create a new layer,* then set the fit size to be a fraction of the zoom factor.

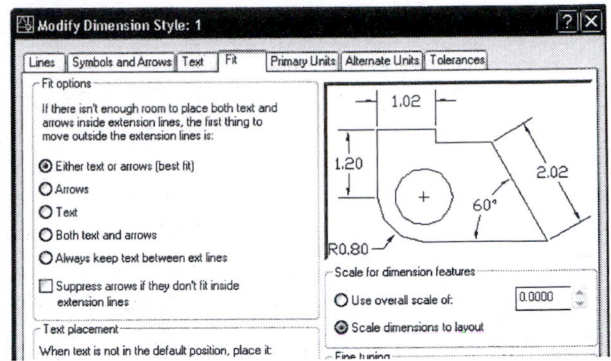

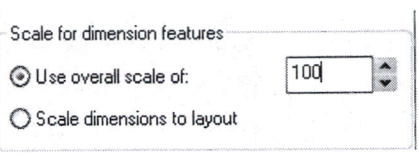

Scale of view	Zoom XP	Fit	Dimensions can be adjusted to
¼ = 1 -0	1/48XP	48	40 – 48
1:100	1/100	100	90 – 110

If your dimensions just do not look right, you can adjust the size of a custom dimension. Changing the zoom factor on a drawing to adjust the automatic paper space size will not help. The scales for a drawing must be standard. A view scale of 1:46 or ¼ = 1 -2 is just not going to work. Flexibility is *only* on the dimension scale factor.

AutoCAD's Template Drawings

AutoCAD has a series of drawing templates that can be used. These title blocks are set up in paper space and will not interfere with your drawing as you develop it. For a list of AutoCAD's templates, use NEW as in Figure 12.6.

Command: **NEW**

The ANSI (American National Standards Institute) templates are imperial, the ISO (International Standards Organization) and JIS are metric. All are professional layouts like the one in Figure 12.7.

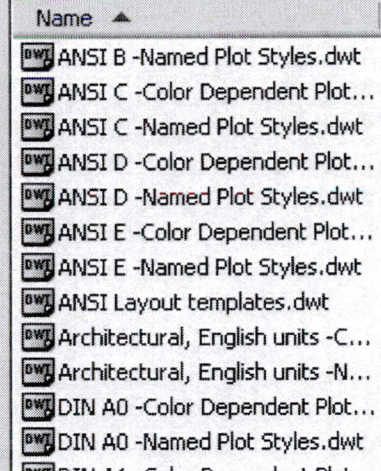

Figure 12.6

Figure 12.7

You can use these, or you can set up your own template file that also has layers and dimension styles. Simply set it up and save it as Mytemplate. Once the bugs are worked out, you can use it every time you start a drawing.

In this example we will take a floor plan of a house that has a circular fireplace and a kitchen, and we will create one drawing sheet with a plan view and the two details described. Take your layout from Chapter 7, or use another floor plan. This drawing will have three scales as follows:

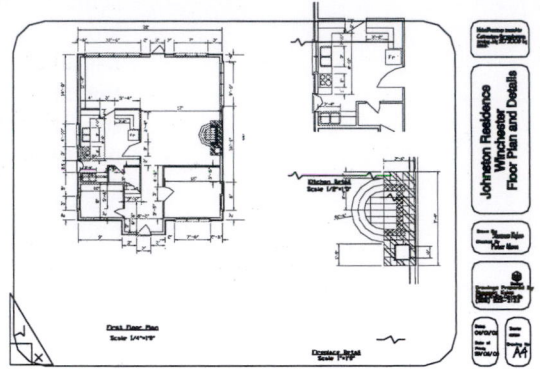

Plan view of house	1/4 = 1 -0
Detail of fireplace	1 = 1 -0
Detail of kitchen	1/2 = 1 -0

Final Drawing

Step 1   *Retrieving the File*

Retrieve the floor plan file on which you want to create a drawing.

If you would like to use an AutoCAD template, open a new file and choose Architectural English Units. Then you won't need to draw a title block. Insert the floor plan or Copy/Paste into model space.

Now create the following layers:

Layer name	Color
DIM1	red
DIM2	yellow
DIM3	green
BORDER	magenta
Pspace	blue
Detail	cyan

Step 2   *Opening Paper Space*

Make sure your BORDER layer is current.

Pick the Layout tab at the bottom of your screen or use the TILEMODE system variable.

```
Command:TILEMODE
Enter new value for
   TILEMODE <1>:0
Regenerating drawing.
```

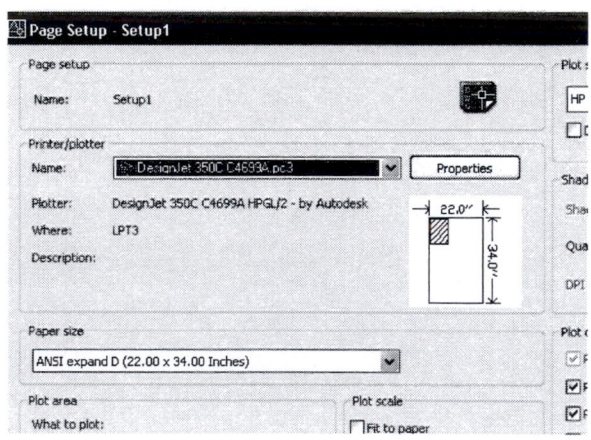

This will get you into a non-tiled space. There will be a view of your model on screen with a magenta border.

Right-click the Layout tab to set the size of your paper. Make a new layout called Floor plan.

Now you want to pick the plotter size that you need for this drawing. We are going to use a 34 x 22 sheet.

If your plotter options do not include a plotter this size, go back to AutoCAD and use the CONFIG command plus Configure a Plotter Wizard to configure a Designjet or similar plotter. Just accept all the defaults. The driver should come with AutoCAD.

Then right-click again on Layout1. Set up your page as above, and be sure to make this layout current on your way out.

Step 3   ***Creating the Views***

Now you can add your views relative to this paper with the command MVIEW.

First, in model space zoom the current view to a scale of 1 to 48.

```
Command:Z
  Specify corner of  window, enter a scale factor (nX or nXP)
    or [All/Center/Dynamic/........Object] <real time>:1/48xp
```

or in paper space pick the border and choose 1/4 = 1 -0 in the properties menu.

You must be in paper space to use the Properties toolbar.

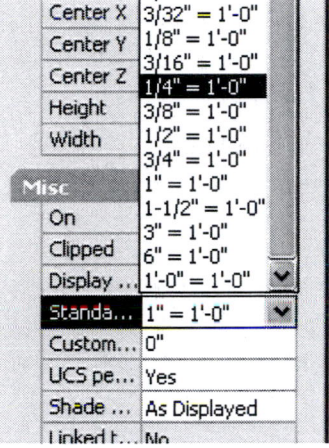

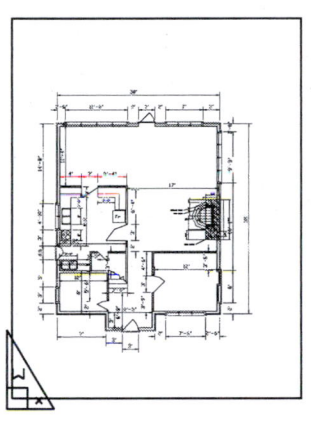

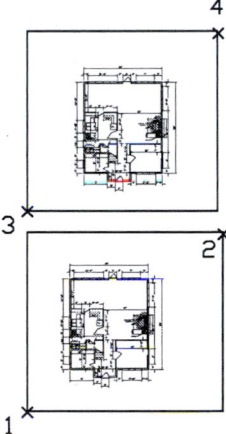

Now adjust the size of this viewport using the grips on the corners.

Go back to paper space mode for this.

Pick up the frame on the corner and drag it into place.

Now add two more views with MVIEW.

```
Command:MVIEW
Specify corner of viewport or
  [ON/OFF/Fit/Hideplot/Lock/Object/
  Polygonal/Restore/2/3/4]:(pick 1)
Specify opposite corner:(pick 2)
Command:MVIEW
Specify corner of viewport or
  [ON/OFF/Fit/Hideplot/Lock/Object/
  Polygonal/Restore/2/3/4]:(pick 3)
Specify opposite corner:(pick 4)
```

Each new view added will have the full floor plan on it.

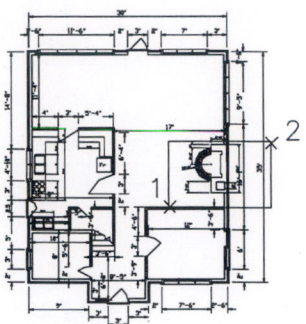

Step 4   *Zooming the Views*

On the lower right view, toggle into model space, then zoom into the views so that you get approximately the fireplace view.

Then use ZOOM with the xp scale to zoom it to 1 =1 -0   This will be a zoom of 1/12XP.

```
Command:Zoom
Specify corner ...<real time>:1/12XP
```

Go back to paper space and move your screen so that you can see the upper right view. Zoom into the kitchen, then zoom it to a scale of ½ = 1 -0 or 1/24XP.

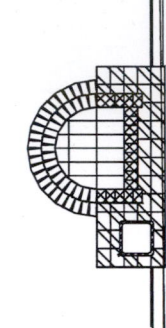

```
Command:Zoom
Specify corner ...<real
    time>:1/24XP
```

The ZOOM command is viewport dependent. You must be in the view in model space to zoom the objects with ZOOM, and in paper space to use the Properties menu.

Step 5   *Dimensioning the Views*

With your kitchen still on screen and zoomed at the correct view, go to the Dimension Styles dialog box and create a new dim layer called Pspace.

Under the Fit tab, change the overall scale to 0. This will size all of the dimensions to the respective views.

Set this dimension style to current.

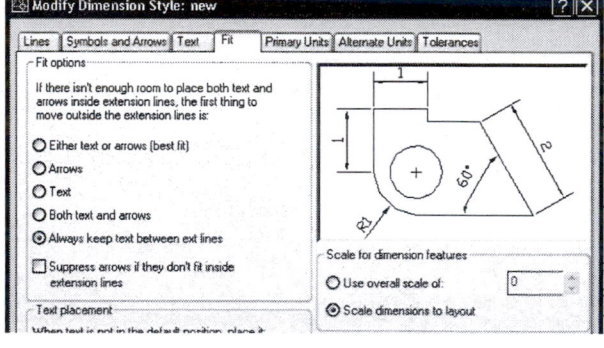

Toggle to paper space to zoom into the view so that you can see it properly.

Once you have the kitchen on your screen return to model space so that you can access the information within the viewport in order to dimension it.

Since this is a drawing done in imperial, make sure that your primary units are set to imperial as well.

In model space, go to the Layer pull-down menu and make DIM1 frozen in the current viewport. Then make DIM3 current.

Use the linear dimension to dimension the room as shown.

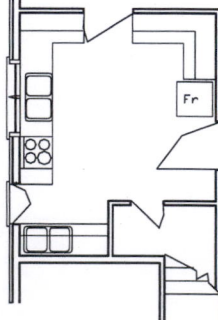

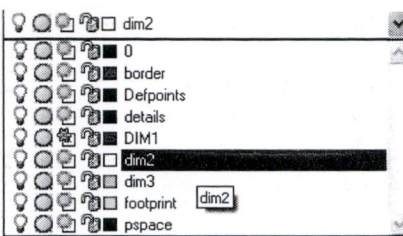

You can use either the Layer dialog box or the Layer pull-down menu to freeze and thaw layers within individual viewports.

This must be done in model space.

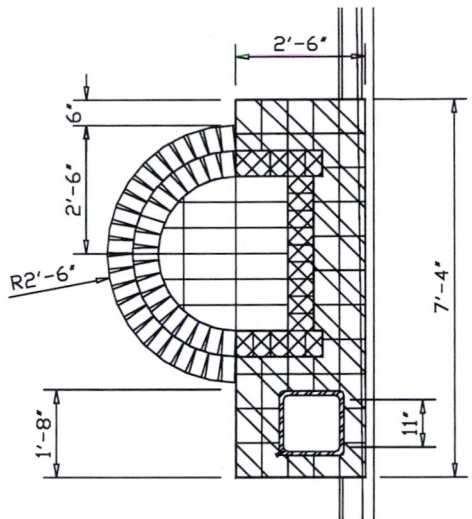

LWT PAPER

Once it is completed, toggle to paper space to pan the drawing over so that you can access the fireplace.

LWT MODEL

Then toggle back to model space, make the DIM1 layer frozen in that viewport, make DIM2 current, and dimension the view as shown.

LWT PAPER

Now go back to paper space and zoom the full paper. Notice that you must make DIM2 and DIM3 frozen in the floor plan view. You can use either the Layer pull-down menu with model space on, or the VPLAYER command in paper space.

LWT PAPER

```
Command:VPLAYER
Enter an option [Freeze/Thaw/Reset/Newfrz/Vpvisdflt:F
Layer(s) to Freeze:DIM3,DIM2
All/Selected/<current>:S
Select objects:(pick the plan view)
Enter an option [Freeze/Thaw/Reset/Newfrz/Vpvisdflt:↵
```

You may notice that the default dimension size sometimes just doesn't work. The numbers look too large and clumsy for the layout.

It is always useful to know how to create a custom-scaled dimension layer for cases like this.

Automatic dimension scaling is OK for many applications, but in this case a dimension scale of 10 may be better than one of 12, which is the default for this window zoom factor.

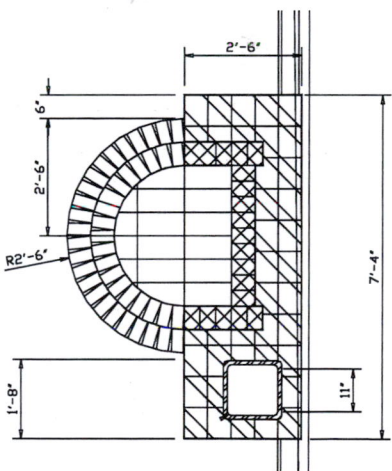

LWT MODEL

Go to the Dimension Style dialog box. Create a new dimension called fireplace. Set the Fit factor for the overall size to 10, then make it current.

Put one dimension on with the new size. If you like it, use Match properties to update the other dimensions.

You should never deflect more than a few units from the default size or the dimensions will look wrong, but sometimes a little 'tweaking' is the difference between a good plot and a great one.

If, when you go to paper space and zoom the whole paper, you can read the dimensions, they are too big.

Step 6   Adding the Notations

Your three views are ready to be labeled. In paper space, make your pspace layer current.

Add the title for each view at a scale of ¼ lettering. The TEXT command is easier here than MTEXT. The toggle for underlining is %%U.

Remember that the view titles are always placed under the view.

In addition, when the borders are frozen you will need break lines to show that the wall continues past the borders of the frames. AutoCAD drawings must maintain the same protocols as manual drawings.

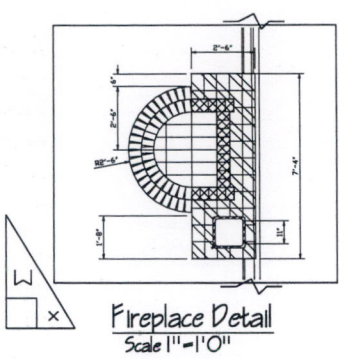

Fireplace Detail
Scale 1"=1'0"

Step 7   Placing the Title Block

The dimensioning layers are frozen within certain views and not others. In paper space you want to do a global freeze on the border layer.

With your border layer frozen, you often see that the layout could use a bit of minor adjustment to make it look better.

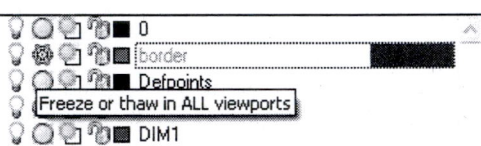

Bring the borders back. Move them around if necessary, go into the views to tighten up some of the dimensions, and move the view titles around so that there is room for a title block.

Finally, if you have not used the AutoCAD architectural template, add a title block. Your drawing is now ready to plot.

Use the PLOT command and plot to a scale factor of 1:1.

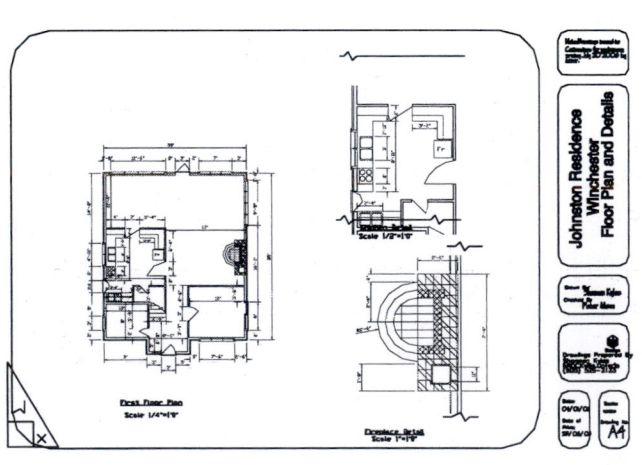

Drawing using custom title block

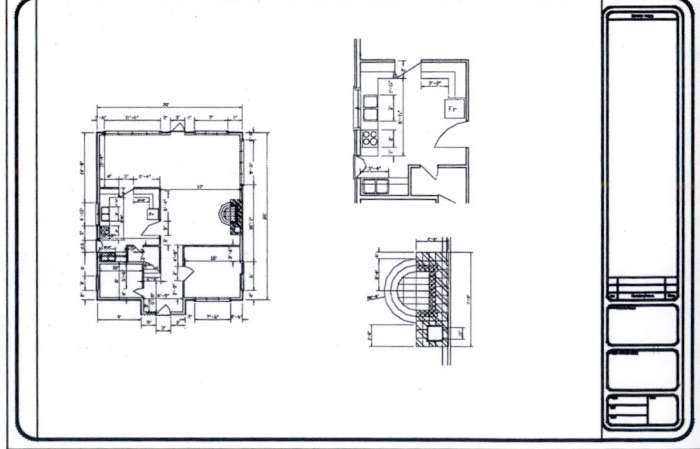

Drawing using AutoCAD architectural template

In this example we will draw two footings and show how they are assembled on a footing layout. The footing layout will be at a scale of 1:500. Footing A will be at a scale of 1:50. Footing B will be at a scale of 1:25.

Be sure to start in a metric drawing. aca-diso.dwt.

This will be plotted at 297 × 210 in landscape orientation. We will add our own title block.

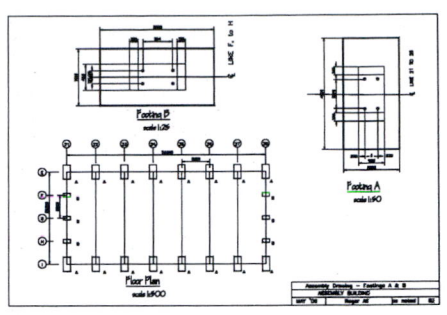

Step 1

First create the following layers

Layer name	color
DIM1	red
DIM2	yellow
DIM3	green
Border	magenta
Footprint	blue
Pspace	blue
Detail	cyan

Draw the footings. Make the footprint (darker lines) in the Footprint layer, and the other details of each view in the Detail layer. Do not dimension the footings yet.

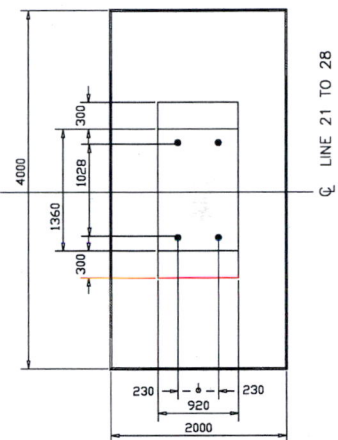

Footing A

Footing B is half the size of footing A. Clearly COPY and SCALE will make the drawing a lot simpler.

The footings will be assembled on a layout grid. On the detail views you want to show the outside dimension or footprint of each footing (the darker line) as well as the pin and upper level details.

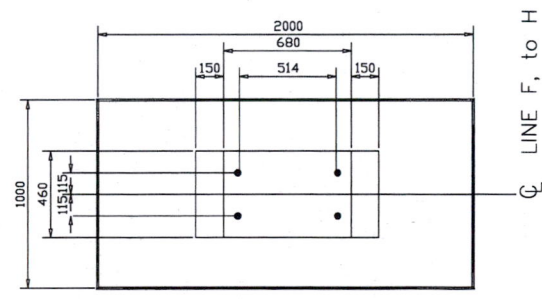

Footing B

On drawings of large construction projects, the footings are often laid out in bays or grids.

Step 2 *The Layout*

On Layer DIM1 draw in a grid of eight vertical lines spaced at 8000 units each and five horizontal lines at 6500 units each.

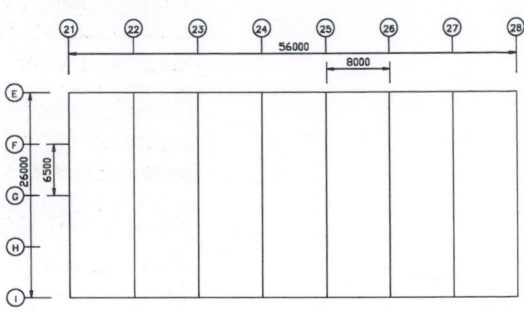

Draw bubbles along the top and side corresponding to this grid. Put in the letters and numbers shown.

Layout Grid

Use COPY to place the footings on the grid intersections.

Use the grid lines to place the footings centered on the grid lines. ARRAY will also work.

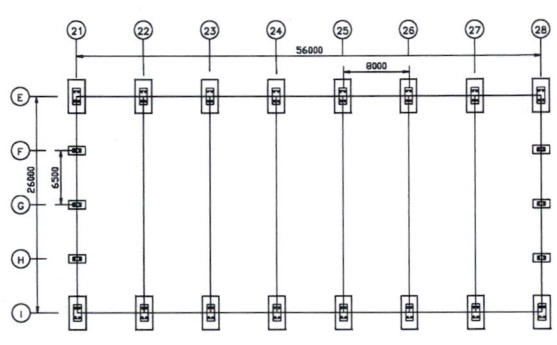

Note that the footings are centered on the grid lines so the base point is quite important.

If you have not done so already, save the drawing at this point.

Layout grid with footings

Step 3 *Opening Paper Space*

Make sure your Border layer is current.

Pick the Layout1 tab at the bottom of your screen or use the TILEMODE system variable.

```
Command:TILEMODE
Enter new value for TILEMODE <1>:0
Regenerating drawing.
```

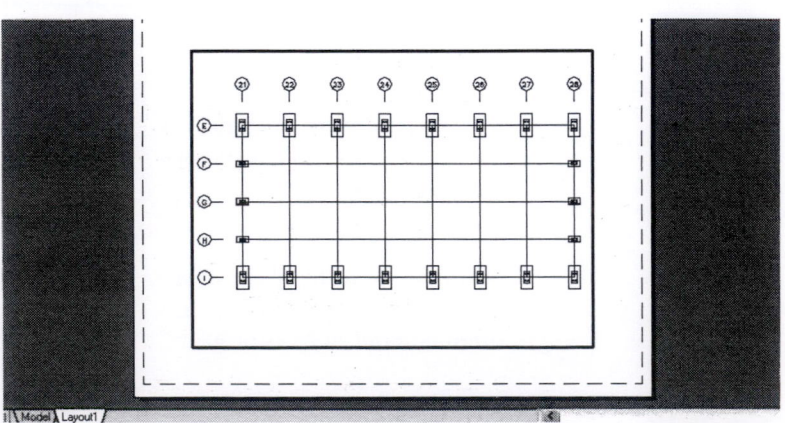

This will get you into a non-tiled space. There will be a view of your model on screen with a magenta border. If the border is not magenta, you are not in the correct layer. Make sure your frame or border is on the Border layer.

Right-click the Layout1 tab to get to the Page Setup Manager.

Make a New layout called Setup.

Now you want to pick the plotter size that you need for this drawing. We are going to use an 8.5×11 sheet or 297 × 210 mm.

Set the plotter to the one you generally use. Pick the paper size.

Make sure that the plotter will plot from 1:1. Then click OK and return to the drawing.

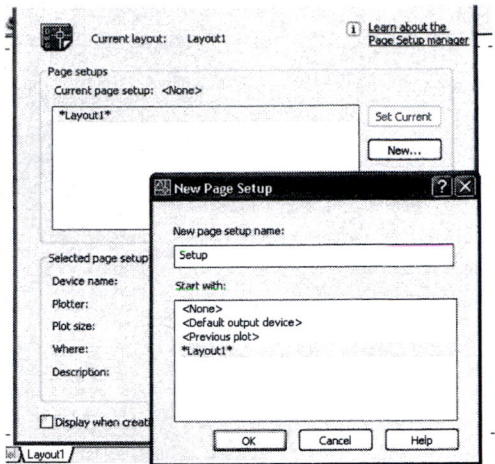

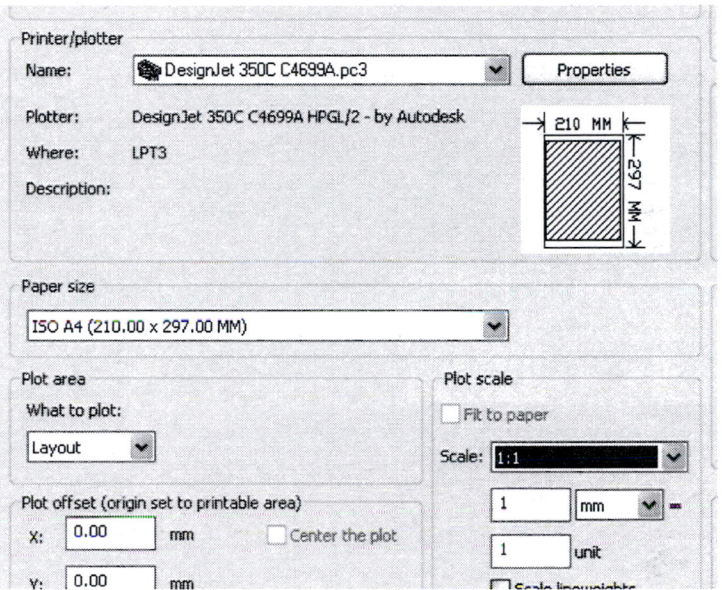

Step 4

Setting up the First Viewport

Now you can add your views relative to this paper using either the Floating Viewports or the command MVIEW. First set up the existing viewport. In model space, ZOOM the current view to a scale of 1 to 500.

```
Command:MS
Command:Z
  Specify corner of  window,
    enter a scale factor (nX
    or nXP)
    or[All/Center/Dynamic/...
    .....Object] <real
    time>:1/500xp
```

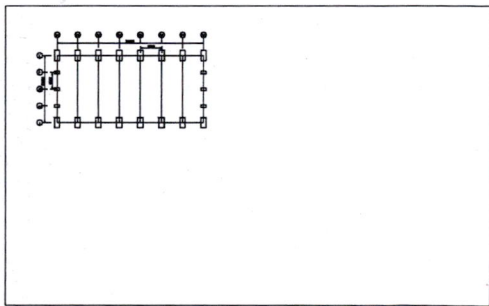

This zooms the view to exactly 1/500th or 1:500.

There is no standard scale for this in the Properties manager. Use the Properties toolbar to set up a custom size of 1-500 or use ZOOM.

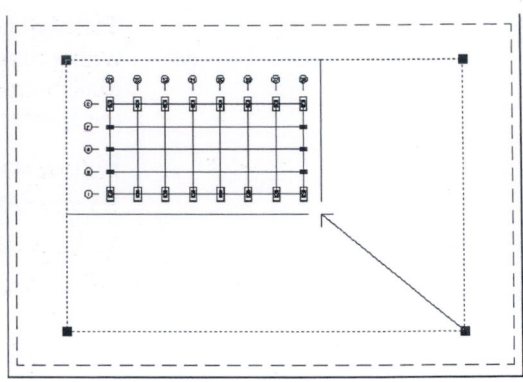

Now adjust the size of this viewport using the grips on the corners. Pick the edge of the frame to highlight the grips, then pick the corner to stretch the frame smaller.

You must be in paper space for this.

Pick up the frame on the corner and drag it into place.

Step 5   *Making New Views*

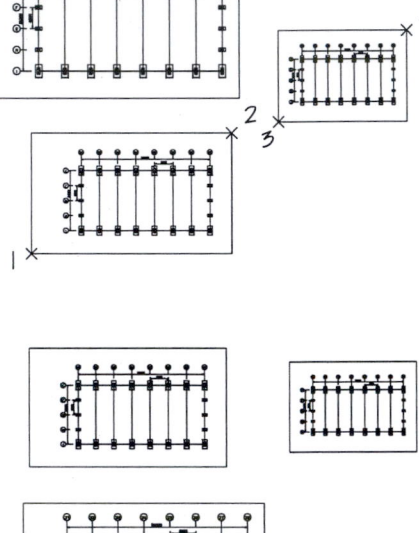

Now add two more views with MVIEW.

```
Command:MVIEW
Specify corner of viewport or
   [ON/OFF/Fit/Hideplot/Lock/Ob
   ject/ Polygonal/Restore
   /2/3/4]:(pick 1)
Specify opposite corner:(pick 2)
Command:MVIEW
Specify corner of viewport or
   [ON/OFF/Fit/Hideplot/Lock/Obje
   ct/ Polygonal/Restore
   /2/3/4]:(pick 3)
Specify opposite corner:(pick 4)
```

It is more practical to have the smaller views on top, so move the views and stretch them until they look like this.

```
Command:M  for move
```

Then pick the view and move it.

Keep the view that is zoomed at 1:500 on the bottom.

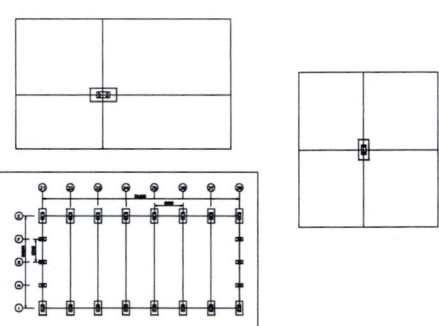

Step 6   *Zooming the Views*

Toggle into model space, then zoom into the footing views so that they are approximately centered in the view.

Then use ZOOM with the XP scale or Properties to zoom them. Footing A should be at a scale of 1:50. Footing B should be at a scale of 1:25.

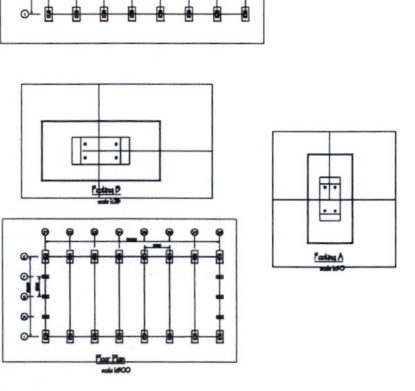

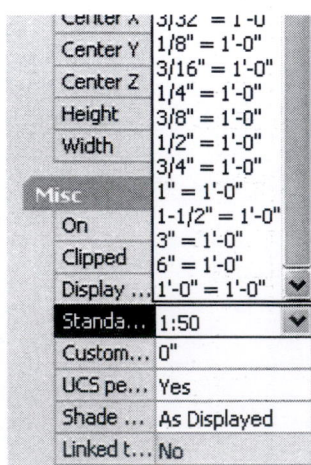

```
Command:Zoom
Specify corner ...<real
   time>:1/50XP
Command:Zoom
Specify corner ...<real time>:1/25XP
```

The ZOOM command is viewport dependent. You must be in model space to zoom the objects using ZOOM, or in paper space to change the scale using the Properties menu shown on the left.

Step 7   **Adding View Titles**

So you don't forget what scale you have zoomed the views to, add the notations.

Make the Pspace layer current, then use the TEXT command. The underline toggle is %%U. Make the title text height 2.5 mm for the title and 2 for the scale.

```
Command:text
Specify start point of text or [Justify/Style]:(pick a point
  beneath the view)
Specify height<10.000>:2.5
Specify rotation angle of text<0>:↵
Enter text:%%UFooting A↵
Command:text
Specify start point of text or [Justify/Style]:↵
Specify height<10.000>:2
Specify rotation angle of text<0>:↵
Enter text:%%UScale 1:50↵
```

Add the titles and scales for Footing B (scale 1:25) and the layout (scale 1:500).

Step 8   **Dimensioning the Views**

Go to the Create Dimension Style dialog box and make a new dimension layer called Pspace.

Under the Fit tab, choose Scale dimensions to layout. This will make all of the dimensions sized to the respective views.

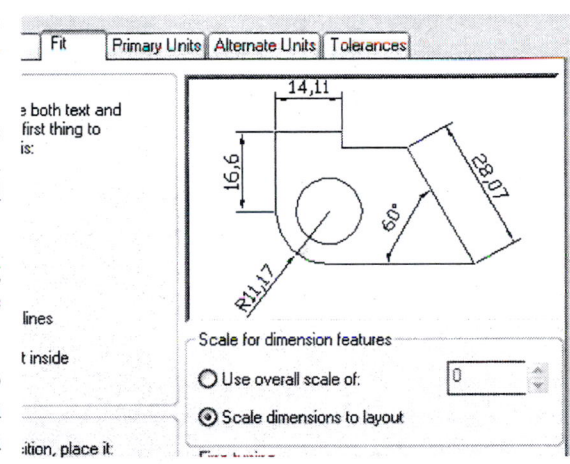

Choose OK to exit this dialog box, and set current to have the Pspace dimension style active.

Toggle to paper space to zoom to the outside of your border so you can see your footing well enough to dimension it.

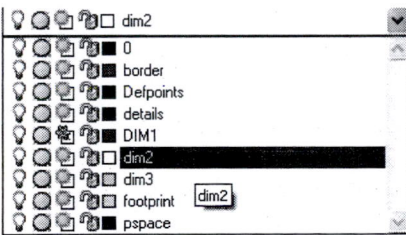

Toggle back to model space to access the information within the viewport in order to dimension it.

In model space, go to the Layer pull-down menu and make DIM1 frozen in the current viewport. Then make DIM2 current.

Use linear dimension to dimension the part as shown.

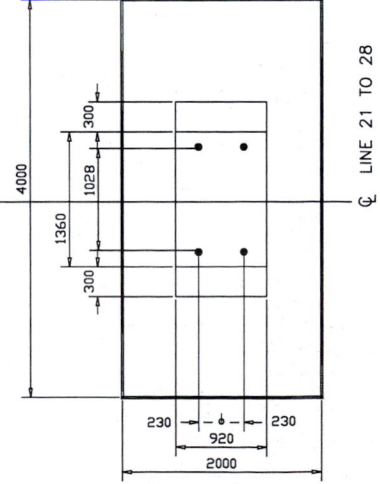

Once it is completed, toggle back to paper space to pan the drawing over so that you can access Footing B.

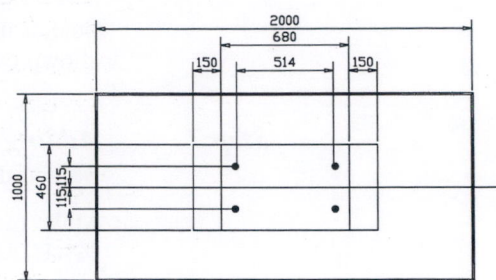

Then toggle back to model space, make the DIM1 layer frozen in that viewport, make DIM3 current, and dimension the view as shown.

Do the same for the layout view. Notice that you must make DIM2 and DIM3 frozen as well. You can also freeze the Detail layer in this view.

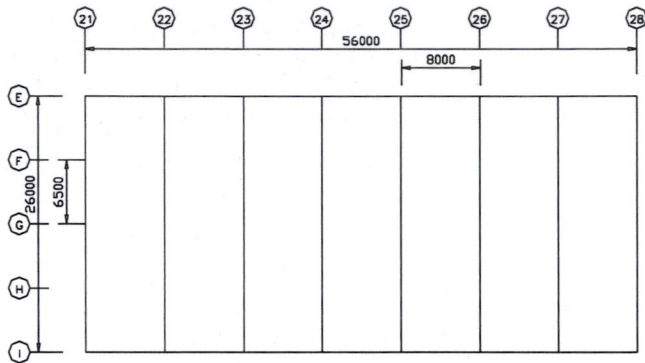

You may also find that you need to freeze the Footprint layer to place the dimensions properly. Once the dimensions are completed, thaw the Footprint layer.

You may notice that when the Footprint layer is turned back on, the default dimension size just doesn't work. The numbers look too large and clumsy for the layout.

It is always useful to know how to create a custom-scaled dimension layer for cases like this. The automatic dimension scaling is OK for many applications, but in this case a dimension scale of 450 is better than one of 500, which is the default for this window zoom factor.

Go to the Dimension Style dialog box. Create a new dimension called Fullview.

Set the Fit factor for the overall size to 450, then make it current. Put one dimension on with the new size. If you like it, use Match properties to update the other dimensions.

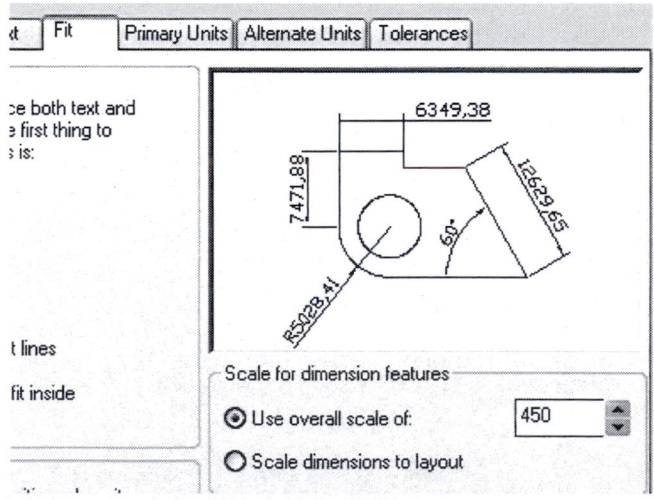

It is a good idea at this point to go into each view and re-zoom at 1/500XP, 1/50XP, and 1/25XP just in case you have zoomed the views while dimensioning.

Thaw the Footprint layer if you have frozen it.

The convention for a repetitive dimension like that of the footing bays is to put in one dimension and assume that the rest are the same.

The footings must be labeled: A beside all A footings, and B beside all B footings.

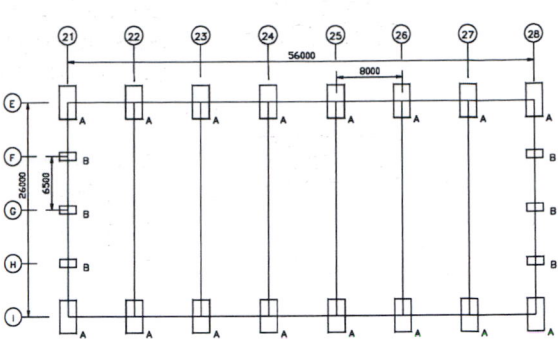

Step 9 *Adding the Title Block*

Your three views are ready, and they are already labeled. Now all you need to do is freeze the Border layer and add a title block.

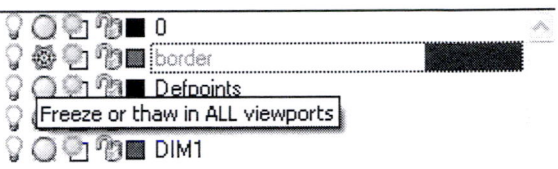

With your Border layer frozen, you often see that the layout could use a bit of minor adjustment to make it look better.

Bring the borders back. Move them around if necessary, go into the views to tighten up some of the dimensions, and move the view titles around so that there is room for a title block.

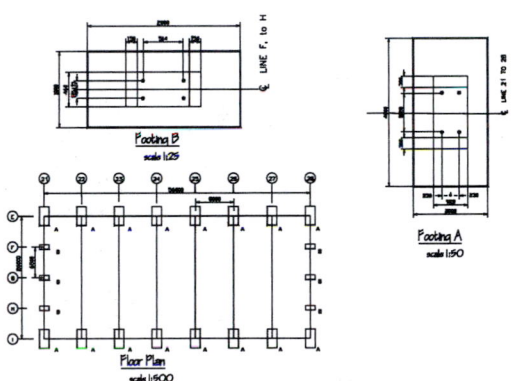

Finally add a title block. Either draw a new one or fill in the one that you opened. Your drawing is now ready to plot.

Use the PLOT command and plot to a scale factor of 1:1.

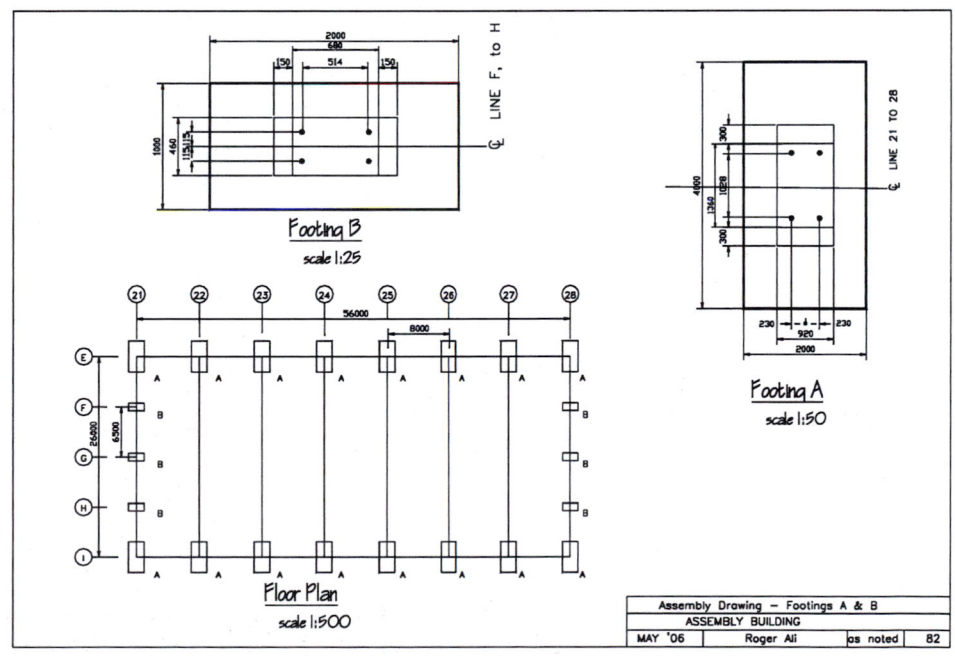

Exercise 12 Practice

Create the fireplace drawing, then open new viewports
in paper space to create the details.

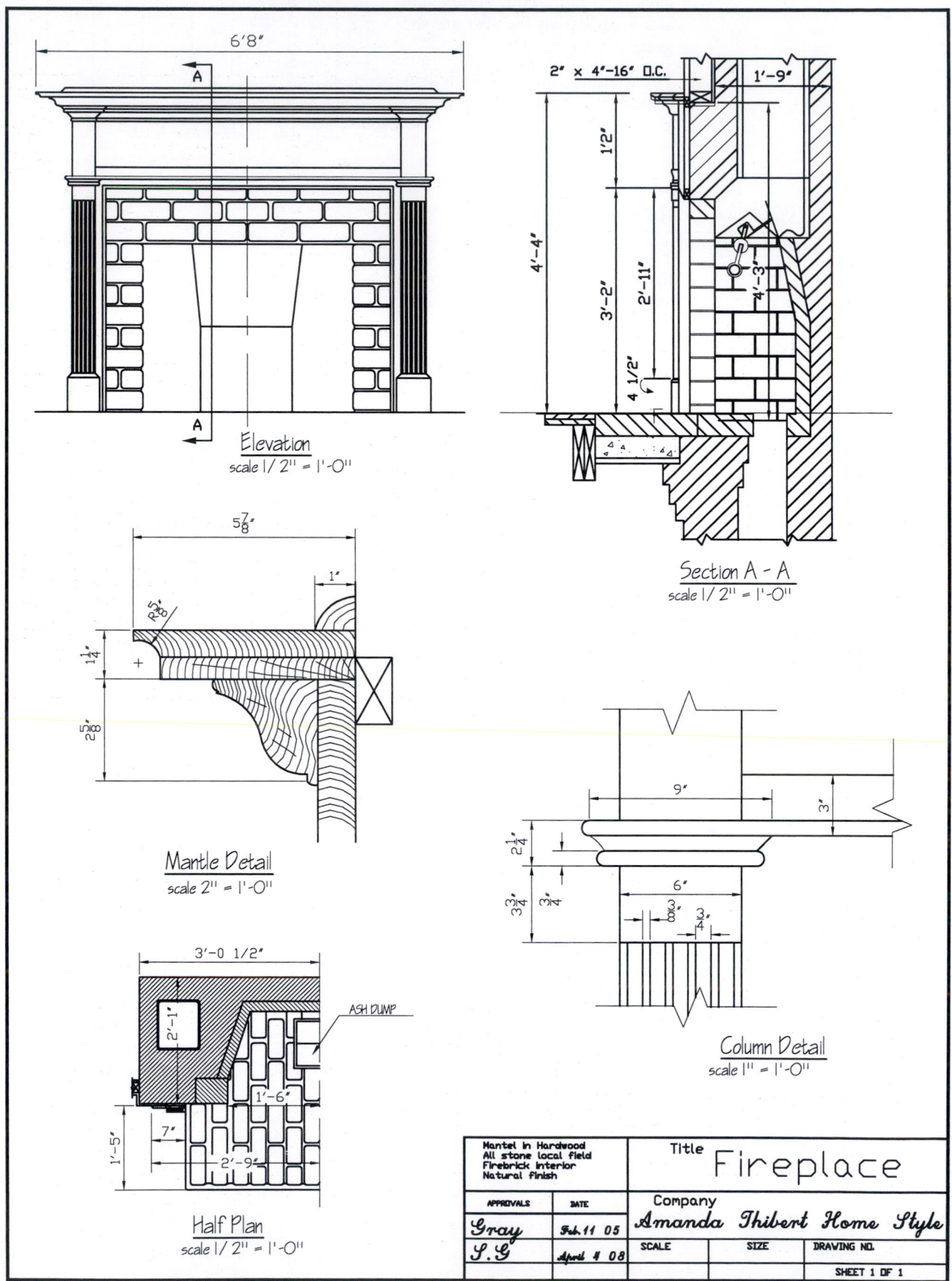

6'8"

A

Elevation
scale 1/2" = 1'-0"

2" x 4'-16' O.C.   1'-9"

1'2"

4'-4"

3'-2"

2'-11"

4'-3"

4 1/2"

Section A - A
scale 1/2" = 1'-0"

5 7/8"

1"

R 5/8"

1 1/4"

2 5/8"

Mantle Detail
scale 2" = 1'-0"

9"   3"

2 1/4"

3 3/4"   3/4"

6"

3/8"   3/4"

Column Detail
scale 1" = 1'-0"

3'-0 1/2"

2'-1"

ASH DUMP

1'-6"

1'-5"

7"

2'-9"

Half Plan
scale 1/2" = 1'-0"

Mantel in Hardwood All stone local field Firebrick interior Natural finish		Title	Fireplace
APPROVALS	DATE	Company	Amanda Thibert Home Style
Gray	Feb. 11 05		
S. G	April 4 08	SCALE	SIZE	DRAWING NO.
				SHEET 1 OF 1

Exercise 12 Architectural

Use the drawings on pages 194 and 195 to compile this commercial building in paper space. Use ISO A1 sheet size.

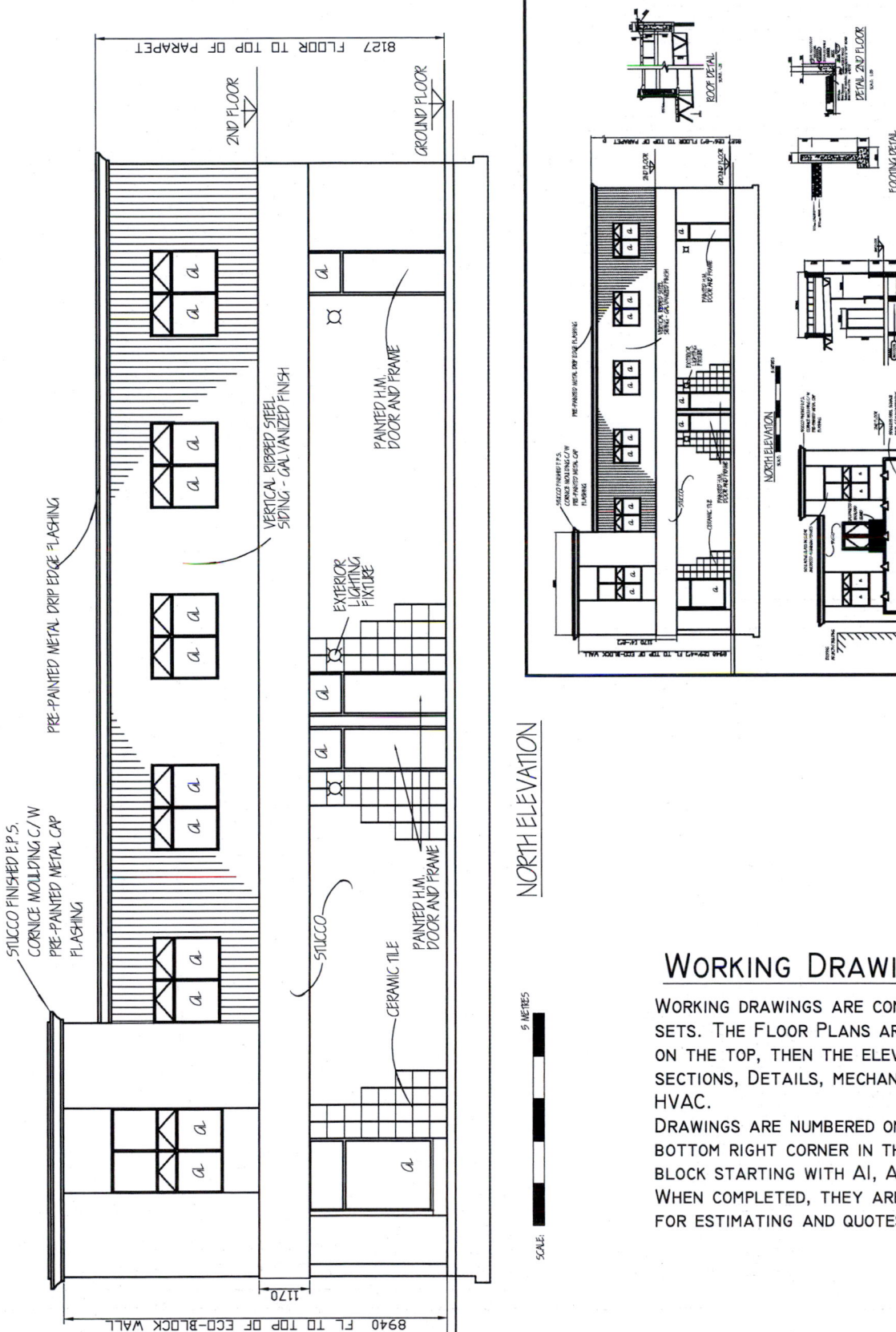

WORKING DRAWINGS

WORKING DRAWINGS ARE COMPILED IN SETS. THE FLOOR PLANS ARE USUALLY ON THE TOP, THEN THE ELEVATIONS, SECTIONS, DETAILS, MECHANICAL, AND HVAC.

DRAWINGS ARE NUMBERED ON THE BOTTOM RIGHT CORNER IN THE TITLE BLOCK STARTING WITH A1, A2, ETC. WHEN COMPLETED, THEY ARE SENT OUT FOR ESTIMATING AND QUOTES.

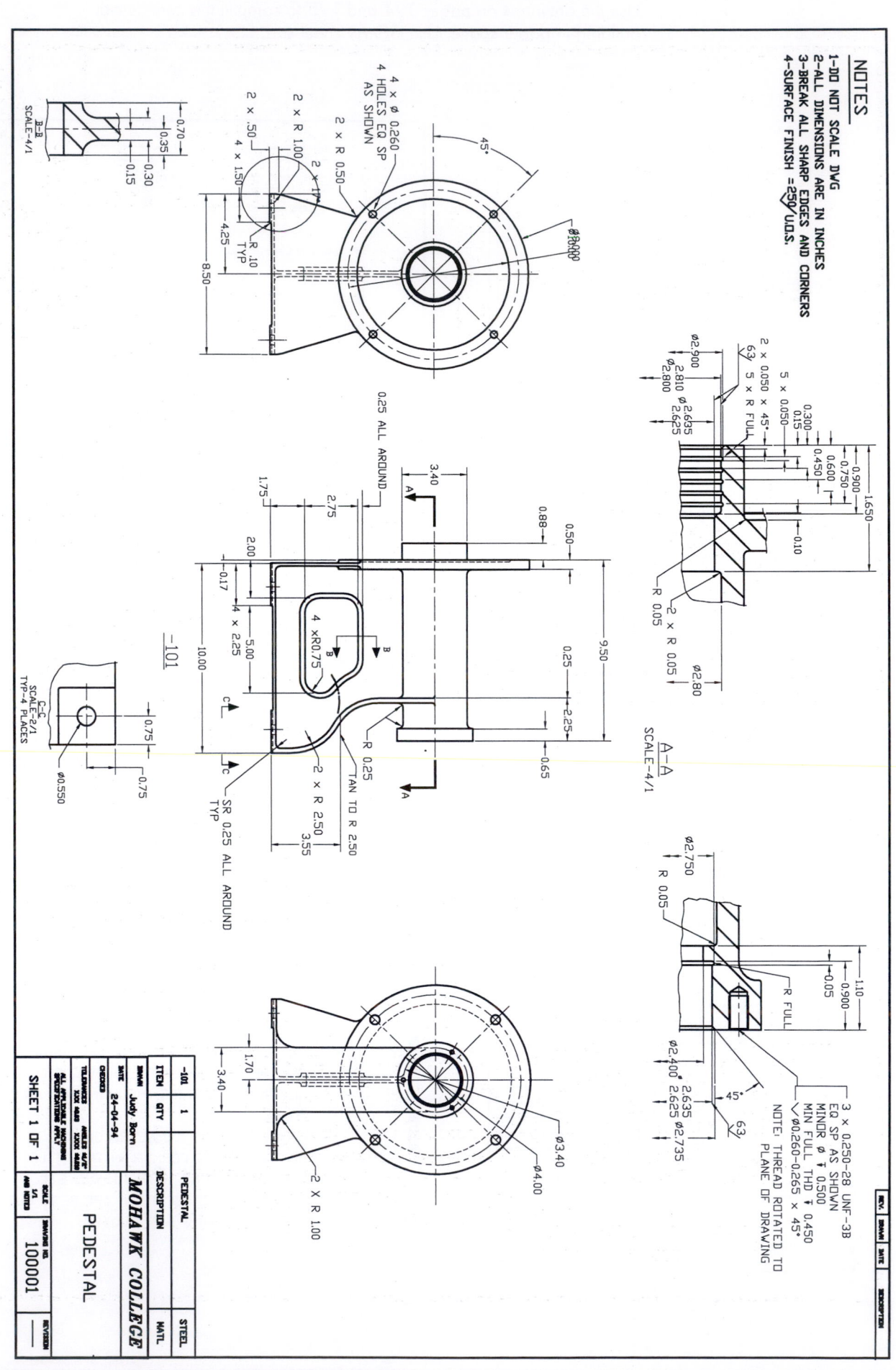

Exercise 12 Wood

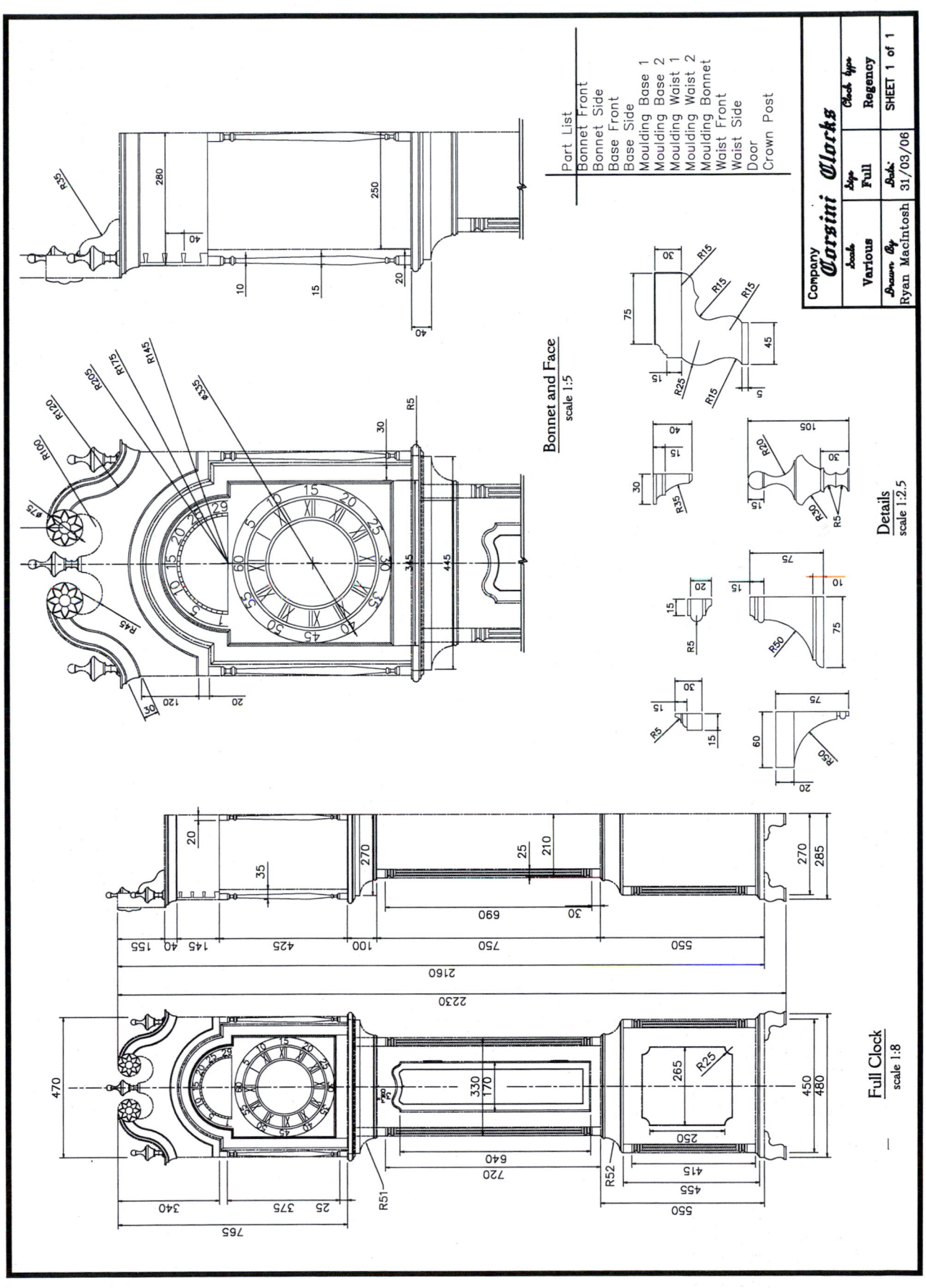

Exercise 12 Challenger

Use paper space to create a detail drawing of Insulated Concrete Forms (ICF).

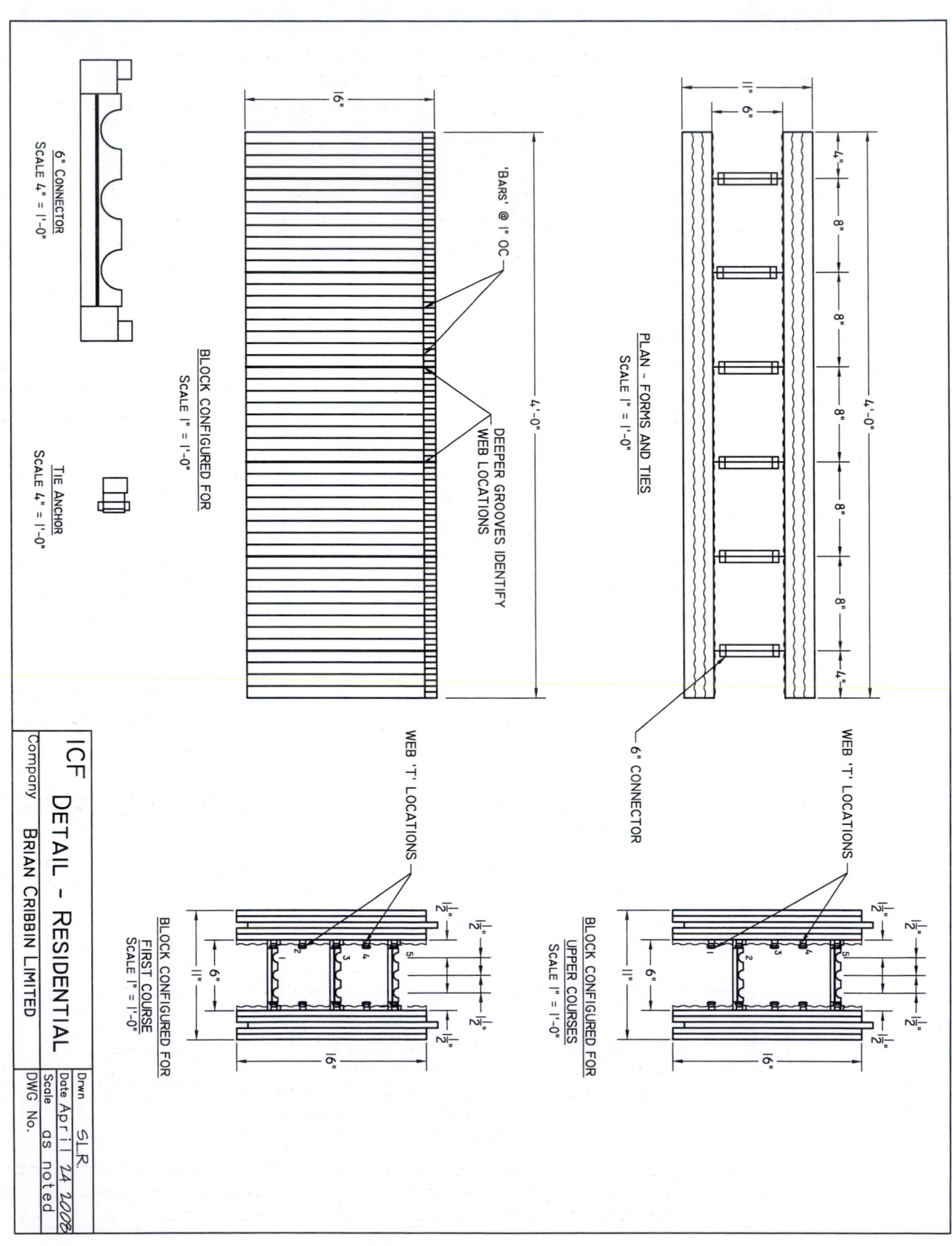

13

POINT, DIVIDE, MEASURE, and Inquiry Commands

On completion of this chapter, you should be able to:

1. Change PDMODE and PDSIZE
2. Use DIVIDE and MEASURE to place points and/or multiple objects where desired within a model
3. Use the SPLINE command
4. Check the parameters and properties of objects within the model
5. Check the overall size of the model
6. Calculate the area of a model with AREA or BOUNDARY
7. Use the SPLINE command.

Points, Point Display and Point Size Options

Points are used in spline generation, in many 3D applications, and as node or reference points which you can snap to or offset from. When you DIVIDE or MEASURE an object, points are used to show the divisions. You can set the style of the point and its size either relative to the screen or in absolute units.

> **Toolbar** From the Draw menu choose
>
> **Pull-down menu** From the Draw menu, Point, then choose single or multiple.

The command line equivalent is POINT.

The default point display is as shown on the icon, simply a one pixel point. In order to make use of points along other objects it is necessary to change the point display.

Point Style

The Point Style dialog box is found under the format menu or by typing in DDPTYPE.

To set the point size or style without using the dialog box, use PDMODE for the style and PDSIZE for the size.

In PDSIZE, a positive number will represent the actual size in drawing units of the point. A negative number is taken as a relative percentage of the screen and a difference in the zoom factor will have no effect on the size of the point display.

First choose whether you want the point to be relative to the screen or in absolute values. AutoCAD stores the point size in the PDSIZE system variable. All points will be added relative to the new size, and all existing points will be updated according to this size on the next regeneration.

Points, either created by MEASURE and DIVIDE or entered using the POINT command, can be accessed with the OSNAP option NODE. This is particularly important when entering blocks at specific points or, in 3D, when finding centers for fillets, etc.

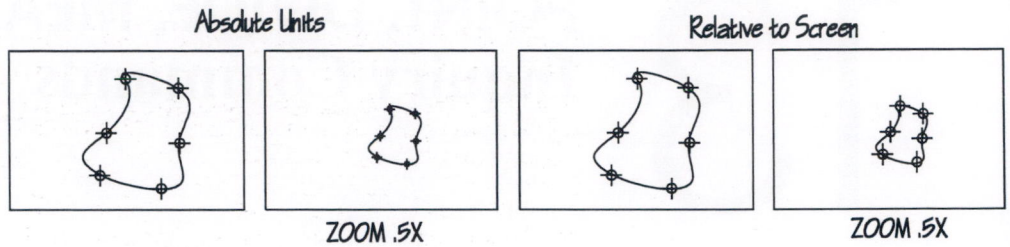

Absolute Units   Relative to Screen

ZOOM .5X   ZOOM .5X

Figure 13.1

The points will then be added in at the points determined by either the DIVIDE or the MEASURE command, as shown in Figure 13.1.

Using DIVIDE and MEASURE

If you have an object or space that needs to be cut into equal pieces or portions, you can use DIVIDE or MEASURE.

DIVIDE will visually divide any linear element– an arc, a circle, a line, or a pline – into a specified number of equal parts.

MEASURE will visually measure a linear element into segments of a specified length.

The DIVIDE Command

DIVIDE places equally spaced point objects or blocks along the length or perimeter of an object.

Toolbar There is no icon available

Pull-down menu From the Draw menu, choose Point, then Divide.

The command line equivalent is DIVIDE.

DIVIDE with Points

Using the Point Style dialog box, set the current point to one that can be easily seen, not the first one.

```
Command:DIVIDE
Select object to divide: (pick the object)
Enter the number of segments or [Block]:6
```

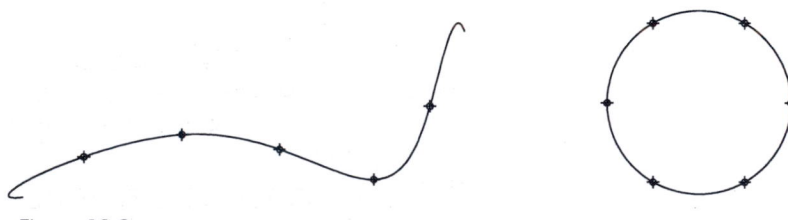

Figure 13.2

The selected object is not altered in any way, but as can be seen in Figure 13.2, there are points in the current style at regular intervals. The points become objects on the current model. To access these points with OSNAP use NODE.

This command is particularly useful for placing lines and arcs along a polyline, as can be seen in the arcs on pages 119 and 120.

DIVIDE Using a BLOCK

DIVIDE can also be used with blocks. In the following example, a mullion block is used to divide a window into equally spaced sections. Both a curved window and a block called mullion are needed.

In Figure 13.3 we have a curved window and a block called mullion, which is the shape of a mullion. The insertion point on the block is the middle of the bottom line.

The command will place this block at regular intervals along the window. The base point of the block is very important. In Figure 13.3a the mullion is placed on the lower arc and rotated, in 13.3b the mullion is placed on the upper arc without rotation.

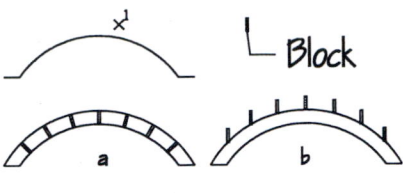

Figure 13.3

```
Command:DIVIDE
Select object to divide:(pick 1)
Enter the number of segments or [Block]:B
Enter name of block to insert:MULLION
Align block with object?[Yes/No]:(Y for a, N for b)
Enter the number of segments:8
```

The MEASURE Command

The MEASURE command is very similar to the DIVIDE command in that it divides a specified object into a series of equal portions. The difference is that the equal portions are given a specific length, and thus there may be a portion of the object left over when the command is finished. Again, the MEASURE command works on lines, arcs, circles, and plines, and again, the markers can be either points or blocks.

> **Toolbar** There is no icon available
> **Pull-down menu** From the Draw menu, choose Point, then Measure.

The command line equivalent is MEASURE.

MEASURE with Points

In the following example, use point display 4, or the vertical line in the dialog box, and change the length to 18m with PDSIZE 18. Then use MEASURE to divide a road illustrated by a pline, Figure 13.4a, into equally spaced lots, Figure 13.4b.

First draw the pline representing the road.

Type in the system variables PDMODE and PDSIZE.

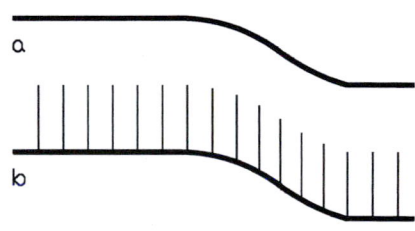

Figure 13.4

```
Command:PDMODE
Select new point mode <1>:4
Command:PDSIZE
Enter new point size< 1.000>:18
Command:MEASURE
Select object to measure:(pick (a)
   Specify length of segment or [Block]:15
```

We now have a road that is divided into equal portions of 15 m lengths with a depth of 18 m each. Notice that the road is created with pline and thus has line and arc segments.

MEASURE Using BLOCKs

In the next example, we will place a block of a toilet along an existing wall. Both a LINE representing a 19' wall and a BLOCK that represents a toilet will be needed.

Create a toilet that has an interior space of 3' × 5,' as in Figure 13.5.

Now BLOCK the toilet, making sure that the insertion base point leaves enough space for a 2″ wall on the back and on the sides.

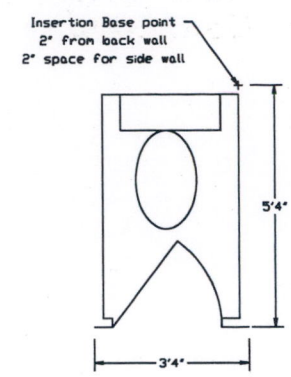

Figure 13.5

Now use MEASURE to place the toilet along the 19' long wall, as in Figure 13.6.

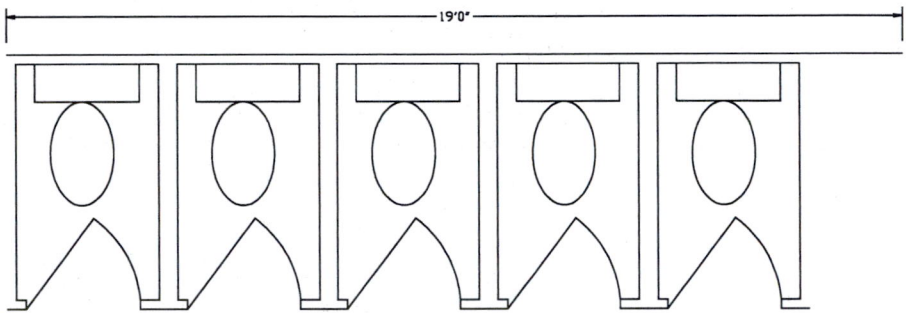

Figure 13.6

```
Command:BLOCK
Block name:TOILET
Insertion base point:corner
Select objects:(pick up all the objects)


Command:MEASURE
Select object to measure:(pick 1) (take the left side of the
   wall)
Specify length of segment or [Block]:B
Enter name of block to insert:TOILET
Align block with object? [Yes/No]:↵
Specify length of segment:1000
```

When creating the block to be used in a MEASURE command, be sure that you have no overlapping items.

If you use points in either the DIVIDE or the MEASURE commands, you will have to change the PDMODE to be able to see the displayed points. Once placed, these points become objects in the file and will be affected by editing commands such as ERASE, MOVE, COPY, etc. If you do not change the PDMODE, these points may be difficult to see.

The SPLINE Command

In Chapter 5 we looked at polylines and how to edit them into splines. There is also a command to create splines or smooth curved lines without accessing the PLINE command.

AutoCAD uses the Nonuniform Rational B-Spline (NURBS) formula to describe the splines entered. A NURBS curve produces a smooth curve between control points; this spline can be either quadratic or cubic.

If you are creating a large drawing with multiple splines for mapping or airfoil design, a drawing containing splines uses less disk space and memory than a drawing with polylines. To access the SPLINE command:

Toolbar From the Draw toolbar, choose

Pull-down menu From the Draw menu, choose Spline.

The command line equivalent is SPLINE.

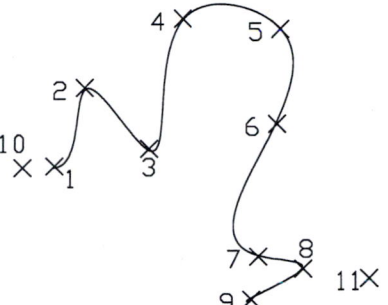

In Figure 13.7 a spline is drawn dynamically as the vertices are being entered.

```
Command: SPLINE
Specify first point or [Object]:
  (pick 1)
Specify next point: (pick 2)
Specify next point or
[Close/Fit tolerance] <start point>:
  (pick 3 through 9)
Specify start tangent: (pick 10)
Specify end tangent: (pick 11)
```

Figure 13.7

Entering Points

Enter points to add additional spline curve segments until you press ↵. Like the LINE and PLINE commands, U for Undo will remove the last entered point.

As the points are entered you can see the spline being created.

Start and End Tangency

The Enter start tangent: prompt specifies the tangency of the spline at the first point; the Enter end tangent: prompt does the same for the end point. You can specify tangency at both ends of the spline, and you can use point, TANgent, or PERpendicular object snaps to make the spline tangent or perpendicular to existing objects.

```
Specify start tangent: (pick 12)
Specify end tangent: (pick 13)
```

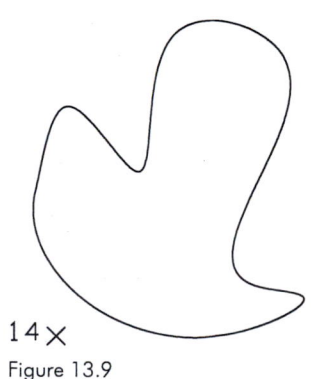

Figure 13.8

Figure 13.9

Points 12 and 13 in Figure 13.8 and 14 in Figure 13.9 are points of continuity for the spline curve.

Close

As in the PLINE command, the Close option defines the last point as coincident with the first and makes it tangent there.

Fit Tolerance

This changes the tolerance for fitting the spline through the points. The number is higher or lower depending on how you want the spline to fit through the points.

Object

This option converts either 2D or 3D polylines into splines.

The SPLINEDIT Command

The SPLINEDIT command edits the spline object.

Toolbar From the Modify menu, choose

Pull-down menu From the Modify menu, choose Object, then Spline.

The command line equivalent is SPLINEDIT.

The options for this edit command are similar to those of PEDIT.

Inquiry Commands

The Inquiry commands are used to see and list the parameters of a model. Objects are put in relative to the origin, 0,0. The size, shape and properties of the objects can be listed with the Inquiry commands.

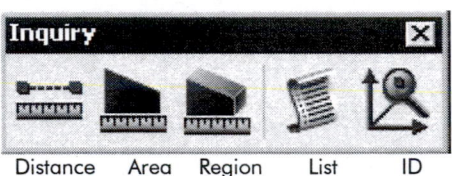

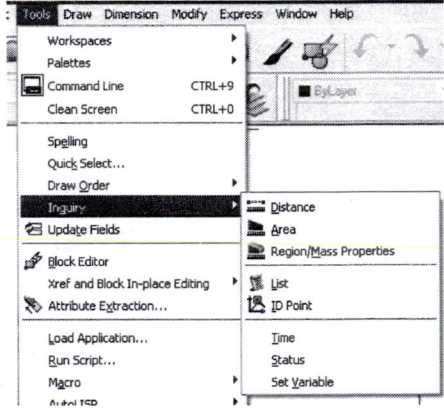

Figure 13.10

The Inquiry commands can be found on the Inquiry toolbar shown above or in the Tools pull-down menu under Inquiry, as shown in Figure 13.10. These commands can also be typed in at the Command: prompt.

AREA	computes the area of a closed polygon
DIST	computes the distance between two points
LIST	lists the position and properties of a specific object or group of objects
ID	identifies the position of a point
STATUS	displays a listing of all the statistics of a file along with other information

The **AREA** Command

AutoCAD offers built-in area computational abilities which also display the perimeter of the object calculated. This can be extremely useful for calculations of plines and irregular shapes.

AREA *Using an Entity*

In Figure 13.11 is a pline that has been fitted with a spline curve, and the area and perimeter are to be calculated. In this case an entity is chosen for the area option.

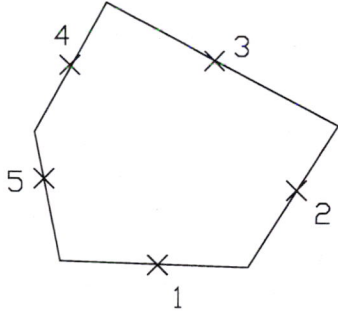

Figure 13.11

```
Command:AREA
Specify first corner point or [Object/Add/Subtract]:O
Select objects:(pick 1)
Area = 16.1434 Perimeter = 19.6824
```

AREA *Using Lines*

When calculating a series of straight lines, the default is to have the shape entered by a series of points as shown in Figure 13.12. OSNAPs are needed for accuracy.

```
Command:AREA
Specify first corner point or [Object/Add/Subtract]:
 END of (pick 1)
Specify next point or press ENTER for total:END of (pick 2)
Specify next point or press ENTER for total:END of (pick 3)
Specify next point or press ENTER for total:END of (pick 4)
Specify next point or press ENTER for total:END of (pick 5)
Specify next point or press ENTER for total:↵
Area = 67.5000 Perimeter = 31.1131
```

Figure 13.12

Subtract and Add with **AREA**

To calculate the net floor area of a bathroom, a drawing of a bathroom will be needed. Draw in a bathroom 3 m × 2 m, as shown in Figure 13.13.

Set OSNAP to ENDpoint and calculate the total floor area.

```
Command:OSNAP
Object snap modes:ENDpoint
```

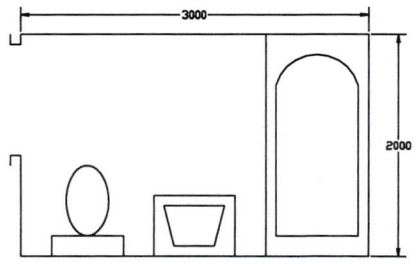

Figure 13.13

LIST can also be used to show the number of objects within a specified window to check that the objects follow good CAD practice. In the following example, we will see how LIST can also be used to tell us if the dimension for an object is real or made up.

In Figure 13.18 there are two associative dimensions listed. To the eye, they look the same. They are identified by being highlighted in the DOT linetype.

Note the difference in the readouts after the LIST command has been used.

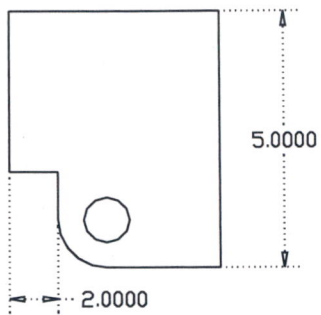

Figure 13.18

Notes

The Properties command has much of the LIST information in a dialog format.

```
Command: LIST
Select objects: (pick the dimension)
                    DIMENSION Layer: dim
                    SPACE: Model Space
                    Handle = 1D14
type: horizontal
1st extension defining point:  X= 2.0000 Y= 4.5000 Z= 0.0000
2nd extension defining point:  X= 4.0000 Y= 3.5000 Z= 0.0000
dimension line defining point: X= 4.0000 Y= 2.5000 Z= 0.0000
default text position X= 3.875 Y= 2.5000 Z= 0.0000
default text
dimension style *UNNAMED

                    DIMENSION Layer: dim
                    SPACE: Model Space
                    Handle = 1D14
type: vertical
1st extension defining point:  X= 6.0000 Y= 8.5000 Z= 0.0000
2nd extension defining point:  X= 6.0000 Y= 3.0000 Z= 0.0000
dimension line defining point: X= 6.5000 Y= 2.5000 Z= 0.0000
default text position X= 6.500 Y= 5.7500 Z= 0.0000
dimension text modifier: 5.0000
dimension style *UNNAMED
```

You can see that the first dimension has been entered properly, and the text for the dimension is the default text. In the second dimension, however, the dimension text has been modified to read 5.0000. If you subtract the second extension defining point from the first in the Y value, you will notice that the actual distance should read 5.5000, but the dimension text has been altered before it was entered.

In addition to checking your own work to be sure that it is all entered correctly and that there are no overlapping items, with the LIST command you can check to see that the dimensions associated with the drawing you have on file are correct and not altered in any way. If the dimensions have been altered, check to see that *all* of the necessary dimensions have been altered.

Another great advantage of LIST is that it enables you to determine if there are overlapping lines when making hatches and dimensions. Use LIST, then Crossing, to see the number of objects overlapping.

The ID Command

The ID function is similar to the LIST function, in that it gives an exact position. The difference is that it gives an exact position of a point rather than an object. This command is used in 3D modeling more frequently than in 2D modeling, to determine the Z depth of items. In 2D this can be useful for finding out the exact position you are looking at on a zoomed screen, or for simply getting your bearings. Keep in mind that if you are picking a point in space, a snap can give a more usable readout.

The location of the point will be relative to the origin or 0,0,0 of the model. If a point in space is chosen, the Z depth will be the current elevation; if an object is chosen with an OSNAP, the actual Z depth of the object will be used.

ID can be used as a reference point for the next point entered using $@$:

```
Command: LINE
From point: @2,0 (starts the line 2 units in X from the ID
    point)
```

There are basically two ways of using ID. The first is to find the parameters of a point on the screen.

```
Command: ID
Specify point: (pick a point)
X = 34.6375    Y = 24.8758    Z = 0.0000
```

This will give the coordinates of a point in space in the defined units. The second is to locate a point by typing in the coordinates.

```
Command: ID
Specify point: 23,4,0
```

The STATUS Command

Another useful command for determining what is happening is STATUS. As you become proficient with AutoCAD, you will find the STATUS command more and more useful, because it displays a listing of all the statistics of a file.

The STATUS command also offers information on memory and the partition on the hard drive where your temporary file or .AC$ file is being stored. If you run out of space in this partition (in a classroom this is often the A: drive), the program will terminate after first saving your file.

STATUS is most often used by beginners to determine if the color setting is overriding the layer color setting.

Enter STATUS at the command prompt and read the status of your file.

Step 1 Set up three layers, for Points, Object, and Dimensions. Make Points current. Then add points.

Make the top center or third point active, and change Size Relative to Screen to 5%.

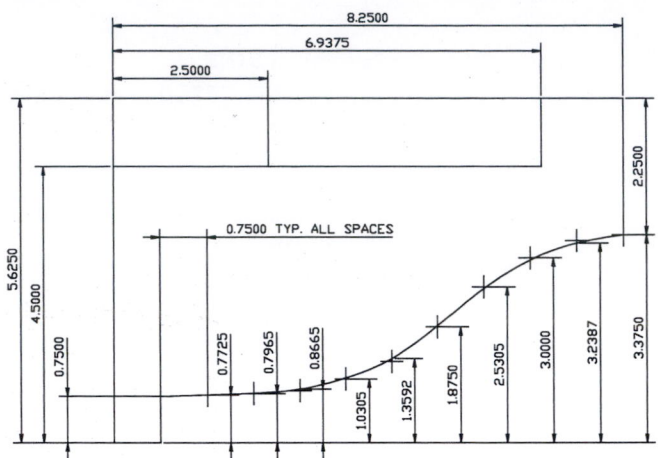

Step 2 Add the points. The lower left corner will be 0,0.

```
Command:POINT
Specify point:.75,.75
Command:POINT
Specify point:1.5,.7725
Command:POINT
Specify point:2.25,.7965
Command:POINT
Specify point:3.0,.8665
Command:POINT
Specify point:3.75,1.0305 etc.
```

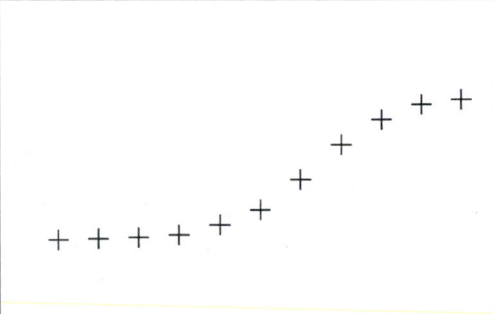

Step 3 Change OSNAP to NODE and use SPLINE to create a spline through all of the points.

```
Command:OSNAP
Object snap modes:NODE
```

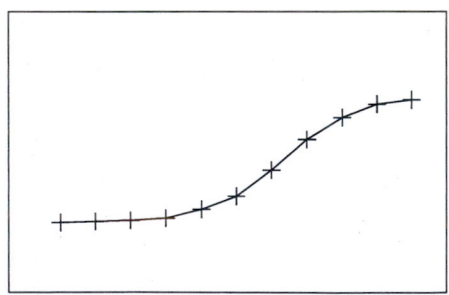

Now use SPLINE to create a spline through the identified points.

```
Command:SPLINE
Specify    first    point    or
   [Object]:(pick the first point)
Specify next point:(pick the next point)
Specify next point or [Close, Fit Tolerance]:(pick the points
   in sequence)
Specify next point or [Close, Fit Tolerance]:(pick the last
   point)
Specify next point or [Close, Fit Tolerance]:↵↵
Specify start tangent:(pick the first point)
Specify end tangent:(pick the last point)
```

Step 4   Using the drawing given at the beginning, add the outside lines.

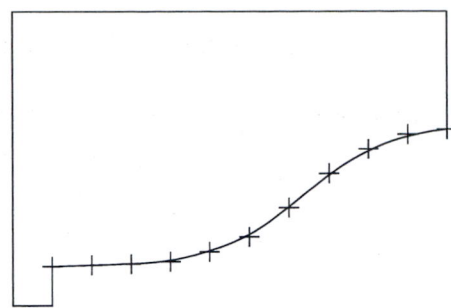

Now use the Boundary option to help determine the area of the top of the part.

Step 5

From the Boundary Creation dialog box, choose Make New Boundary Set.

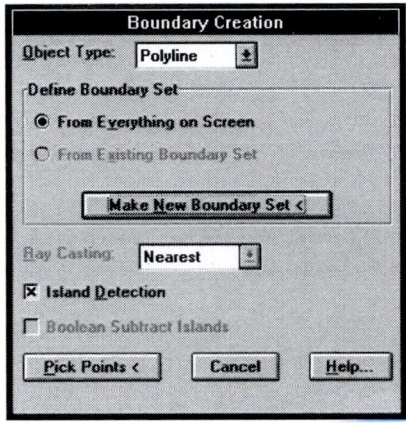

Now pick the lines and the spline.

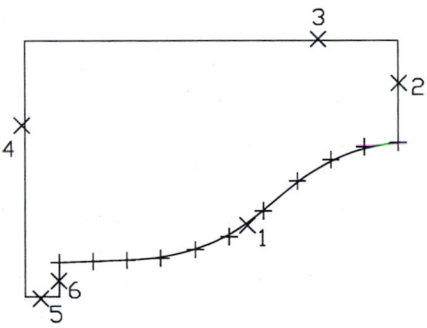

Press ↵ when you are finished and you will return to the Boundary Creation dialog box.

From the dialog box, select Pick Points. You will be prompted to pick the internal point; choose the inside of the part.

A region will be created and you will have the area for the part.

If your system doesn't allow this, enter the AREA command, use the Object option, and L for last object.

Step 6   Now list the properties of the spline and one of the point nodes.

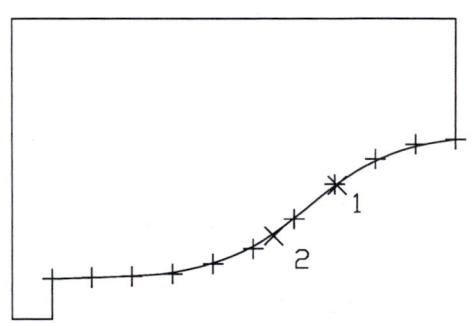

You will get a listing of the properties.

Step 7

Now add some points along the top edge of the part that corresponds to the *X* measurement of all of the points. This is to double-check that all of your points are entered correctly. Use DDPTYPE to change the point style to the vertical line as shown.

```
Command:MEASURE
Select object to
  measure:(pick 1)
Specify length of
  segment or [Block]:.75
```

Use Undo to remove the points once they have been used to double-check the lower points.

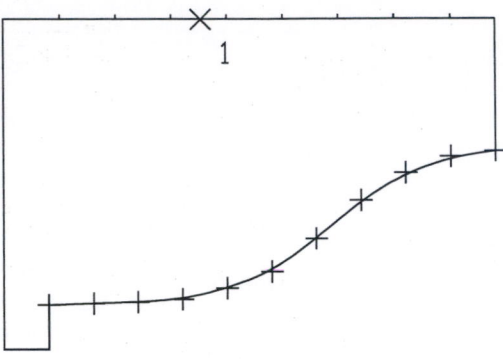

Step 8

Now add the remaining lines and dimensions to this view.

Step 9

Pan the screen over and, using the same method, draw in the front view of the gusset.

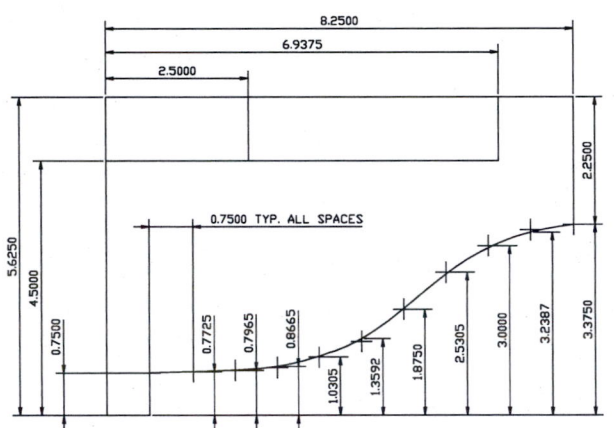

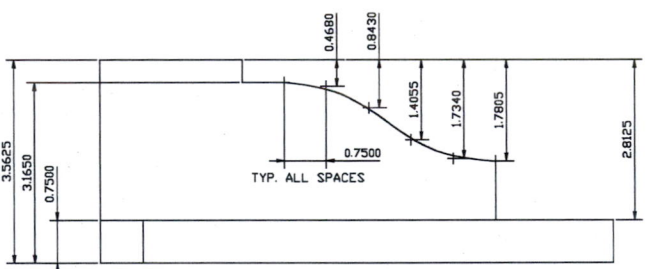

The points will be much easier to enter if you make 0,0 the upper left corner. Move the first view up and out of the way.

When you are done, add a title block and notations to complete the drawing.

Type in the word TIME to see how long it took you to do the drawing.

DIVIDE and MEASURE are used most effectively on PLINEs. In this example we will draw a pseudo four-centered arch.

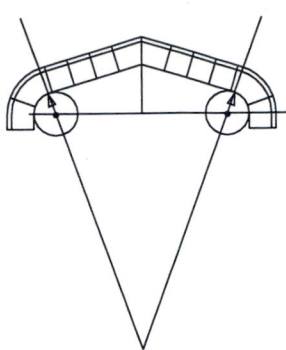

Step 1

First we must determine the span of the arch. Start by setting up some construction lines with the span and the rise of the arch.

We will make the span 5′ or 60 ″, and the rise 14 ″.

Draw in two construction lines as shown.

Then put in two circles, each one centered 6″ from the end of the span line, having a radius of 6″.

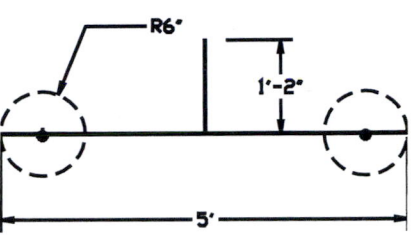

Step 2

Draw lines tangent to the circles terminating at the top of the rise. Trim the circles.

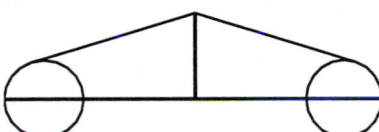
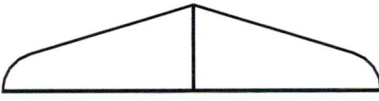

Step 3

Use PEDIT with the Join option to make two plines, one on either side of the rise.

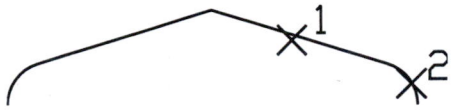

```
Command:PEDIT
Select object:(pick 1)
Object selected is not a polyline
Do you want to turn it into one? <Y>:↵
Enter an option [Close/Join/Width...rve/Ltypegen/Undo]: j
Select objects: (pick 2)
Select objects:↵
1 segments added to polyline
```

Step 4   Use DIVIDE to divide the plines into six equal sections for the stone pieces. First set PDMODE. Under the Format menu, pick Point Style.

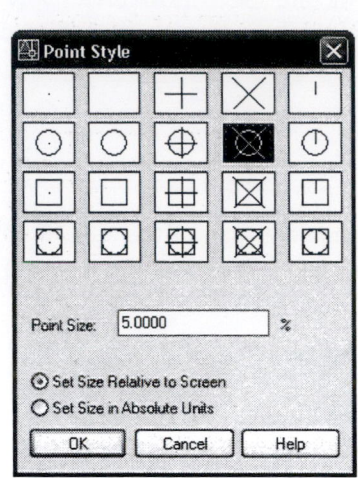

```
Command:DIVIDE
Select object to divide:(pick 1)
Enter the number of segments or
    [Block]:6
```

Your PLINE should have six points on it. Do the same with the other side.

Step 5   Now OFFSET the PLINEs first by 6″, then by 7″.

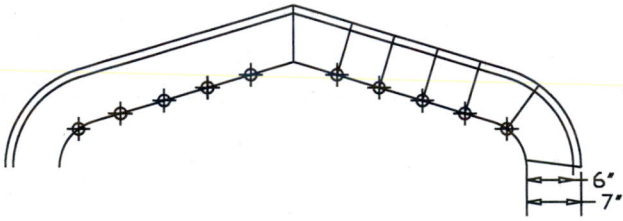

Step 6   Use FILLET with a radius of 0 to close the top off. Then use LINE with the OSNAP NODE and PERpendicular to make the lines for the stone.

The hidden lines and arrows show the method of calculation.

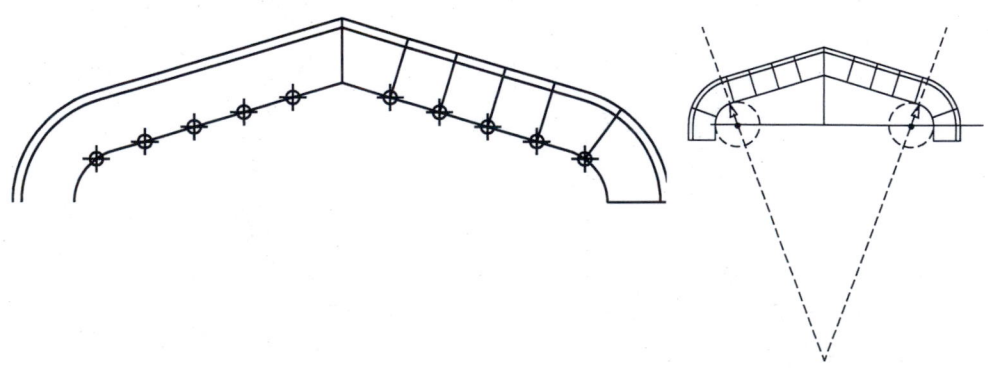

Exercise 13 Practice

Create the simple floor plan shown below.
Use DIVIDE to divide the glass wall and gliding doors into equal portions.
Use MEASURE with BLOCK to create the decorative brick wall.

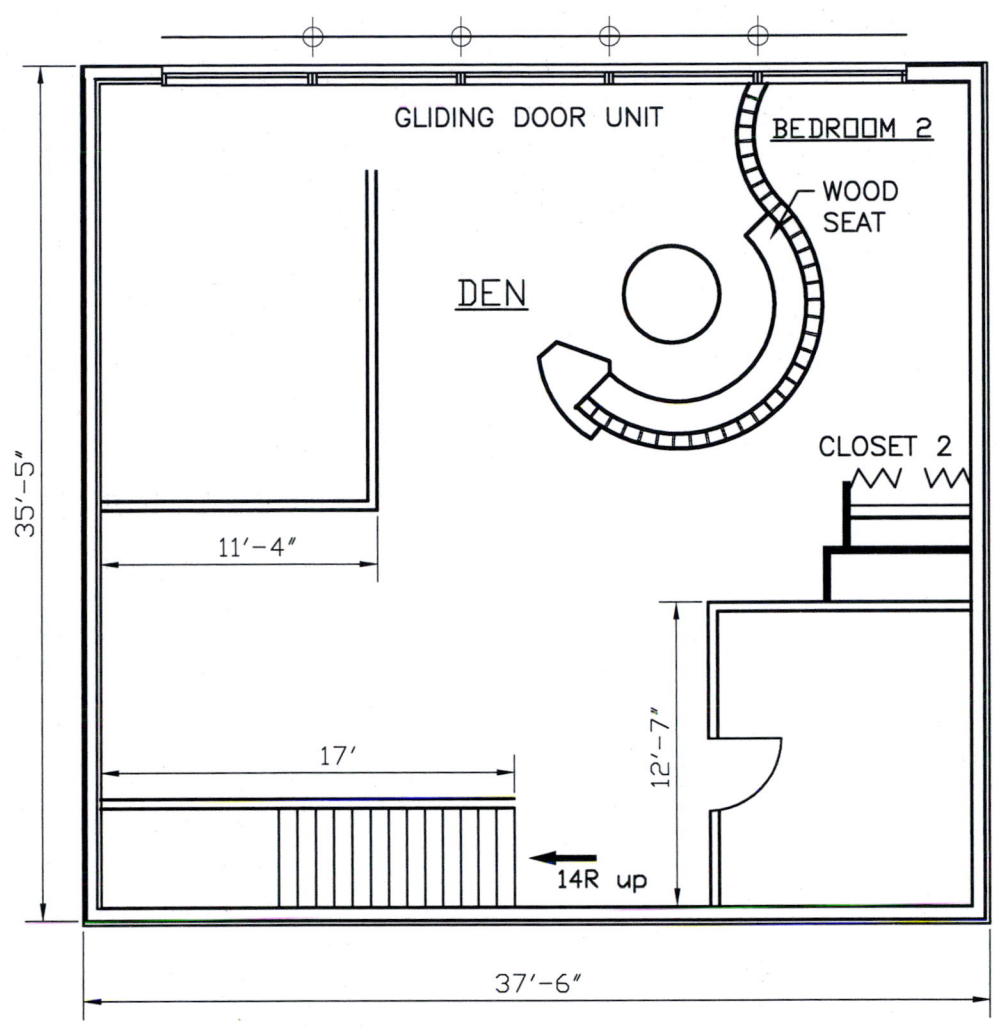

GLIDING DOOR UNIT

BEDROOM 2

WOOD
SEAT

DEN

CLOSET 2

35'-5"

11'-4"

17'

12'-7"

14R up

37'-6"

FLOOR PLAN

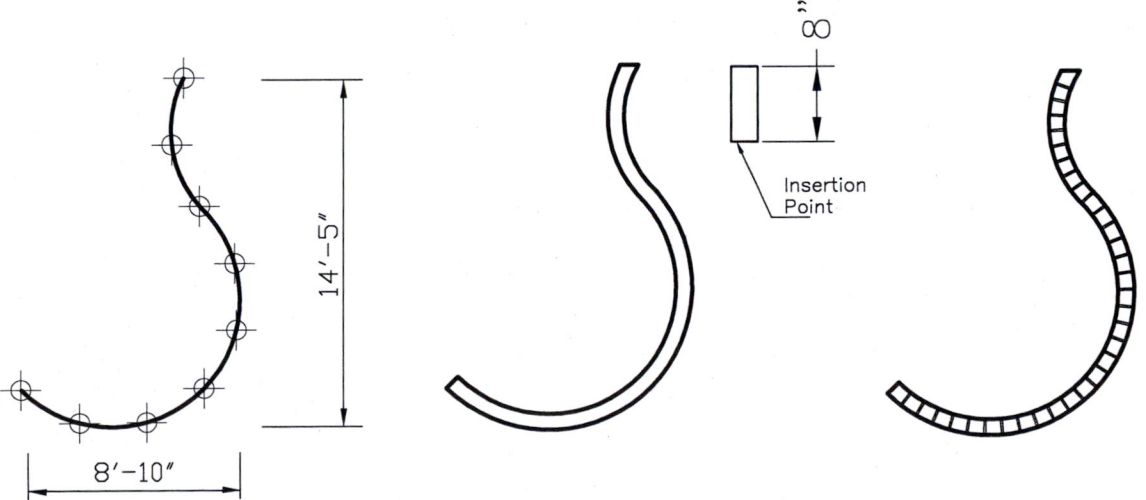

14'-5"

8'-10"

8"

Insertion
Point

DECORATIVE WALL

Exercise 13 Architectural

This steel roof framing plan is very difficult without DIVIDE.
To place the 'marks' make a dimension style without extension or dimension lines.

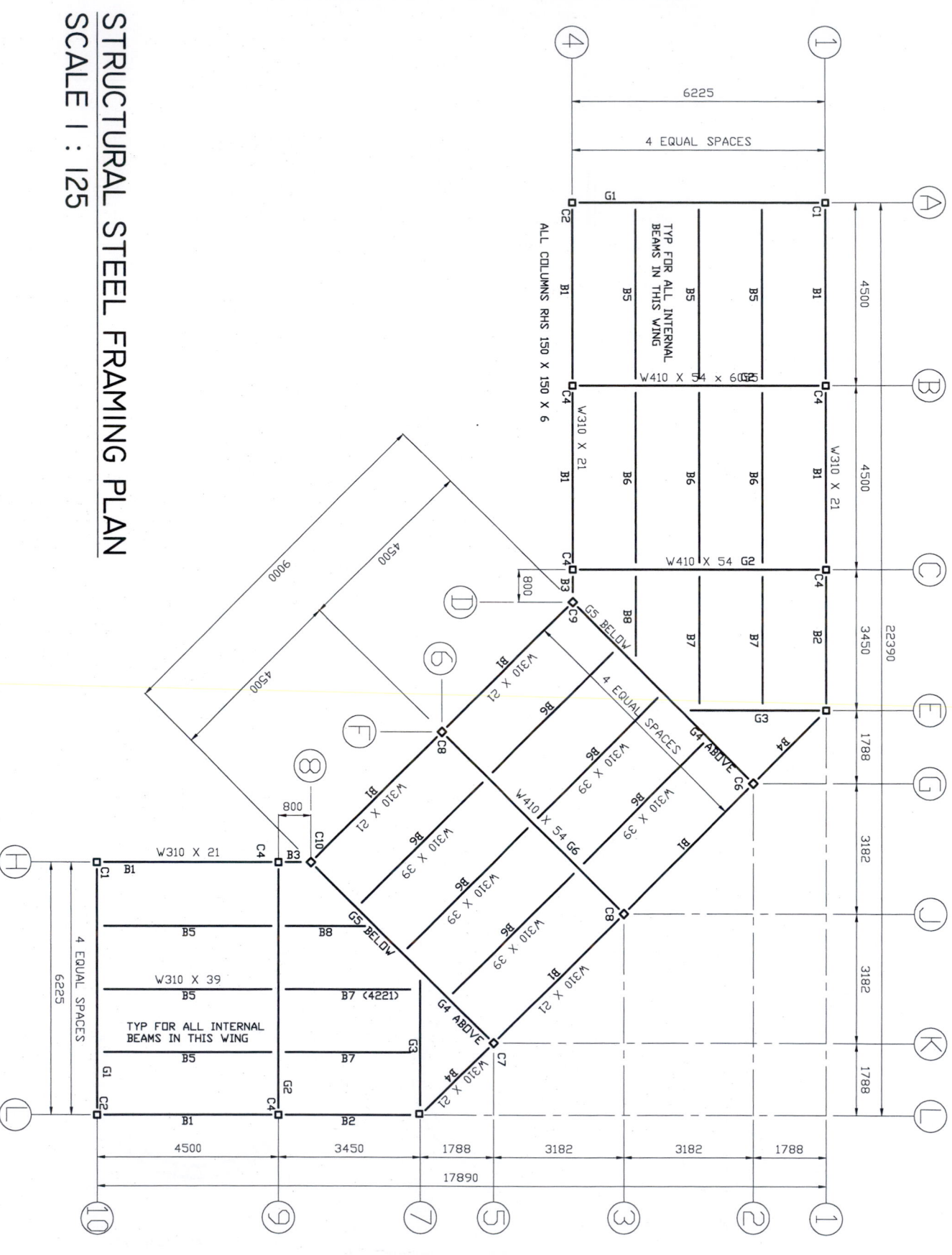

STRUCTURAL STEEL FRAMING PLAN
SCALE 1 : 125

Exercise 13 Mechanical

Use DIVIDE and MEASURE to produce 13A.
Use AREA to find the various surface area of 13B.
Use PDSIZE to display the nodes at the correct distance.

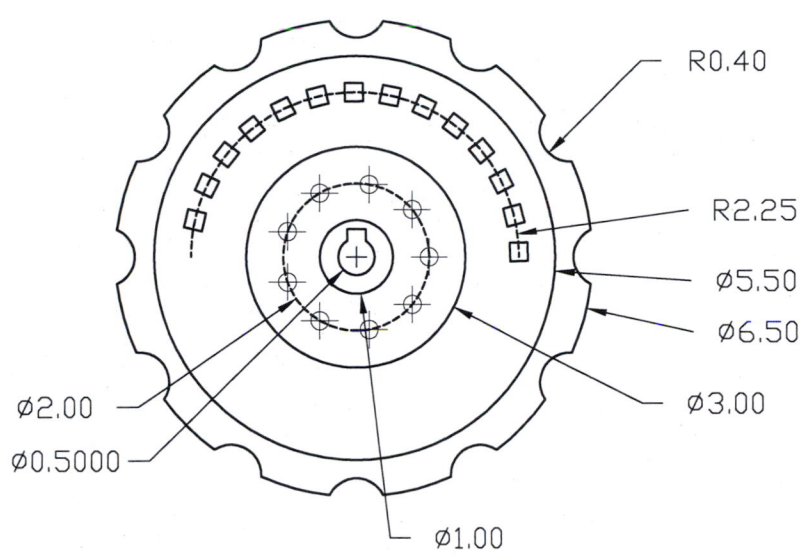

R0.40

R2.25

Ø5.50

Ø6.50

Ø3.00

Ø2.00

Ø0.5000

Ø1.00

13A

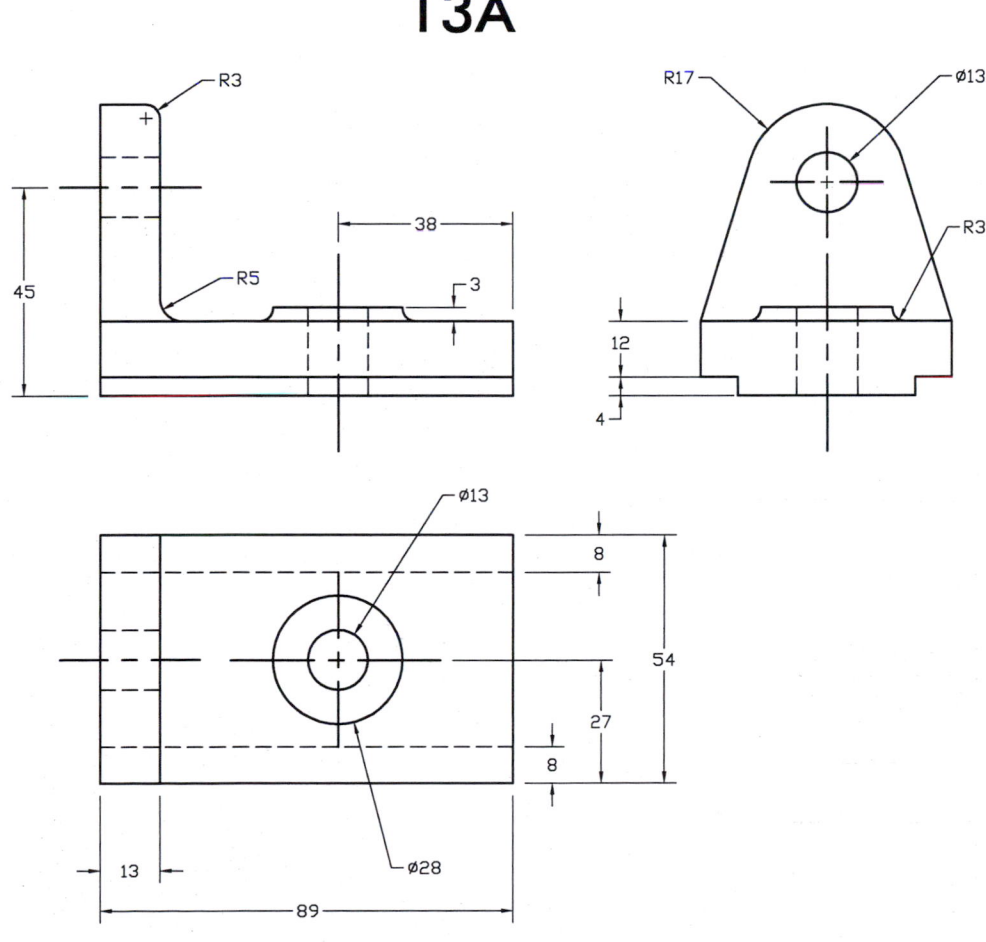

13B

Exercise 13 Challenger

Find the area of the exterior surface.

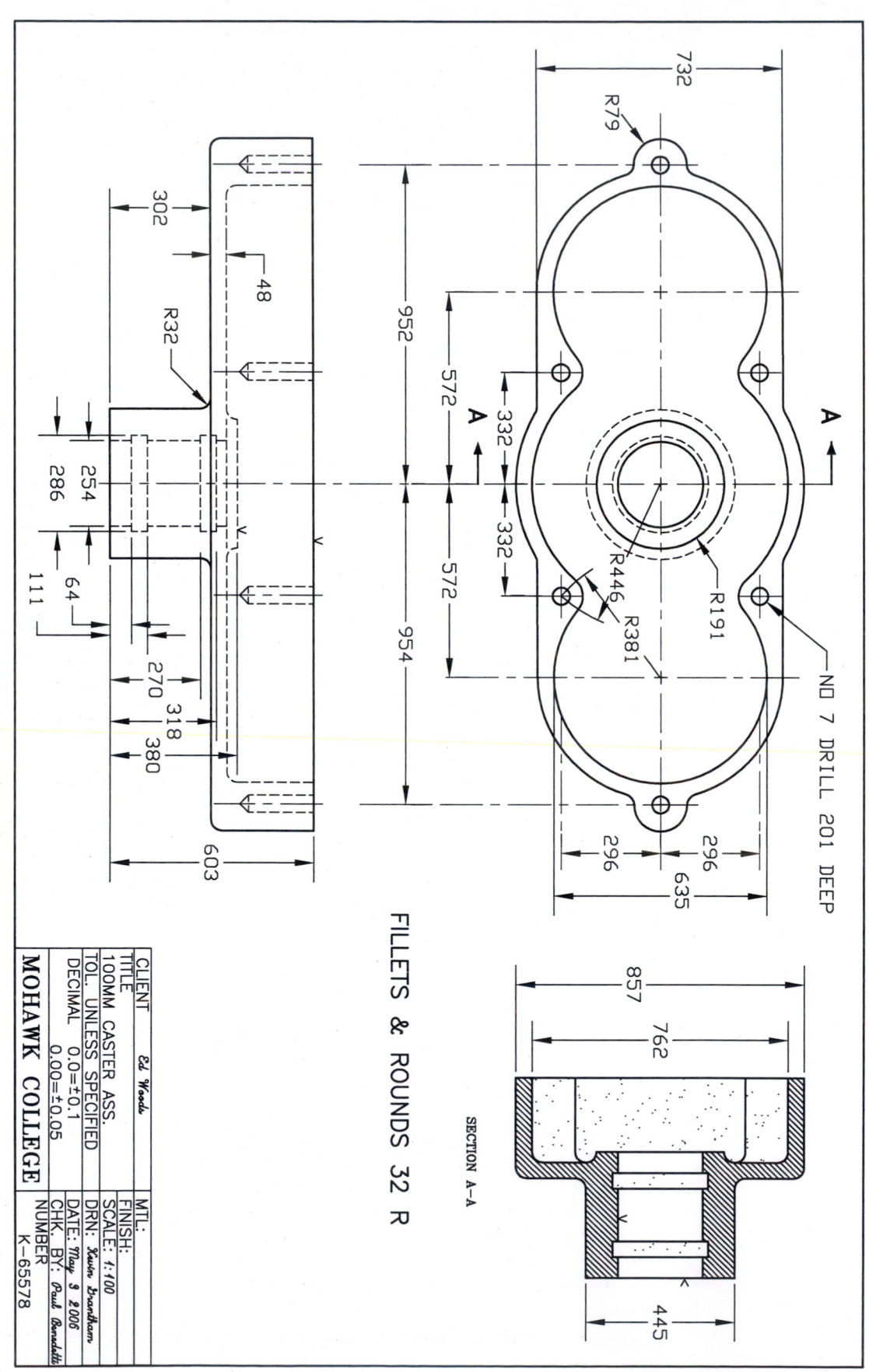

14 Attributes

On completion of this chapter, you should be able to:

1. Define a series of attributes
2. Block the attributes
3. Insert the attributes onto a drawing
4. Use the ATTribute DIAlog boxes
5. Create an attribute table.

Introduction

In addition to creating geometry and drawings, AutoCAD allows for the generation of non-graphic information which can be accessed in the form of bills of materials, schedules, parts lists, and other data that is cross-referenced on the drawing. This non-graphic intelligence is called an *attribute*. Attributes provide a label or tag that lets you attach text or other data to a block. This attribute information can then be downloaded onto your file in a text format.

Attributes can also be used to generate templates for fill-in-the-blanks situations such as on drawing notations and title blocks. This information can then be used by your system manager to track files and drawings.

Attributes for Title Blocks and Notations

Attributes are commonly used to fill in relevant information on title blocks and other drawing notations. All of the sizes for the inserted text are determined at the time the attribute is defined and blocked, as are the placement of the text and the lettering font. All the user needs to do is add the missing information for the customer, name of part, date, name of designer, etc.

In the title block in Figure 14.1, all the information relative to the current drawing (in italics) has been added using attributes.

Figure 14.1

```
Command:-INSERT
Enter block name or [?]:TITLE
Specify insertion point or [Scale
  /X/Y/Z/Rotate/PScale/PX/PY/PZ/P
  Rotate]:0,0
Enter X scale factor,specify opposite corner,or
  [Corner/XYZ]:↵
Enter Y scale factor <use X scale factor>:↵
Specify rotation angle <0>:↵ (the same as block inserts)
Customer name <Skylab Industries>:↵
Name of Part<>:Hook Link
Enter the current date (Month/Day/Year):March 17 2006
Drawn by <B. Challis>:↵
Scale <1:5>:↵
Drawing no. DFC-####<>:DFC-1003
```

Remember that the angle brackets contain defaults, e.g. <Skylab Industries>.

Defining the Attributes

To define attributes, use the ATTDEF command. In a title block the text font for the block titles (Name, Date, Drawn By) should be different from the text font for the current information. Before creating the attribute definitions, load the new text font.

To define the attribute use -ATTDEF or the Attribute dialog box.

The *ATTDEF* Command

The ATTDEF command allows you to define an attribute. We will create the attribute definition for the date. Access the ATTDEF command through the command line.

> **Pull-down menu** Under the Draw menu choose Block then Define Attributes.

The command line equivalent is ATTDEF for the dialog box and -ATTDEF for the command string.

Figure 14.2 illustrates a corner of a title block. The word Scale is added in text to identify the information within that box. The tag Scale will then be added with the ATTDEF command. See the Attribute Definition dialog box on the next page as well.

```
Command:-ATTDEF
Current attribute modes -- Invisible-N, Constant-N,
  Verify-N, Preset-N
Enter an option to change [Invisible/Constant/Verify/Preset]
  <done>:⏎
```

Change the modes if you want the attribute to be constant, invisible, preset, or verified. In this case no change is necessary.

The *attribute tag* is a one-word summary of the subject of the attribute. In this case, you want the user to enter the scale of the drawing.

```
Enter attribute tag name:Scale
```

The attribute prompt is what is actually going to appear on the command line. This prompt asks you for the required information, in this case the scale, and it should offer any further information that may be required in order to enter that information.

```
Enter attribute prompt:Enter scale of drawing
```

The default attribute value is what the user will usually want to use. If most of your drawings are done with a scale of 1:50, that is the scale you should use as a default.

```
Enter default attribute value:1:50
```

Now that the you have identified what your attribute is going to say, you need to place it on the title block. ATTDEF works like the TEXT command.

```
Specify start point of text or [Justify/Style]:J

Align/Center/Fit/Middle/Right/TL/TC/TR/ML/MC/MR/BL/BC/BR:R
Right: (pick the right to have your text right justified)
Specify height <0.1800>:.25
Specify rotation angle <0>:⏎
```

Notes

Use -INSERT and -ATTDEF to avoid the dialog boxes.

Text in the Standard font was used to define the text for the title block. The tag for the scale attribute is in the Italic Complex Font. This is where the current scale will appear. The tag shows up as shown in Figure 14.2 until the attributes are blocked.

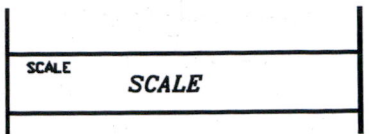

Figure 14.2

Using the Define ATTribute DIAlog Box

In the Attribute Definition dialog box you can set the modes, tag information, prompt, location, and text style as described above. When Insertion Point is chosen, the dialog box will disappear until the point is chosen, then reappear once the selection is made.

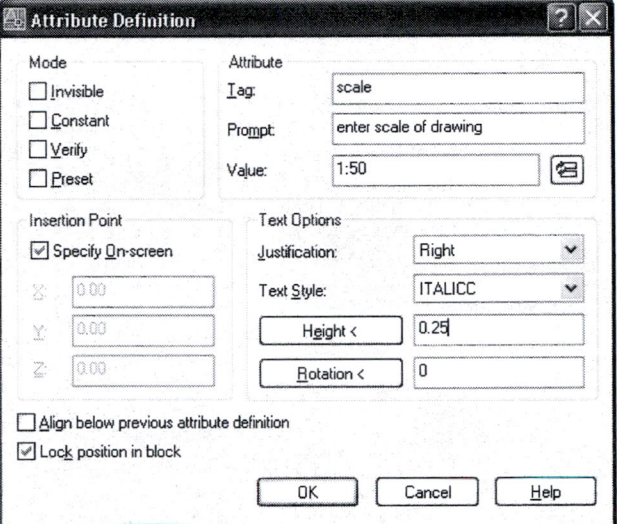

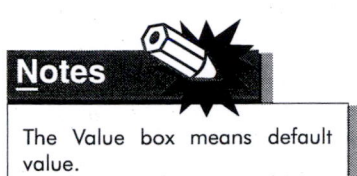

When using the dialog boxes you need to be sure to enter a point for the reference point (Specify On-screen). If not, your attribute could end up anywhere.

Once the attribute definitions are all entered, the attributes can be blocked and then inserted.

Editing Attribute Definitions

If you have made an error in defining the attribute you can change the attribute definition with the CHANGE command or with the Edit Attribute Definition dialog box.

> **Pull-down menu** From the Modify menu, choose Object, then Text.

The command line equivalents are ED, DDEDIT or CHANGE.

DDEDIT and ED give you a dialog box where you can change the tag, prompt, and default values, etc. The CHANGE command does the same without the dialog box.

```
Command:CHANGE
Select objects:(pick the definition you wish to change)
Select objects:↵
Command:DDEDIT
<Select a TEXT or ATTDEF object>/Undo:(pick the ATTDEF)
```

The Edit Attribute Definition dialog box will offer you three input boxes to change the tag, prompt, or default value. Note how the default value is here called the Default.

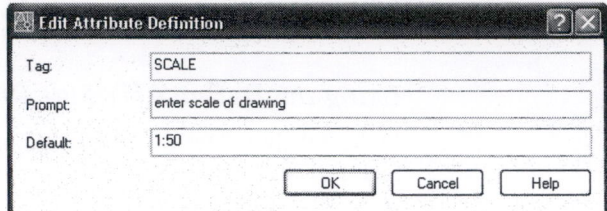

Press OK to complete the change.

BLOCKing the Attributes

Once the attributes are all defined, you must BLOCK the information. Pick the attribute definitions in the order that you want your prompts to appear by picking each attribute in order. If you use Window, you may be prompted in the reverse order.

Inserting the Attributed Blocks

Having created the attributed block, now use INSERT to place it on your drawing. Once inserted, the block will contain the values for each attribute in that particular instance or application.

If you use -INSERT you will be prompted for the information on the command line. If you use INSERT, you will get the Insert dialog box. You can also use the ATTDIA dialog box for inserting attributes. If this dialog box does not appear, toggle ATTDIA on.

```
Command:ATTDIA
Enter new value for ATTDIA<0>:1
```

Changing Attribute Definitions

While inserting the block, you may notice some errors in the block definitions. To change the attribute definitions at this point, use EXPLODE and then CHANGE or DDEDIT. When you EXPLODE the attributed block, it will revert back to the tag format.

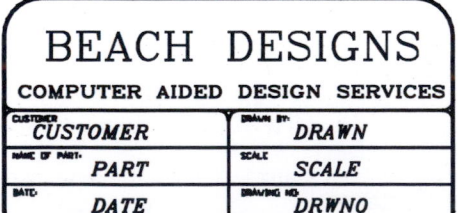

Attribute definition with tags

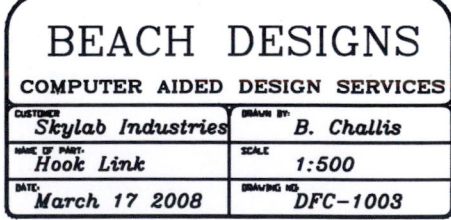

Attributed block instance

Figure 14.3

In Figure 14.3 the graphics with attribute tags before they are all blocked are shown on the left, and the inserted block with the instance values on the right. If you want to change any of the default values once they have been entered, use EXPLODE then CHANGE or DDEDIT.

Displaying Attributes

When attributes are added to objects, they can be made universally invisible by setting the Invisible mode in the ATTDEF command, or you can make them invisible later using the ATTDISP command.

The *ATTDISP* Command

ATTDISP controls the display of attributes. It is only available on the command line.

```
Command:ATTDISP
Enter attribute visibility setting
  [Normal/ON/OFF]<Normal>:OFF
```

Where: ON = all attributes visible

    OFF = all attributes invisible

   Normal = normal visibility, set individually

Attribute Modes

If you want to INSERT the attributed blocks without having to turn the display off, you can set the attribute mode to invisible. You can set these as follows to make attributed block insertion easier:

INVISIBLE   makes all attributes invisible

CONSTANT   makes an attribute uneditable; for example, the president's desk is always the president's desk even though the president may change

VERIFY   allows you to take a final look at what you have entered prior to having it added to the drawing

PRESET   is used when creating attributes that will always have the same value

Creating Attributes for Data Extraction

Attributes can be used both for title blocks and drawing notations, and for occasions where the information can be downloaded to a price list or bill of materials. The attributes are all defined in the same way, and the extractions take place once the attributed blocks are all inserted.

When attributes are used for drawing notations they are almost always visible on the drawing. When they are used for bill of materials, parts list or schedules, the attributes are usually invisible on the drawing but shown on the Attribute Extract Table, as explained on page 267.

Attributes are always associated with blocks. If you want an attributed block, first create the geometry for the final block, if there is any, then add the attributes. You will need to have lines and text as well as attributes in a title block, but you will not need geometry in an attributed block that is meant, for example, to extract room colors.

Tutorial 14b gives an example of attributes used for extraction to a Bill of Materials.

The concepts for attributes are relatively difficult for non-computer users. The only way to see how they are used is with practice. There are three tutorials at the end of this chapter starting on page 268. If you do all of them, you should understand better how attributes work.

Editing Attributes

There are two ways of editing attributes: one at a time or in groups. For one at a time use ATTEDIT or the Enhanced Attribute Editor.

Toolbar From the Modify II toolbar, choose

You will be prompted to choose the block to edit. Then you will get this screen.

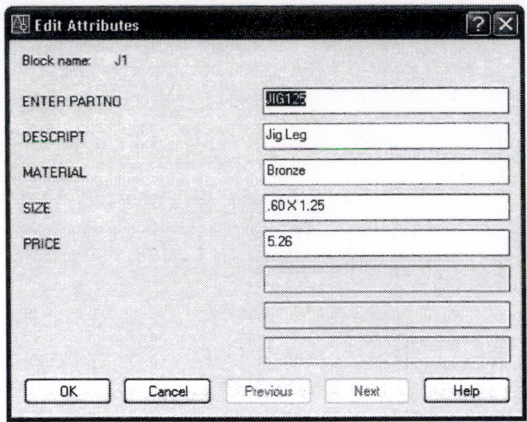

You can change the value of the attribute on this tab, the size and quality of the text on the next, and the properties (layer, color, linetype) on the last. Make changes then click OK.

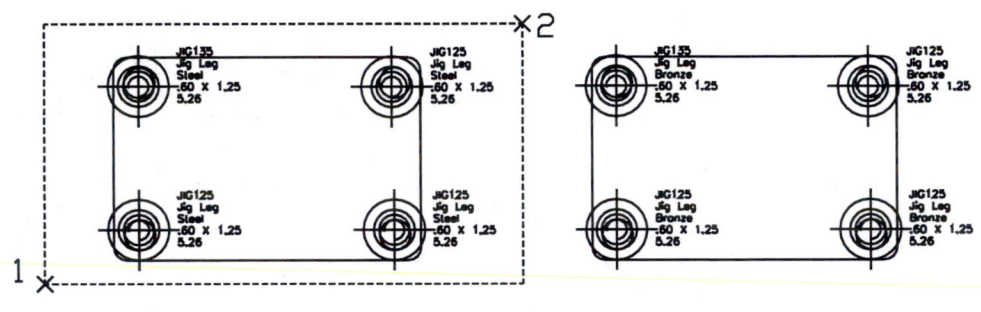

Before ATTEDIT   After ATTEDIT

Figure 14.4

```
Command:-ATTEDIT
Edit attributes one at a time? [Yes/No]<Y>:N
Performing global editing of attribute values
Edit only attributes visible on screen? [Yes/No]<Y>:↵
Enter block name specification <*>:JIGLEG
Enter attribute tag specification <*>:Material
Enter attribute value specification <*>:↵
Select Attributes:(pick 1 pick 2)
Enter string to change:Steel
Enter new string value:Bronze
```

Figure 14.4 show a global change on a number of attributes. The value is case sensitive, i.e. Steel will work, STEEL won't. Make sure that you have typed the information in correctly or you will have problems.

If you need to change a lot of the information on an attributed block, another way of doing this is to EXPLODE the block, redefine it, then re-block it under the same name. All of the block instances will be updated.

Data Extraction

Now that the data have been entered correctly, you can have the file printed out onto a spreadsheet or materials list by using the command ATTEXT. This operation does not change the drawing in any way, but takes the attribute data to applications such as Excel, Lotus 1-2-3, or your favorite word processor. You can either export the data, or have it written immediately on your drawing with EATTEXT.

ATTEXT and Template Files

To extract data with ATTEXT you must create a template file in .txt; Wordpad or Notepad are the easiest to use. The template file might look like this:

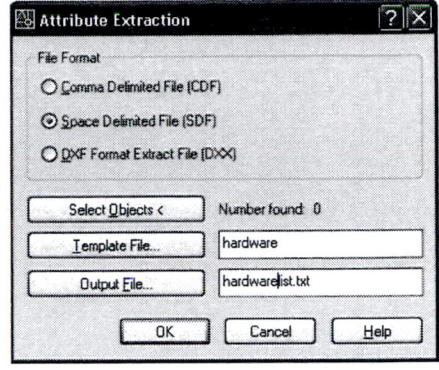

```
PARTNO   C008000
DESCRIPT C015000
MATERIAL C008000
SIZE C008000
PRICE C008000
```

The template file is set up with the tag plus the code:

```
PARTNO   C008000
```

Where: **PARTNO** = TAG
     C = character
     008 = 8 characters maximum in the jig name (C**008**000)
     000 = number of decimal places (none are needed for a character field)

Once the template file is created, invoke the ATTEXT command and you will get the Attribute Extraction dialog box. The template file is the one you have just created. The extract file is the one you are trying to make. View the extract file in Wordpad or Notepad.

EATTEXT

If what you are after is a compiled listing of your attributes on screen, use the EATTEXT command. This will walk you through the process of extracting all of your attribute data and placing it on your screen. This is a new command with Release 2006. In Release 2008 it has become much more complicated. Simply remove all references to objects other than attributes when prommpted and it will work beautifully.

You will be prompted for such things as a table heading, in this case the heading Jig Assembly was chosen. Figure 14.5 shows an extraction from quite a large drawing. The lists can be numbered according to any of the tags.

hardware							
Quantity	DESCRIPT	MATERIAL	Name	PARTNO	PRICE	PRODNO	SIZE
1	Jig Leg	Bronze	J1	JIG135	5.26		.60 X 1.25
1		brass	jigleg			none	
2	Jig Leg	Steel	J1	JIG135	5.26		.60 X 1.25
4			1				
4	Jig Leg	Bronze	J1	JIG125	5.26		.60 X 1.25
4	Stand. Bushing	Steel	B1	BU135	3.29		.125 X .55
12	Stand. Bushing	Steel	B1	BU125	3.29		.125 X .55
19	Jig Leg	Steel	J1	JIG125	5.26		.60 X 1.25
31			2				

Figure 14.5

Step 1

Use LINE and PLINE to create the design for a standard title block. Use TEXT to enter the headings for each area. Use imperial or metric sizes.

Your title block should be in a layer called title. If it is on Layer 0, it will pick up the linetype, layer and color of the layer current in the receiving file when it is inserted.

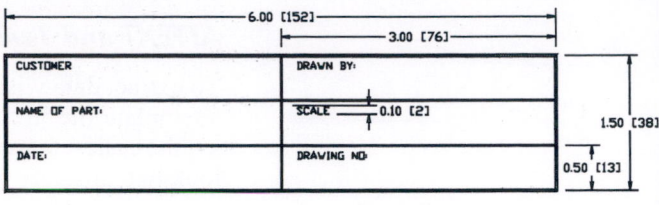

Step 2

Add the attribute definition for the date.

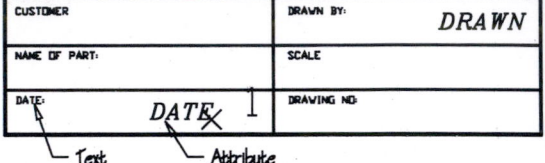

First use the STYLE command to set a new text style for the attributes. Use the typed-in version of ATTDEF for this one.

```
Command:STYLE (set a new text style with the name attributes
    using a different font)
Command:-ATTDEF
Current attribute modes:Invisible-N Constant-N Verify-V
    Preset-P
Enter an option to change [Invisible/Constant/Verify/Preset]
    <done>:↵
Enter attribute tag name:Date
Enter attribute prompt:Enter today's date
Enter default attribute value:
Justify/Style/<start point>:J
Align/Center/Fit/Middle/Right/TL/TC/TR/ML/MC/MR/BL/BC/BR:R
Right point: (pick 1)
Specify height <5.0000>:.15
Specify rotation angle <0>:↵
```

Step 3

Now enter the attribute for the designer using the Attribute Definition dialog box.

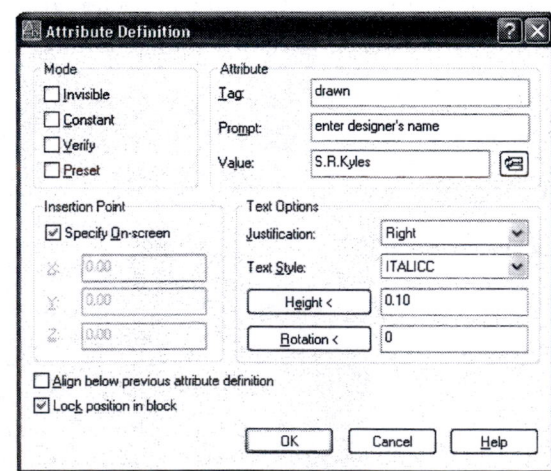

Use your own name as the default.

If your attribute is not where you want it, the Insertion Point area cannot have been set up right. Use Undo to undo this command, and try again being sure to specify the insertion point on screen by first choosing Insertion point from the dialog box.

Step 4 Use either the command line or the dialog box method of defining the other attributes so that each area has an attribute tag.

Once these are in, the title block can be inserted onto other drawings.

Step 5 Title blocks are meant to be used on many different drawings. Save this file under the name of Title1 just to see if it works. Be sure you know what directory it is going to. When in doubt use the A: drive or your flash drive.

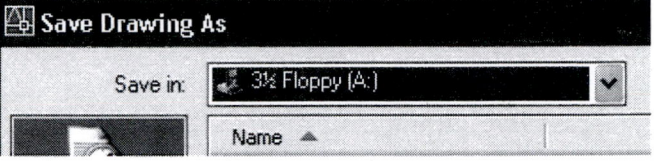

```
Command: SAVE

TITLE1
```

Step 6 Open a new file and test the block out. Use either the dialog box or the command line form of the INSERT command. Keep the Title1 file open as well.

```
Command: New
Command: -INSERT
Enter block name or [?]: TITLE
Specify insertion point or [Scale/X.../PRotate]: 0,0
Enter X scale factor, specify opposite corner, or
    [Corner/XYZ]: ↵
Enter Y scale factor <use X scale factor>: ↵
Specify rotation angle <0>: ↵
Customer name <Skylab Industries>: ↵
Name of Part <>: Hook Link
Enter the current date (DD/MM/YY): 17/03/06
Drawn by <S.R.Kyles>: Bill Challis
Scale <1:50>: ↵
Drawing no. DFC-####<>: DFC-1003
```

With the ATTDIA system variable on, you will get this dialog box to let you see all of your attributes. To change the defaults, simply type over them.

Turn on ATTDIA as shown.

```
Command: ATTDIA
Enter new value for
    ATTDIA<0>: 1
```
Click OK to place them on screen.

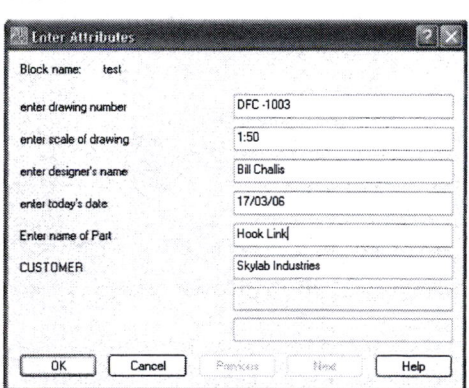

Step 7 A good way of checking your attributed block is to have a fellow student insert the block and fill in the details of the attributes without your help. If he/she has any problem understanding the prompts, go back and edit them. While you may understand your prompts now, the fact that someone else doesn't understand them may indicate that you won't either after a few days.

Other areas you may want to adjust are the attribute text sizes or justifications.

Go to the file Title1. Use CHANGE, DDEDIT , or the Properties menu shown on the left to make necessary changes.

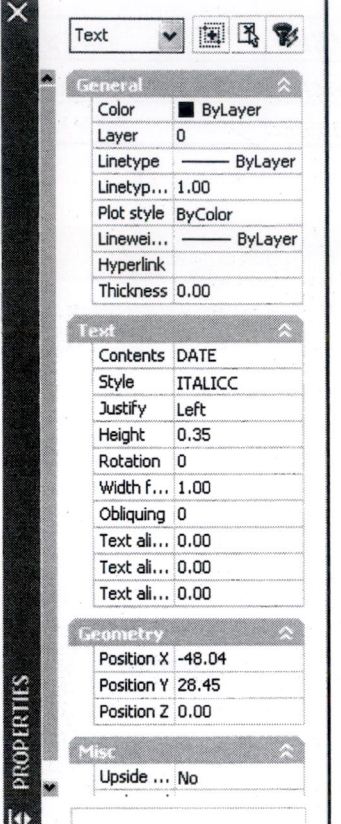

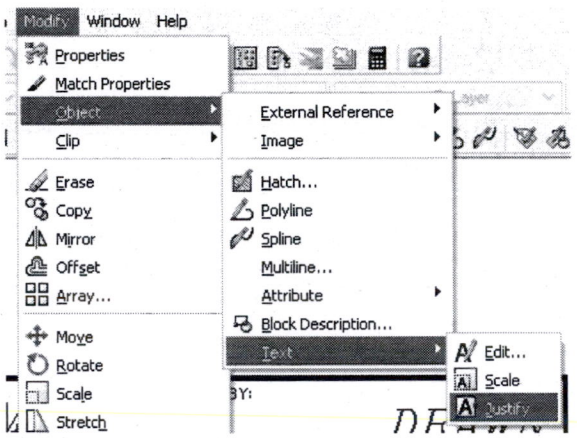

For the justification of attributes such as in Skylab Industries and DFC - 1003, shown below, neither CHANGE nor DDEDIT will help. Use the Properties bar under the Modify menu shown on the left or

```
Command:-justifytext
Select objects: (pick the text)
Select objects:↵
```

Once changed, save the file again and reinsert it.

Step 8

If you have updated the file, resaved it, and reinserted it and the same problems still exist, erase the Title1 block, then use PURGE to erase it from memory. Sometimes, even if you have updated the file, it will bring back the old file instead of the new one. Purge is a great way to clean up your files of unwanted data.

```
Command:PURGE
```

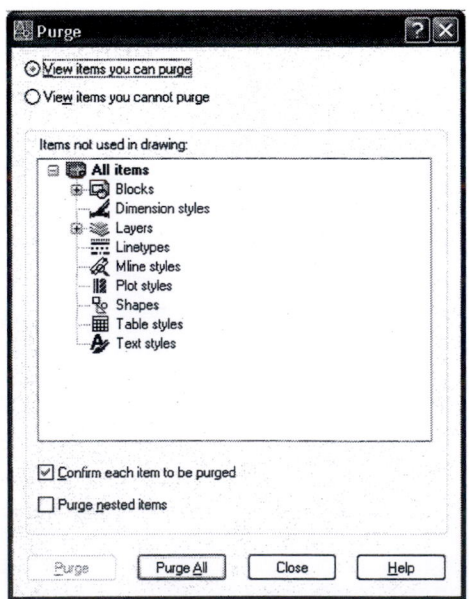

In this tutorial, you will create a jig with a bushing and jigleg. In a manufacturing environment, it is usual to have items such as bushings purchased rather than made on-site to save both time and money. If you were to create a drawing of this part using these items, you could also have the ordering information stored as attributes with the block before it is inserted into the drawing.

Step 1

Draw up these very simple parts in imperial measures on different areas of your screen. Make separate layers for the bushing and the jigleg.

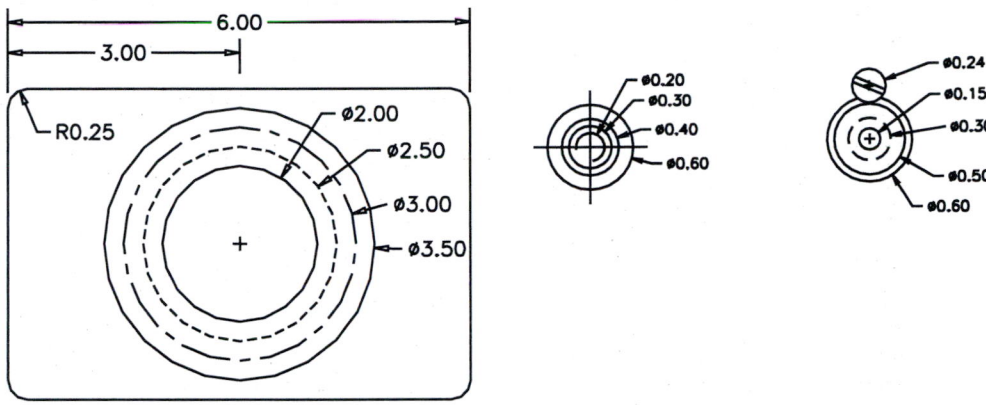

Step 2

If you wanted to create a bill of materials for these items, the information could include:

- Part Number

- Description

- Material

- Size

- Price

Then you would add the quantity and order the parts. The information listed above will be added as attributes to the block before it is blocked and then extracted later.

The words in this illustration are attribute tags, not text. No text is needed in this exercise.

PARTNO
DESCRIPT
MATERIAL
SIZE
PRICE

Add the attributes for the part using these default factors:

Tag	Prompt	Default Value
Partno	Enter Part Number	JIG125
Descript	Enter Part Description	Jigleg
Material	Enter Material	Steel
Size	Enter Part Size	.60 × 1.25
Price	Enter Current Price	5.26

Enter the information for the jig with the -ATTDEF command. Use the Attribute Definition dialog box for the bushing.

```
Command:-ATTDEF
Current attribute modes:Invisible-N Constant-N Verify-V
  Preset-P
Enter an option to change [Invisible/Constant/Verify/Preset]
  <done>:↵
Enter attribute tag name:PARTNO
Enter attribute prompt:Enter Part Number
Enter default attribute value:JIG125
Justify/Style/<start point>:(pick a spot beside the part)
Specify height <5.0000>:.03
Specify rotation angle <0>:↵
```

When entering the next line for the attribute, the attribute definition will line up with the last entered line if you press ↵ at the text justification line or simply choose OK on the dialog box. The attribute definitions will line up perfectly.

```
Command:-ATTDEF
Current attribute modes:Invisible-N ... Preset-P
Enter an option to change [Invisible.../Preset] <done>:↵
Enter attribute tag name:DESCRIPT
Enter attribute prompt:Enter Part Description
Enter default attribute value:JIGLEG
Justify/Style/<start point>:↵
Command: ↵
ATTDEF
 Current attribute modes:Invisible-N ... Preset-P
Enter an option to change [Invisible...Preset] <done>:↵
Enter attribute tag:MATERIAL
Enter attribute prompt:Enter Material
Enter default attribute value:Steel
Justify/Style/<start point>:↵
```

Continue with the final two definitions until all five are complete. It's simply a matter of typing.

Use ↵ at the Justify text prompt to line the attribute up with the previous definition. Also use ↵ to bring back the previous command to make the attribute definition more rapid.

Step 3   When all of the information has been added, create a block of the part with a new name to help you identify it as an attributed block.

```
Command:-BLOCK
Block name (or ?):ATTJIGLG
Insertion base point:CENter of (pick 1)
Select objects:(pick 2, 3)
```

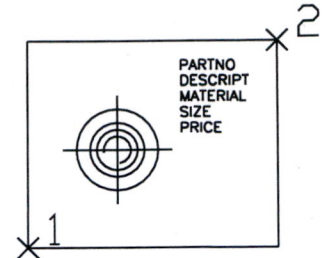

Remember that the block will disappear.

Step 4   Do the same with the bushing using the dialog box. Use the values listed below.

Tag	Prompt	Default Value
Partno	Enter Part Number	BU125
Descript	Enter Part Description	Stand. Bushing
Material	Enter Material	Steel
Size	Enter Part Size	.125 × .55
Price	Enter Current Price	3.29

Use the dialog box by either typing in ATTDEF or on the Draw menu, choose Block then Define Attributes.

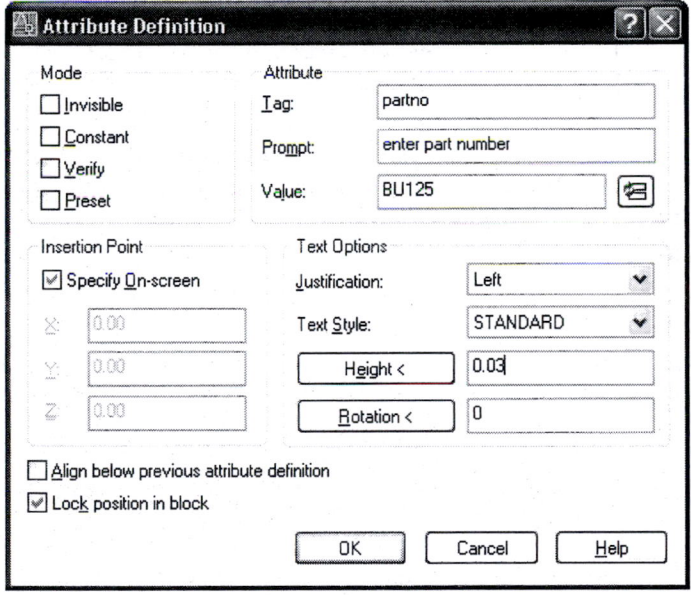

After the first definition is entered you can pick the 'Align under previous definition' button to have the attribute definitions line up.

Your part should look like this.

Step 5   Use the BLOCK command with the dialog box to save the geometry and attributes as a block.

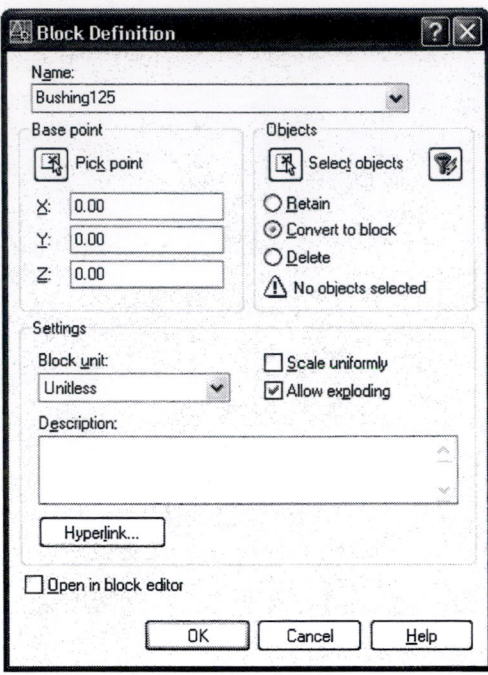

The important thing to remember with this is to pick the insertion base point. You are not prompted for it, so often people forget to pick it. The default 0,0 will make placing the bushings on the part quite difficult.

Step 6   Use ZOOM All to get the original file back onto the screen. Add four circles for the jiglegs at 0.50 from the edge of the part in each corner. Add four circles at the QUADrant of the center line circle on the inside. Now insert four jiglegs and four bushings onto the file.

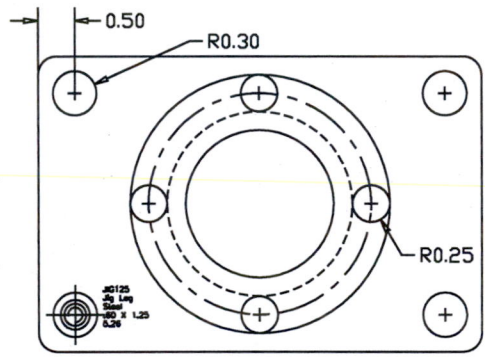

```
Command:-INSERT
Block name (or ?):ATTJIGLG
Insertion point:CEN of (pick 1)
X scale factor <1>/Corner/XYZ:↵
Y scale factor (default = X):↵
Rotation angle <0>:↵
Enter Part Number <JIG125>:¿
Enter Part Description< Jigleg>:↵
Enter Material <Steel>:↵
Enter Part Size <.60 x 1.25>:↵
Enter Current Price <5.26>:↵
```

To get all four of these jiglegs placed, you could insert the object three more times or you could copy the existing block.

Step 7   Use the Attribute Dialog box to enter the final four parts.

INSERT the four bushings, making sure to rotate each one by 90 degrees so they line up correctly.

    Command: **INSERT**

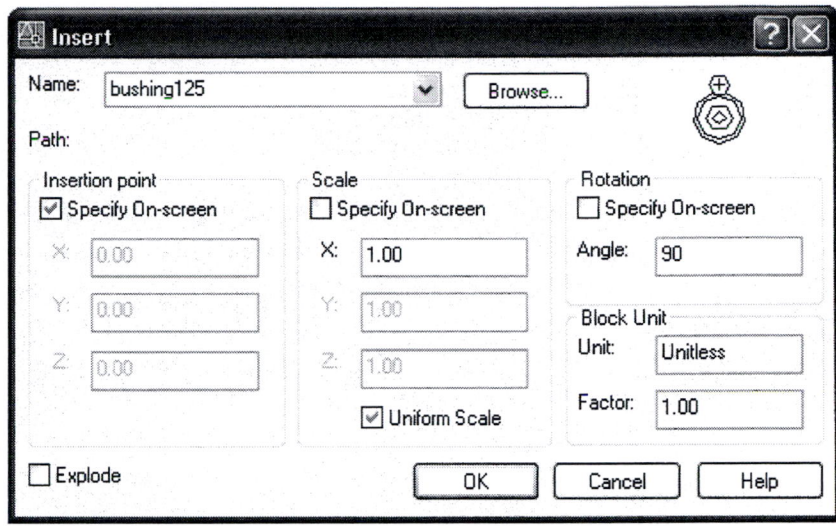

When all four bushing blocks have been entered, your file will look like this:

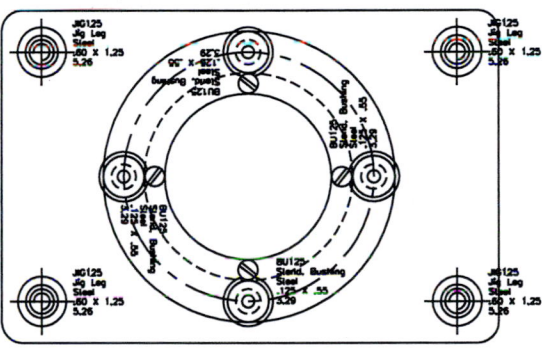

Notice that when you insert the bushings, the attributes will be rotated along with the geometry. This is of no importance because the attributes will not be part of an overall drawing. In an electrical drawing, however, the attributes must be visible and legible; in this case, you should create attributed blocks for both directions. When attributes are added to parts in mechanical or architectural applications, they are not usually printed or plotted.

Step 8   If you were just making a drawing, it would never look like this. Use ATTDISP to turn all of the attributes off.

    Command: **ATTDISP**
    Normal/On/Off <current value>: **OFF**

Step 9   Finally create an attribute table containing the Bill of Materials. Use EATTEXT and simply follow the menus.

    Command: **EATTEXT**

In this Tutorial we will create attributes for the footings in Tutorial 12b page 229.

Step 1   Open Tutorial 12b and save it as Tutorial 14c. Make one copy of footing A and one of Footing B outside the grid.

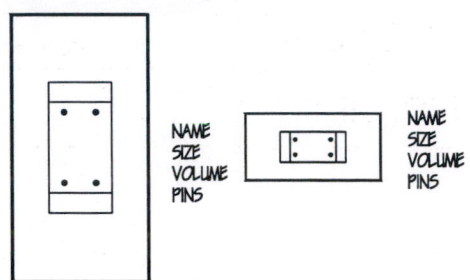

Use the ATTDEF command to define attributes with the same defaults as shown below.

Footing A

Tag	Prompt	Default Value
Name	Enter Footing Designation	Footing A
Size	Enter Size	4000×2000×2300
Volume	Enter Volume	7 cubic meters
Pins	Enter Pin description	4 × 250 projection

Footing B

Tag	Prompt	Default Value
Name	Enter Footing Designation	Footing B
Size	Enter Size	2000×1000×1300
Volume	Enter Volume	3.5 cubic meters
Pins	Enter Pin description	4 × 125 projection

Step 2   Block the footings and insert them into the grid layout as they were before.

Step 3   Use ATTDISP to turn the Attribute Display off.

Step 4   Use EATTEXT to get a list of the footings on the part placed on the drawing. Simply open the dialog box, use Next for every entry, add the title 'Footings' where required. In Release 2008, you will nedd to remove any information not related to the attributes.

Footings					
Quantity	Name	NAME(1)	PINS	SIZE	VOLUME
6	footingb	Footing A	4 x 250 projection	4000 x 2000 x 2300	7 cubic meters
16	footinga	Footing A	4 x 250 projection	4000 x 2000 x 2300	7 cubic meters

Exercise 14 Architectural

First create this title block with attributes as shown. Save the file.
Draw the two sample windows shown. Add the attributes. BLOCK them.
Draw the simple layout shown. INSERT the attributed windows.
Use EATTEXT to get a window schedule. Insert the attributed title block.

Title:	Drawn By:		Scale:	Revision:
TITLE	DRAWN BY		SCALE	REV
Project:	Checked By:		Date:	Drw #
PROJECT	CHECKED BY		DATE	DRW.#

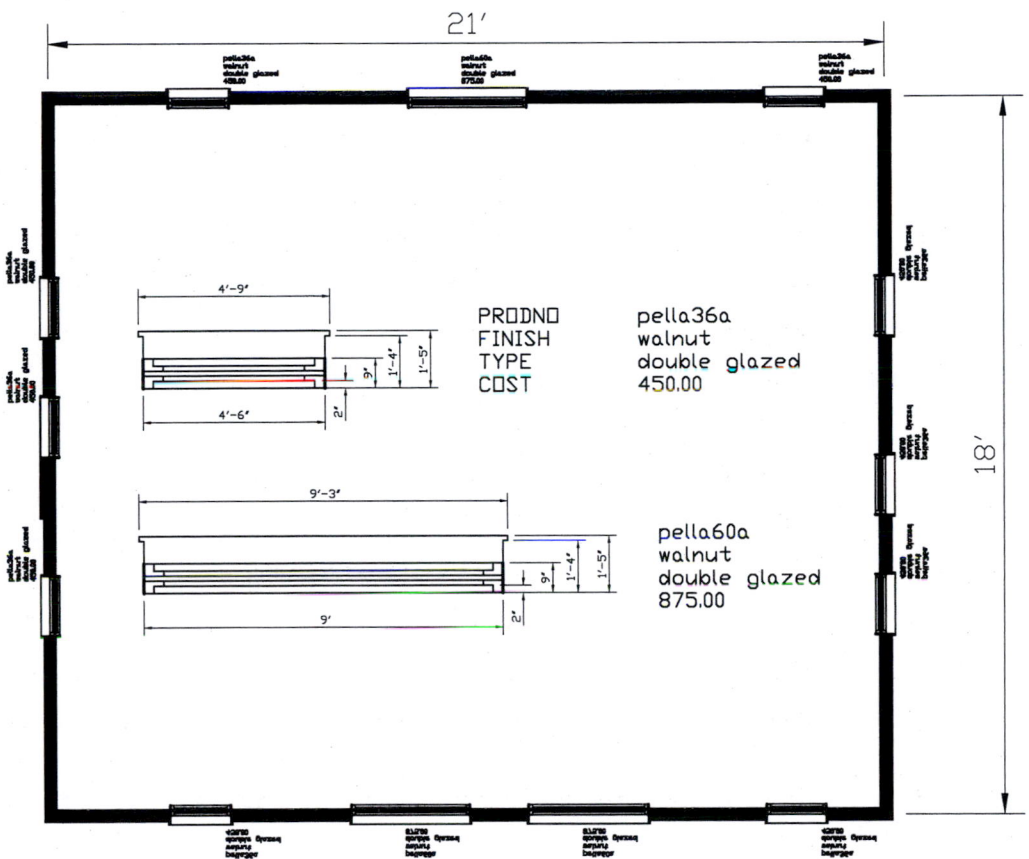

PRODNO   pella36a
FINISH   walnut
TYPE   double glazed
COST   450.00

pella60a
walnut
double glazed
875.00

Second Floor Window Schedule					
Quantity	COST	FINISH	Name	PRODNO	TYPE
3	875.00	walnut	window1	pella60a	double glazed
10	450.00	walnut	window1	pella36a	double glazed

Title:	Drawn By:		Scale:	Revision:
WINDOW PLAN	JAMIE JONES		NOTED	5
Project:	Checked By:		Date:	Drw #
WESTOVER HOUSE	WENDY BENEDETTI		01/04/08	A2

Exercise 14 Challenger

Draw these desks, the dimensions are on page 67.
Attribute as shown. BLOCK them and then make the room layout.
Create a Furniture Schedule using EATTEXT.

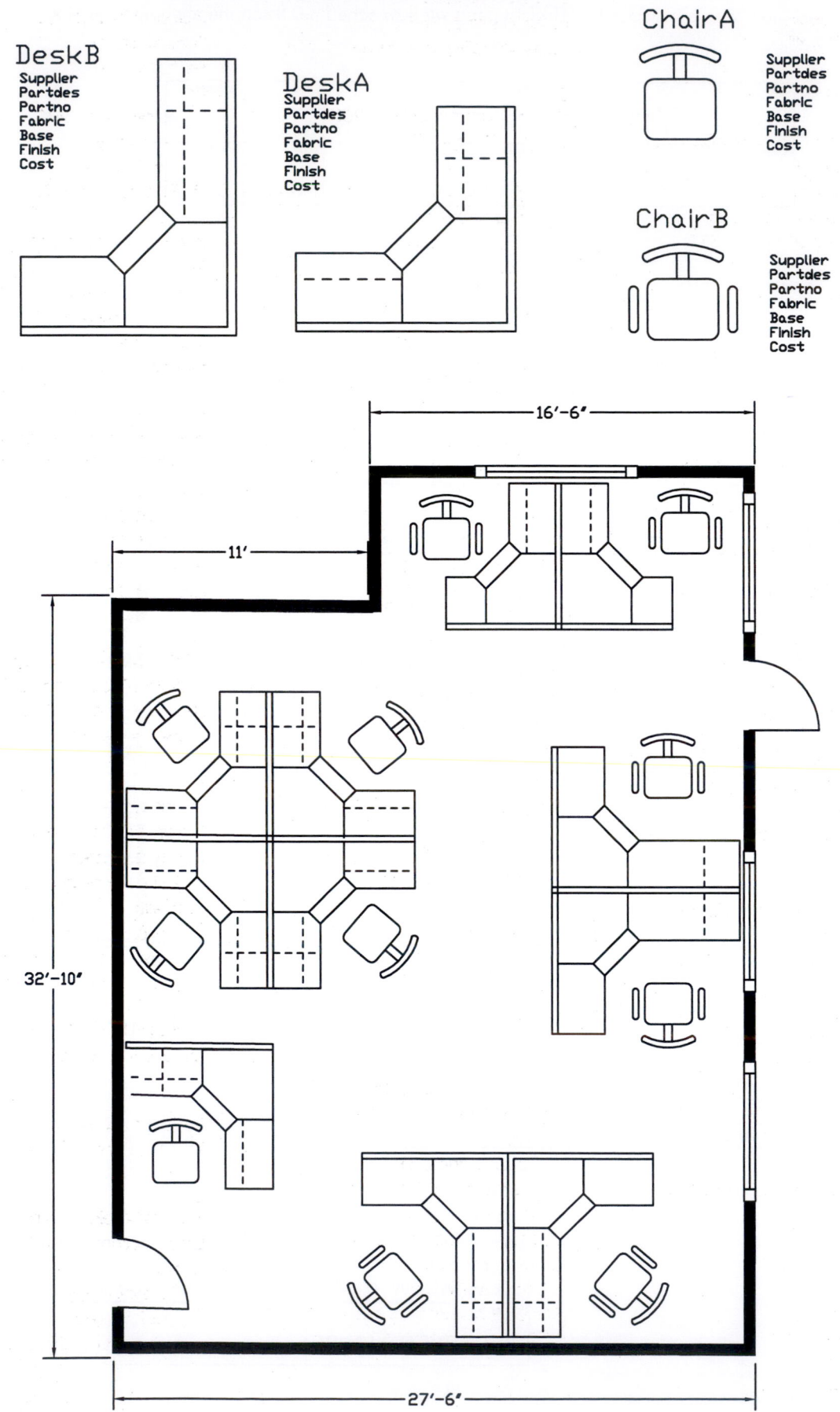

Index

Drawings